Lecture Notes in Physics

Edited by J. Ehlers, München, K. Hepp, Zürich
R. Kippenhahn, München, H. A. Weidenmüller, Heidelberg
and J. Zittartz, Köln
Managing Editor: W. Beiglböck, Heidelberg

85

Applied Inverse Problems

Lectures presented at the RCP 264
"Etude Interdisciplinaire des Problèmes Inverses",
sponsored by the Centre National de la
Recherche Scientifique

Edited by P. C. Sabatier

Springer-Verlag
Berlin Heidelberg New York 1978

Editor
Pierre C. Sabatier
Université des Sciences
et Techniques du Languedoc
Laboratoire de Physique Mathematique
Place Eugen Batallion
F-34060 Montpellier

Library of Congress Cataloging in Publication Data

France. Centre national de la recherche scientifique. Recherche coopérative sur programme 264.
Applied inverse problems.

(Lecture notes in physics ; 85)
Bibliography: p.
Includes index.
1. Inverse problems (Differential equations) --Addresses, essays, lectures. 2. Mathematical physics--Addresses, essays, lectures. 3. Geophysics--Mathematics--Addresses, essays, lectures. I. Sabatier, Pierre Célestin, 1935- II. Title. III. Series.
QC20.7.D5F73 1978 530.1'5'535 78-26104

ISBN 3-540-09094-0 Springer-Verlag Berlin Heidelberg New York
ISBN 0-387-09094-0 Springer-Verlag New York Heidelberg Berlin

Printing and binding: Beltz Offsetdruck, Hemsbach/Bergstr.
2153/3140-543210

TABLE OF CONTENTS

INTRODUCTION TO APPLIED INVERSE PROBLEMS

Pierre C. Sabatier
Département de Physique Mathématique
Université des Sciences et Techniques du Languedoc
34060 MONTPELLIER CEDEX, FRANCE

Summary

The centers of interest in applied inverse problems are presented, together with a short guide to the literature reviewing the subject and an introduction to the lectures contained in the present book.

Contents

1 - Introduction

Inverse Problems began in Physics as soon as a physicist tried to fit experimental results into a theoretical model. But the first theoretical models were given by so simple rules that there was hardly a problem. Think for example to the Boyle-Mariotte law ($pV = RT$). A very few number of experimental results (e.g one measure of p, V, T), were enough to determine the parameters (R), of the theoretical curve. If other experimental results were available either they fitted the theoretical curve inside the range of possible errors, or they did not, and then it was legitimate to question the validity of the model. The Inverse Problem reduced to only one of its aspects : determining the range of errors for the (often unique) parameters (here R) - this so called "error calculus" was considered as a necessary (but not exciting) work for the experimental physicist. A new aspect of Inverse Problems appeared in 1826 with Abel's problem : in a vertical plane, we define a profile $s(x)$, where s is the curved abscissa, as a function of the altitude x ; for any massive point moving in the gravity field and starting from a point of altitude, say, a, we measure its time of arrival $\tau(a)$ at zero altitude. The problem is to infer $s(x)$ from $\tau(a)$. It is solved by the Abel's integral :

$$s(x) = \pi^{-1} \int_0^x \frac{\tau(a)\, da}{\sqrt{x-a}} \qquad (1.1)$$

Thus, one has here to construct the parameter from the experimental result, and deriving this inverse mapping is the first aspect of the inverse problem. Needless to say, the problem of errors is still present.

In Abel's problem, the "direct" problem $(s \to \tau)$ is ruled by the formula

$$\tau(a) = \int_0^a \frac{ds(x)}{\sqrt{a-x}} \qquad (1.2)$$

Remarkably, a similar formula controls many inverse problems. Let us study the corresponding error calculus. If s_i is the result of a measurement of $s(x_i)$, it is "natural" to appraise the overall error on the measurements by the distance

$$d(\{s_i\}, s) = \mathrm{Sup}\ |\ s_i - s(x_i)\ | \qquad (1.3)$$

Let $\tau_e(x)$ a "parameter" which exactly fits the experimental point. It can be constructed for instance by choosing an interpolation of $s(x_i)$ and using (1.2). It is "natural" to characterize the shift between $\pi_e(x)$ and the exact parameter

$\tau(x)$ by the distance

$$d(\tau_e, \tau) = \sup_a |\tau_e(a) - \tau(a)| \qquad (1.4)$$

Now suppose we know that the error (1.3) on the measurements of s is small. The surprising point is that this does not guarantee that the error (1.4) on the parameter is small. Unfortunately this remains true no matter the number of measurements. Even if this number is ∞, or if one knows that the distance $d(s_e(x), s(x))$, defined like in (1.4), is small, one still cannot guarantee that $d(\tau_e, \tau)$ is small. This means that the mapping $s \to \tau$, as defined by (1.2), is not continuous for distances defined like in (1.3) or (1.4).

Of course, from the physical point of view, lack of stability is very unpleasant a case, and the XIXth century physicists did not think it could exist in a real physical problem. But Hadamard, showed that it could. (Cauchy's problem is a classical example). Since that, many other (worse) situations have been found.

Actually, the only <u>essential</u> point in a physical problem is that the mapping $\mathfrak{m}$ of a set $\mathcal{C}$ of theoretical parameters into a set $\mathcal{E}$ of "results" be correctly defined (<u>direct problem</u>). To solve the inverse problem is

(1) extend the set $\mathcal{E}$ so that it contain altogether computed results and results of measurements.

(2) for any element e of $\mathcal{E}$, show the existence (or not) of a non void subset of $\mathcal{C}$ that $\mathfrak{m}$ maps into e (<u>existence question</u>)

(3) define mappings of $\mathcal{E}$ into $\mathcal{C}$ (<u>construction question</u>)

(4) $\mathcal{C}$ and $\mathcal{E}$ being given structures of metric spaces, study the <u>problem of errors</u>. In the simple case where the question (3) is solved by well defined mappings of $\mathcal{E}$ into $\mathcal{C}$, this problem reduces to checking the continuity of these mappings - the stability question. But most inverse problems are partly overdetermined. Thus a number of <u>"possible" experimental results</u> (i.e. the sets of points in $\mathcal{E}$ that are consistent with error bars) do not correspond to any point in $\mathcal{C}$. It follows that both the definition of errors and the choice of solutions necessitate a special formulation, which is consistent with physical intuition and belongs to the so called techniques of "<u>regularization</u>".

In the case of (at least partly) underdetermined inverse problems, three other questions are more specially related to the physicist's sensibility.

(5) are all the inverse mappings $\mathcal{E} \to \mathcal{C}$ possible to classify by physical criteria (classification question)

(6) are the elements of $\mathcal{C}$ corresponding to an element of $\mathcal{E}$ close to each other ? (question of approximation, or of ambiguities extent)

(7) what extensions of $\mathcal{E}$ (in other words, what new experiments) and of $\mathcal{C}$ (in other words, what refinements of the model) can make $\mathfrak{M}$ a bicontinuous bijection.

From this exposure one can see that inverse problems are really important applied problems, with interdisciplinary characteristics. As so, it is not surprising that they have been studied by many mathematicians and physicists. Unfortunately for the subject, there is a seducing naive way to deal with experimental results. In this way, which is the general fitting technique, one reduces $\mathcal{C}$ to a subset depending on a very few parameters, which are determined by trial and error, using the direct problem. Although these techniques always ignore intrinsic ambiguities and most error questions, it is only when they fail that physicists remember that studying inverse problems is the only complete way of analyzing experimental results. It follows that almost all studies of inverse problems up to 1967 were only devoted to theoretical aspects (construction of inverse mappings and identification of ambiguities). The most interesting problems during this period were inverse problems in spectral theory and in quantum scattering theory. They yielded remarkable ways of constructing 2nd order self-adjoint differential operators from asymptotic information (Gelfand Levitan, Marchenko).

But in the late sixties, the needs for studies of applied inverse problems increased in several fields, and particularly geophysics. It led in 1967 to a simple but very interesting method of analysis for linearized inverse problems (Backus and Gilbert). It also led to a large number of studies of Abel equations, coming in various soundings. Meanwhile, practical applications of inverse problems were also tried in other fields (e.g. in scattering theory by the author). It is the great virtue of L. Colin to have understood this situation, and gathered in 1971 scientists of so many fields in the first meeting on inverse problems. From that time on, many (smaller) meetings on this subject were held in the world. In particular, every year from 1972, I tried myself to gather during a couple of days european specialists of these subjects, with the help of the Centre National de la Recherche Scientifique (Meetings of the RCP 264 : "Etude interdisciplinaire des Problèmes inverses"). On the whole, we succeeded to keep the interdisciplinary character of inverse problems, every speaker trying to be understood by all attendants. In the last years, "applied" aspects, and "aspects relevant to a complete analysis", in the form we described above, became more and more important, more and more refined. It suffices to see the list of lectures titles, which is given at the end of the book, to see this evolution in our meetings. In fact it is a general evolution, and I had predicted it (without special merit), in the NASA meeting of 1971.

In the present book we have decided to publish a set of lectures given in our last meetings and which have the properties.

(a) they are of interest in several fields.

(b) they are devoted to applied inverse problems.

(c) they gather much information that is not so well-known. Besides, we publish here a (much smaller) set of lectures which give particular applications (or attempts of applications) of inverse problems that usually are considered as purely theoretical, and two interesting theoretical extensions.

As so, this book covers topics or methods that

(a) are new when compared to the applied problems treated in the proceedings of the 1971 NASA meeting : Mathematics of Profile Inversion (L. Colin ed.)

(b) are complementary (as regards applications) to this reference, as well as to the recent book "Inverse Problems of Quantum Scattering Theory" by K. Chadan and P.C. Sabatier.

(c) are complementary (as regards the methods), to the review papers that were available in 1977 on geophysical inverse problems.

So as to make this book altogether compact and useful

(a) we avoid redundancy with existing treatises; for instance we do not give lectures on the special topics (e.g. inversion of the radiative transfer equation) that are connected with external geophysics and astrophysics, although these topics were widely covered by our meetings (see the list of titles at the end of the book). The reason is that an exhaustive report on the subject is now prepared by the International Union of Geodesy and Geophysics.

(b) instead, we have decided to give, in this introductory chapter, a short guide both to the lectures of this book and to some review papers or reports that we know in the litterature on interesting samples of applied inverse problems. This guide is given on the way, without systematic recollection.

To begin with, we give below the references pertaining to this introduction, and some general treatises on applied inverse problems. The section 2 is a collection of chosen inverse problems. The section 3 is a survey of the methods of obtaining solutions and the corresponding problems, when errors are not taken into account. The section 4 is a short introduction to quasisolutions and regularization. In the section 5, we present side applications of inverse problems, and we end this introduction by some remarks.

References of the section 1.

Abel Résolution d'un Problème de Mécanique. Journal für die reine und angewandte Mathematik Bd.1 Berlin 1826.

Hadamard I. Sur les Problèmes aux Dérivées Partielles et leur Signification Physique. Bull. Univ. Princeton 13 , 1902.

Tichonov A., Arsénine V. Méthodes de Résolution de Problèmes Mal Posés (French Translation from Russian) MIR Publ. 1976.

Lavrentiev M.M. Some Improperly Posed Problems of Mathematical Physics Springer Tracts in Natural Philosophy 1967.

Agranovich Z.S. and Marchenko V.A. The Inverse Problem of Scattering Theory (English transl.) Gordon and Breach New York 1963.

Chadan K. and Sabatier P.C. Inverse Problems in Quantum Scattering Theory. Springer-Verlag New York Heidelberg Berlin 1977.

Colin L. (Ed.) Mathematics of Profile Inversion NASA Technical Memorandum NASA TMX - 62 150.

Fymat A.L. (Ed.) Inversion Methods in Atmospheric Radiation Research - prepared for International Union of Geodesy and Geophysics. Jet Propulsion Lab.CALTEC Pasadena 1977.

2 - A Samples of applied inverse problems

2.1 Linear problems

We do not think it necessary to define a linear problem. Gravity and magnetic interpretation are the most important problems that are (apparently) linear. In fact, the existence of positivity and other bounds impose constraints such that it is not true that any linear combination of two solutions still is a solution. Other "linear" inverse problems are obtained by "linearizing" a non linear problem. One knows the method. Suppose one has to solve in $\mathcal{C}$ the equation

$$\mathfrak{m}(x) = y \qquad (x \in c \ , \ y \in \mathcal{E}) \tag{2.1}$$

where $\mathcal{C}$ and $\mathcal{E}$ are subsets of normed spaces $\bar{\mathcal{C}}$ and $\bar{\mathcal{E}}$. Suppose one knows x_o in $\mathcal{C}$ such that $\mathfrak{m}(x_o) = y_o$ is in $\mathcal{E}$ (and "reasonably close to y"). Suppose one knows a subsets $\mathcal{C}^*$ of $\bar{\mathcal{C}}$ and a mapping $\mathfrak{m}_o$ of $\mathcal{C}^*$ into $\bar{\mathcal{E}}$ that defines a subset $\mathcal{E}^* = \mathfrak{m}_o(\mathcal{C}^*)$ of $\bar{\mathcal{E}}$, and that has a (generalized) inverse, in the sense that there exists a mapping $\mathfrak{m}_o^{-1}$ of $\mathcal{E}^*$ into $\mathcal{C}^*$ which for any vector v of $\mathcal{C}^*$ verifies

$$\mathfrak{m}_o^{-1} \ \mathfrak{m}_o \ v = v \tag{2.2}$$

Now, one can seek a solution of (2.1) as the limit x, if it exists (i.e. with $x - x_o$ in $\mathcal{C}^*$), of the successive approximations to the solution of

$$x = x_o + \mathfrak{m}_o^{-1} \left(y - y_o - \mathfrak{n} \left(x_o \ ; \ x\right) \right) \tag{2.3}$$

where

$$\mathfrak{n} \left(x_o \ , \ x\right) = \mathfrak{m}(x) - \mathfrak{m}(x_o) - \mathfrak{m}_o \left(x - x_o\right) \tag{2.4}$$

If the mappings $\mathfrak{m}_o$ and $\mathfrak{m}_o^{-1}$ are linear, the zero-order approximation simply is the linear problem defined by the mapping $\mathfrak{m}_o$. The question of obtaining $\mathfrak{m}_o^{-1}$ from $\mathfrak{m}_o$ is called the linearized inverse problem of $\mathfrak{m}$ in the neighbourhood of $(\mathfrak{m}\ x_o = y_o)$. Clearly conditions of validity and convergence of the successive approximations algorithm, together with the a priori constraints on x, will reduce the possible choices of y and values of x to sets $\underline{\mathcal{E}}$, respectively $\underline{\mathcal{C}}$, which are contained in, but usually much smaller than , the intersection of $\mathcal{C}$ and the translation of $\mathcal{C}^*$ by x_o, respectively the intersection of $\mathcal{E}$ and the translation of $\mathcal{E}^*$ by y_o.

$\underline{\mathfrak{E}}$ and $\underline{\mathfrak{C}}$ not only depend on x_o, y_o, on the a priori constraints, and on $\mathfrak{M}$, but also on $\mathfrak{M}_o$ and the chosen inverse $\mathfrak{M}_o^{-1}$.

Nevertheless, the result is so often relevant in physical cases that one can hardly exaggerate the importance of linear inverse problems for applications. The lectures by V. Courtillot et al. , D. Jackson, M. Cuer, M. Bertero et al. in the present book deal with linear or linearized inverse problems. We shall meet also linearized inverse problems and the corresponding references with the non-linear problems from which they have been obtained and with the studies of corresponding methods of solution, as given in § 3.2.

Finally, we refer the reader who likes to get a more general geeling of these problems, from the point of view of a physicist, to the following recent reviews :

R.L. Parker Understanding Inverse Theory. Ann. Rev. Earth Planet.Sci. 1977, p.35-64.

R.L. Parker Linear Inference and Underparametrized Models. Rev. Geoph. Space Phys. <u>15</u>, 446-456 (1977)

P.C. Sabatier On Geophysical Inverse Problems and Constraints. J. Geophys. <u>43</u>, 115-137 (1977).

and for linear inverse problems in potential theory to

G. Anger Uniquely Determined Mass Distributions in Inverse Problems. In Veröffentlichungen des Zentralinstitutes für Physik der Erde <u>52</u>, 2, Postdam 1977.

2.2. Spectral problems

Let $\mathfrak{D}$ be a linear operator, mapping into itself a convenient space of functions, $\mathfrak{H}$ The set of values of λ at which $(\mathfrak{D}-\lambda)$ is not inversible is defined as the spectrum of $\mathfrak{D}$. The inverse spectral problem is obtaining $\mathfrak{D}$ from its spectrum or from informations directly related with it. Now there are two extreme cases.

<u>1st case</u> $\mathfrak{D}$ is the sum of a well- known differential operator $\mathfrak{D}_o$ and an unknown function V , completed by known boundary conditions on a known surface :

$$\mathfrak{D} = \mathfrak{D}_o + V \qquad \text{(Fixed boundary)} \qquad (2.5)$$

The problem is to determine V . It has been very much studied, particularly when $\mathfrak{D}$

is a self-adjoint operator, and still more often when it is the Laplacean operator. Most papers only deal with theoretical problems. Yet, this problem has applications, particularly to the normal modes problem of the Earth. We do not give any lecture on this subject in the present book (although some papers were presented in the meetings of the RCP 264). Actually, I do not know any overall review of inverse spectral problems in this first case. But there are many papers giving good partial reviews. I like to quote the following ones, whose list of references, when they are gathered and completed by those of § (3.4), almost cover the subject in the case of a differential 2nd order self-adjoint operator.

B.M. Levitan and M.G. Gasymov Determination of a Differential Equation by Two of its Spectra. Russian Math. Surveys 19, 1-63 (1964).

V. Barcilon Well-Posed Inverse Eigenvalue Problems. Geophys. J. R. Astr. Soc. 42, 375-383 (1975).

O.H. Hald Inverse Eigenvalue Problems for Layered Media. Comm. Pure and Appl.Math. 30, 69-94 (1977)

For a periodic function $V(x)$, one can see

E. Trubowitz The Inverse Problem for Periodic Potentials. Comm. Pure Appl. Math. 30, 321-337 (1977).

The references by Barcilon and Hald are devoted to "applicable" problems. In fact, real treatments of applied problems, up to now, are done by linearizing them and using the technique of § 3.2 . See for example (Earth Inverse Problem)

F. Gilbert Inverse Problems for the Earth's Normal Modes, in Mathematical Problems in the Geophysical Sciences Tome 2 W.M. Reid, Editor. American Mathematical Society 1971

For related problems, see also the lectures by Morel in the present book.

2nd case $\mathcal{D}$ is a well known differential operator, e.g. the Laplacean operator. What is unknown is the surface on which the boundary conditions are imposed.

$$\mathcal{D} = \mathcal{D}_o \qquad \text{(unknown boundary)} \tag{2.6}$$

A good example is the tambourine, with an homogeneous membrane, a known tension, but an unknown shape. The problem has been remarkably surveyed by M. Kac in a famous lecture :

M. Kac Can One Hear the Shape of a Drum. Am. Math. Monthly 73, II, 1-23 (1966).

Real applications exist (e.g. constructing a resonant cavity from its pure tones). See the references of the lecture by Céa in the present book.

2.3. Problems of quantum scattering theory

Again we deal with the operator $\mathcal{D}_0 + V$, where $\mathcal{D}_0$ is the Laplacean operator, V is a multiplication operator called the potential. Boundary conditions are given at ∞. They fix the incoming flux. The scattered flux is characterized by an operator called the S-matrix, which in principle can be determined from the experimental results. In the inverse problem one determines V either from S or directly from the experimental results.

Theoretical aspects and applications have been reviewed in the book by Chadan & Sabatier (quoted in § 1.1). An interesting review lecture is lacking in this reference, and we like to quote it here :

Faddeyev L.D. Inverse Problem of Quantum Scattering Theory II. J. Sov.Math. Vol. 5, 334-396 (1976).

In the present book, the lecture by Pelosi give a practical application of matrix methods in the inverse problem at fixed energy, the one by Coudray gives an attempt to apply Loeffel's method to the same problem with complex potentials.

The inverse problems of quantum scattering theory correspond in some sense to a generalization to an infinite domain of the inverse spectral problems in the 1st case (2.5) which has been given above. This analogy can also be used in the reverse way - see

Sabatier P.C. Spectral and Scattering Inverse Problems, to be published in the Journal of Mathematical Physics.

2.4. Continuous scattering problems

The scattering of elastic waves by continuous media, that of optical waves by media with varying index, yield inverse problems whose mathematical form is very similar to that of the previous ones. Here again, one has a self adjoint operator, $\mathfrak{D}$, depending on an unknown function, and one makes measurements at fixed boundaries. For stationnary processes, again these measurements are those of flux and intensity. But the most important processes often are not stationnary. The time of arrival of specially identified parts of the signal, in certain cases phases or amplitudes, are then the most important measurements. These three dimensional problems sometimes are studied in particular frames, corresponding to ideal soundings, and reducing the number of dimensions. However, in most analyses, it is usually necessary to make preliminary approximations of $\mathfrak{m}$. For this reason, there are hardly any review papers of these problems in a general framework. In all cases, an approximation method is imposed, and the lecture by Kennett, which we give in the present book, is not an exception.

2.5. Scattering by a finite object

This problems is related to the 2nd case of the spectral problem like the previous one are related to the 1st case. One has to determine a boundary surface from scattering experiments, of elastic or electromagnetic waves. A standard reference of the direct problem is the well-known book

P.D. Lax and R.S. Phillips Scattering Theory. Ac. Press 1967

Interesting reviews of the inverse problem can be found in the "Mathematics of Profile Inversion" quoted in § 1.1.

2.6. Miscellaneous

The previous samples of applied inverse problems were but a part of the subject. We should not forget, although we shall not give anything more than quotations the transport, heat transfer, and radiative transfer inverse problems, the ones that are connected with various atmospheric soundings, the construction of object from their projections, the determination of images through randomly fluctuating media, not to forget shape recognition, electrocardiography, etc. Some mathematical problems and methods which come with these problems are reviewed in § 3 and § 4. For other information, let us give a few quotations of papers with important references lists.

Alain Fymat - quoted in § 1.

K.T. Smith et al. Practical and Mathematical aspects of the Problem of Reconstructing

Objects from Radiographs. Bull.Amer·Math.Soc. 83, 1227-1270, (1977), and Invited address at the Far West Sectional Meeting of the American Mathematical Society, Monterey Ca 1975.

L.A. Shepp and J.B. Kruskal Computerized tomography : The new Medical X-Ray Technology. The American Mathematical Monthly, Volume 85, Number 6, p.p. 420-439, (1978).

R.H.T. Bates Imaging Through Randomly Fluctuating Media. Proc. I.E.E.E. 65, 138-143 (1977)

Dyson F.I. Photon Noise and Atmospheric Noise in Active Optical Systems. Journ.Opt. Soc. Am. 65, 551-558 (1976).

G. Anger Some Remarks on Inverse Problems in Differential Equations - Rostocker Mathematisches Kolloquium Rostock (1978).

R. Bellman, B. Kashef and R. Vasuderan, The Inverse Problem of estimating Heart Parameters from Cardiograms. Mathem. Biosciences 19, 221-130 (1974).

A. Bamberger, G. Chavent and P. Lailly Etude mathématique et numérique d'un problème inverse pour l'équation des ondes à une dimension. Centre de Mathématiques Appliquées de l'Ecole Polytechnique, Rapport n° 14 (1977).

D. Colton Integral Operators and Inverse Problems in Scattering Theory. In Function Theoretic Methods for Partial Differential Equations. Lecture Notes in Mathematics Vol. 561 Springer Verlag 1976.

M.M. Lavrentiev, V.C. Romanov and V.G. Vasiliev Multidimensional Inverse Problems for Differential Equations Springer Verlag 1970.

R.G. Newton Inverse Problems in Physics SIAM Rev. 12, 346-356 (1970).

A. Bjorck (ed.) Symposium in Mathematical and Numerical Analysis of Inverse and Ill-Posed Problems Dep. of Mathematics, Linköping University, Linköping, Sweden (1977).

J.L. Lions Contrôle optimal de systèmes gouvernés par des équations aux dérivées partielles Dunod 1968.

H. Grosse and A. Martin Theory of the Inverse Problem for confining Potentials Preprint CERN 1978.

Section 3 - Dealing with solutions

Here we assume that the results are perfect (no error). If it is the case, overdetermination in an inverse problem is trivial : either the results are consistent or they are not. If they are, the redundant ones always can be suppressed. If they are not, there is no solution, unless the set of admissible results is arbitrarily reduced. Hence we are led here to survey only problems in which any element e of $\mathcal{E}$ corresponds at least to one element of c . The questions (3), (5), (6), (7) of § 1 are of interest as well as the stability question, but only in the sense of continuous (?) inverse mappings. Thus constructing inverse mappings and classifying them is the largest part in the job. It is done either by exact methods or by approximate ones, among which we only survey the linearized methods, the ray methods and the numerical methods.

.1 Exact methods

We deal with the samples of problems that are surveyed in § 2, except the linear ones, which of course are to be studied in § (3.2). All these problems are controlled by an operator $\mathcal{D}$ which in most cases is linear and self adjoint. Typically, an exact method is made of two steps :

(a) constructing from the experimental results a function - say - s, that is associated with the spectrum of $\mathcal{D}$.

(b) constructing $\mathcal{D}$ from this function s .

Typically, the first step is not difficult. But it is very important because the ambiguities in solutions and in certain cases the lack of stability, come in at this step. The ambiguities are due to the fact that experimental results do not determine completely the function s . But since their relations with s in usually very simple, it can be used to classify the equivalent solutions by means of the lacking information - for example, in the inverse problems of § (2.3), the bound states and the associated parameters yield a natural classification.

There has been recently an original attempt to classify the solutions of problems (2.2) by extremal properties (V. Barcilon Ideal Solution of an Inverse Normal Mode Problem with Finite Spectral Data to be published in Geoph. J.R.S.). I think that this attempt should be generalized. It is narrowly related with the discrete approaches discussed by Turchetti in the present book.

The lack of stability, when it comes, is due to a necessary extrapolation or analytic continuation of the results. The presence of this step in the Gelfand, Levitan, Jost, Kohn methods but not in the Marchenko method makes the second one more interesting for physical applications. Problems connected with these continuations are studied in the present book in the lecture by Atkinson, whereas an approach like Turchetti's may yield a way to squeeze them.

The second step of the construction can be approached in several ways.

(1) The key to methods like Gelfand-Levitan's, Jost-Kohn's, Marchenko's, etc is the existence of a transformation operator, which has the remarkable property of yielding the data which correspond to a value of parameters from those corresponding to another one. Besides, the trace of the operator readily yields the new value of parameter. These methods apply to the problems of the " 1st case " studied in § 2. The references we gave in § 2 can be used to understand the methods. Some new points about applications can be found in the present book in the lectures by Pelosi and by Coudray. Applications to solving non linear differential equations will also be seen in § 5. But in our general lectures, we prefer to give very general, although non-standard, ways to construct exact methods. They can be found in the lectures by Cornille and by Karlsson. From time to time, one sees in the litterature attempts to apply Gelfand-Levitan's or Marchenko's methods to problems in which they are unusual. One can read for instance in the list of lectures to our meetings a relevant one by Weidelt on Earth electromagnetic problems and one by Lefoeuvre on the problems of sound scattering. A favourable point in these applications is that the problem of bound states is ruled out by positivity constraints (see my paper quoted in § 2.1), and this certainly is favourable to the practical use of these exact methods. If the function s can itself be obtained from the experimental result without making necessary an analytic continuation, the whole method has every reason to be practicable, even for experimental results. This is the case in acoustic applications, but not in electromagnetic ones But I have to say that up to now, all the algorithms that are used for the corresponding computations seem to me rather naive, except the attempts by Lambert, Corbella and Thome (Nucl. Phys. B. 90, 267-284), which uses Pade approximants and should be adapted to less specialized problems than they did.

(2) A different way to approach inverse problems was initiated by Jost and Kohn, and developped by Moses, and more recently by Prosser. They propose an algorithm with reproduces the operator $\mathcal{D} + q$ essentially be inverting the Born series of $(\mathcal{D}+ q)^{-1}$.

This method formally applies to all problems that we studied in § 2 but it is not very efficient, and all these algorithms converge only in certain cases. We do not give here any example of applications.

(3) Special approaches to the problems in which a boundary surface must be determined have been given : see in particular the lecture by Céa and its references.

3.2 Linearized methods

The method that is described in § (2.1) easily can be used to linearize problems of the form (2.5) as soon as there exists a value V_0 for which one knows the inverse of the "unperturbed operator" $(\mathcal{D} + V_0)$, with the fixed boundary conditions. This inverse is the well- known Green's operator $(\mathcal{D} + V_0)^{-1}$. It is an easy trick to show then that the solution of the perturbed problem is also the solution of the integral equation

$$\psi = \psi_0 + (\mathcal{D} + V_0)^{-1} (V_0 - V) \psi \tag{3.1}$$

Linearizing the inverse problem is reducing (3.1) to its "Born approximation"

$$\psi = \psi_0 + (\mathcal{D} + V_0)^{-1} (V_0 - V) \psi_0 \tag{3.2}$$

Hence the inverse problem reduces to a Fredholm equation of first kind relating the "unknown parameter" V to the "result" ψ. A similar model holds in most linearized problems and can be analysed on the general form

$$e(x) = \int_\Omega K(x,y)\, c(y)\, dy \quad (e \in \mathcal{E} \; ; \; c \in \mathcal{C}) \tag{3.3}$$

All the "difficulties" in handling this equation come from the obvious fact that $\mathcal{E}$ must be the image of c by the operator K (or included in it), and from the less obvious fact that the distances or norms which are chosen in $\mathcal{C}$ and $\mathcal{E}$ have to be chosen together very carefully if the mapping and its inverse are to be continuous. To understand these points on a general example, assume that $K(x,y)$ is symmetric and belongs to $L_2(\Omega \times \Omega)$. Let $\varphi_k(x)$ be the associated orthonormal function. corresponding to the eigenvalue γ_k (so that $\int_\Omega K(x,y)\, \varphi_k(y)\, dy = \varphi_k(x)$ and $K(x,y) = \ell.i.m. \sum_1^\infty \gamma_k\, \varphi_k(x)\, \varphi_k(y)$). Let e_k be the Fourier coefficient of $e(x)$. Then the Fourier coefficient c_k of $c(x)$ should satisfy

$$c_k = e_k \, \gamma_k^{-1} \tag{3.4}$$

Now, the eigenvalues γ_k go to zero as $k \to \infty$, and very rapidly when $K(x,y)$ is very regular. If the quadratic norm is chosen in $\mathcal{C}$, the direct mapping $\mathcal{C} \to \mathcal{E}$ is continuous. The inverse one is continuous only if $\|e\| \to 0$ implies $\|c\| \to 0$, i.e. $\sum_1^{\infty} e_k^2 \gamma_k^{-2} \to 0$. Hence we see that usually the norm in $\mathcal{E}$ must be much stronger than the one in $\mathcal{C}$, and $\mathcal{E}$ should be much narrower than $L_2(\Omega)$. If not, the space $\mathcal{C}$ has to be increased to be complete for very weak norms and it contains elements that may be far from the "functions" in usual sense. Thus, with many kernels K, a natural choice of the normed space $\mathcal{E}$ leads to a choice of $\mathcal{C}$ that includes "weak solutions" that may be unacceptable from the point of view of a physicist. But in our case of perfect results, the problem reduces to deciding the proper definitions of $\mathcal{C}$ and $\mathcal{E}$. We refer the reader to the treatise by M.M. Lavrentiev quoted in § 1.

Another interesting case, and experimentally relevant, is the one where $e(x)$ is known only at a few points x_i. The problem is then strongly underdetermined. It is customary (from Backus and Gilbert) to study it in the form

$$e_i = \int_\Omega G_i(x) \; c(x) \, dx \quad (i = 1,\ldots N, \; c \in \mathcal{C}) \tag{3.5}$$

One first has to cheek that the $G_i(x)$ are linearly independent. If not, either the e_i's are consistent and then one can eliminate the redundant equations or they are not and then the problem has no solution. Thus we can assume that the G_i's are linearly independent. Assume also they belong to $L_2(\Omega)$, (a weak assumption). Their linear combinations span a subspace $G_2(\Omega)$, and there is only one solution of (3.5) in this subspace-say $c_o(x)$. All the other solutions are given by the formula

$$c(x) = c_o(x) + c_\perp(x) \tag{3.6}$$

where $c_\perp$ is any element of $L_2(\Omega)$ that is orthogonal to $G_2(\Omega)$. It is easy to see that the solution c_o is the least-norm solution in $L_2(\Omega)$. From the mathematical point of view, the problem is completely solved at this point. The physicist would like in addition a physical classification of the infinite ambiguities represented here by the $c_\perp(x)$'s, ie by the orthogonal complement of G_2. One may think for instance to classify them by means of their smoothness. But this is very difficult to succeed. On the contrary, it is very naturally done in the Backus-Gilbert method, (references quoted in § 4), in which solutions or quasisolutions are generated in a very simple way. Linear problems with exact data are not studied in the present book. The references that are given in § (2.1) can be consulted for additionnal information.

3.3. Ray methods.

The problems we surveyed in § 2 were wave propagation problems. When the frequency increases in such a way that the average variations of the parameters are small on a wave length, the wave propagation looks more and more like optical rays propagation. Justifying the approximation again can be done by means of equations (2.1) - (2.4), where the mapping $\mathfrak{m}_o$ is not necessarily linear. Dealing with the direct problem in this way often yields inverse problems that lead to the Abel's equation we saw in § (1.1). They were studied for a very long time, in many different fields. The recent studies put more and more emphasis on the ambiguities that come in, for instance, if the mapping $\mathfrak{m}$ is defined like in (1.1), but with a parameter s that is not necessarily a monotone function. Each domain in which s is not monotone can be "seen" only in part, and a very interesting question is to find extremal bounds for the hidden parts. The reader can find a relatively long references list for these studies in the case of scattering problems in our book quoted in § (1.1). In the present book, we asked Dr Kennett to present the advances of this problem in geophysics because it is certainly the field in which the ray approach is the most important.

3.4. Numerical methods

Everything could fit this very vague title. Here we mean methods which are specially conceived for computers and for dealing with large classes of inverse problems. We give three lectures on the subject (all in French). The first one, by Céa, deals with all kinds of problems of finding a "best shape". Hence it can include for instance all problems of § 2 that correspond to the second case of § 2.2 , ie those in which one looks for an unknown boundary surface. It can also deal with many problems of design, and probably it is the most "applied" lecture in the present book. The second lecture, by Morel, deals with the inverse spectral problem of § 2.2, 1st case, in the case where matrices in $\mathbb{R}_N$ (or in ℓ_2) define the operator. In a satellite lecture at the end of the book, it is given an algorithm that has practically been used to solve these problems. Needless to say, hints on numerical methods are met in many other lectures in this book, but they are narrowly connected with the problem of errors, and we shall see them in § 4

It is curious to notice that purely numerical methods only try to obtain one solution -

the one which minimizes a certain "cost function". The problem of classifying the equivalent solutions, which is essential in a physical treatment of inverse problem, usually is out of their scope and capabilities.

3.5. Miscellaneous methods

Problems that are connected with numerical methods are discretization problems. One can find hints on their specific aspects for inverse problems in the lectures by Morel and by Turchetti. Some tricks to squeeze the non linearity of particular problems can also be found in the lectures on special geophysical applications by Jobert and Cisternas and by Barthès and Vasseur. Finally, let us notice that it is very often of first importance in certain cases to reduce the inverse problems to their simplest form, by taking into account all the simplifications that can come from physical symmetries or similarities. This is studied in the lecture by Feix et al.

Section 4 - Dealing with quasisolutions and errors.

Many experimental physicists, many engineers, like to deal with problems that are overdetermined by excess of measurements. They trust a model. They gather as many observed results as they can, even those poorly measured. The model parameters are then determined by some balanced - fitting method (e.g. "least-square determination"), with a statistical evaluation of errors. Such a policy assumes a very strong faith altogether in the model and in simple rules of statistics, since the overdetermination implies that many observed results may be redundant or contradictory.

On the other hand, the presentation of inverse problems we used for well-posed and for underdetermined cases, which is still convenient for overdetermined problems when errors are forgotten, becomes more cumbersome for this common type of overdetermined problems.

Example Let us be given a simple physical model leading to a linear relation $y = a x + b$ between the physical quantity y and the space coordinate x, with an a priori constrainst (e.g. the graph of y must lie outside of a certain forbidden region $\mathcal{D}$ in the (x,y) plane). Suppose there is a number of measurements $y_i (x_i)$, with "error bars" Δ_i, that we shall first understand in a naive way, for the sake of simplicity. A temptative definition of $\mathcal{E}$ would be the set of couples of points or two components vectors that are consistent with observations. But any $y(x_i)$ between $y_i(x_i) - \frac{1}{2} \Delta_i$ and $y_i (x_i) + \frac{1}{2} \Delta_i$ is consistent with the observations. Thus two measurements yield an infinity of couples. However, certain of them are not consistent with other measurements, or with the a priori constraints in C. They should be eliminated. Let us now consider a third measurement. It enables us to define with the first two measurements two other set of couples. There should be couples, in different sets, belonging to the same straight line (if there was not, the problem would have no solution). One must put a rule to make mathematically identical these equivalent couples because, if not, they would give redundant elements in $\mathcal{E}$. On the other hand, if we give up our too much naive understanding of errors, we have to take into account that the coincidence of many couples along a given straight line increases the "likeliness" of this line as a correct physical model. But how to put this information in $\mathcal{E}$? Obviously, this definition of $\mathcal{E}$ is not very good.

We have to try another definition of $\mathcal{E}$. If there are N measurements, we can represent them by N-components vectors, and $\mathcal{E} \subset R_N$ or C_N. This definition is not bad and will be used. But one should never forget three points. (a) the overdetermination, which was obviously due only to the exess of measurements, is now built in the definition of $\mathcal{E}$. Thus the image $\mathfrak{m}(C)$ of the set of parameters will play a central role in all studies (b) this definition of $\mathcal{E}$ depends on the number N of measurements, a rather unpleasant situation (c) the representation of errors should be made carefully. Actually this last remark is never taken into account and one usually defines a distance in $\mathcal{E}$ just like if all errors were uncorrelated.

To take into account the second remark, we are led to keep in our analysis the following policy. As long as possible, instead of describing the model by the mapping $\mathfrak{m}$ of C into $\mathcal{E}$, we shall describe it be means of several mappings $\mathfrak{m}_i$ into open sets $\mathcal{E}_i$, each mapping corresponding to one (one type of) measurement. For example, in the case which is treated above, the measurements of y at one point x_i correspond to one mapping $\mathfrak{m}_i$. In each set $\mathcal{E}_i$, we call $\underline{\mathcal{E}}_i$ the smallest open subset that contains the value given by the measurement and those which are allowed by error appraisals(e.g. the interval $y_i(x_i) - \frac{1}{2}\Delta_i$, $y_i(x_i) + \frac{1}{2}\Delta_i$ in the example).

Now, if we keep along with our naive understanding of error bars, we define a true solution as any point in C whose i^{th} measurement lies in $\underline{\mathcal{E}}_i$, $i = 1,\dots N$. Let d_i be the distance in $\mathcal{E}_i$, a true solution c cancels the function

$$D_o = \sum_{i=1}^{N} d_i\,(\underline{\mathcal{E}}_i\ ,\ \mathfrak{m}_i\,(c)\,) \tag{4.1}$$

When $\underline{\mathcal{E}}_i$ simply is an openball, centered at e_i , radius $\frac{1}{2}\Delta_i$, $d_i\,(\underline{\mathcal{E}}_i\,,\mathfrak{m}_i(c)\,)$ vanishes when $d_i\,(e_i\,,\mathfrak{m}_i(c))$ is smaller than $\frac{1}{2}\Delta_i$. Suppose we introduce a smoother (e.g. probabilistic) understanding of errors. We are led to characterize a "good fit " of the measurements by the value of a "cost function" D_1 that is very small when $d_i\,(e_i\,,\mathfrak{m}_i\,(c))$ is smaller than $\frac{1}{2}\Delta_i$ for all i's , and is essentially similar with D_o when they are larger for all i's . One could think for instance to

$$D_1 = \sum_{i=1}^{N} w_i\, p_i\, d_i\ \ (e_i\ ,\ \mathfrak{m}_i(c)) \tag{4.2}$$

with

$$p_i = (2\, d_i/\,\Delta_i)\ (\,1 + 4\, d_i^{\,2}\,/\,\Delta_i^{\,2}\,)^{-\frac{1}{2}} \tag{4.3}$$

or any other choice with the same qualitative properties. The w_i's are convenient numbers called the weights. One could also think, for euclidean distances, to the cost function

$$D'_2 = [\sum_{i=1}^{N} w_i \, p_i^2 \, d_i^2 (e_i , \mathfrak{M}_i(c))]^{\frac{1}{2}} \qquad (4.4)$$

Using a cost function has several consequences

(a) one is led to define a "best fit" as "the" parameter c which minimizes the cost function. It is then often preferred to set $p_i = 1$ in cost functions like (4.2) or (4.4), obtaining

$$D_2 = [\sum_{i=1}^{N} w_i \, d_i^2 (e_i , \mathfrak{M}_i(c))]^{\frac{1}{2}} \qquad (4.5)$$

because this form is more pleasant for numerical calculations (non vanishing gradient at the minimum). In many problems of interest, conditions are such that the minimizing parameter c exists and is unique. Thus, if we compare with the analysis above, where there are many possible true solutions (in the "hard" sense of error bounds), the best fit is only one of them. One can expect that it shifts to another one if new measurements are taken into account. But if the problem with perfect measurements has an unique solution, one expects that the shift is small, and tends to zero for large N . Things may be different for a problem that is intrinsically underdetermined (i e with many equivalent solutions, even with perfect measurements).

(b) When there is no true solution, the best fit usually remains defined. The corresponding parameter c is the one which yields results that are the closest to the observed ones. Following several autors, we use the word "quasisolution" to denote the minimizing parameter.

(c) So as to plainly understand the interest of quasisolutions in the problems that are overdetermined only by excess of measurements, with unavoidable errors, we come back to the definition of $\mathcal{E}$ as a set of N-vectors, on which we use D_2 as a distance and we assume that the set of parameters is compact. Then let us quote the following theorem (due to Tichonov).If the equation $\mathfrak{M} c = e$ has on the compact $\mathcal{C}$ one solution and one only for any e in $\mathfrak{M}\mathcal{C}$, and if there exists for any e a unique point Pe in $\mathfrak{M}\mathcal{C}$ such that $d_{\mathcal{E}} (e , \mathfrak{M}\mathcal{C}) = d_{\mathcal{E}} (Pe , \mathfrak{M}\mathcal{C})$, then the quasisolution of $\mathfrak{M} c = e$ is unique and depends continuously on e.

Hence we see that even for these overdetermined problems, the method of quasisolution yields results, in cases of interest, that are essentially those of a well-posed problems. But what happens if the problem is also intrinsically underdetermined ? Clearly we need to have a control on the choice of solutions, because if we do not, this freedom may result in instabilities. The most simple way to do it is to design a cost function that contains also the distance between the parameter and a reference parameter - e.g. 0 if $\mathcal{C}$ is a normed space - or any a priori value c_o of c which is very likely because of other informations. A quasisolution will be have to minimize for instance

$$D_2(\gamma) = \sum_{i=1}^{N} w_i \, d_i^2 \, (e_i \, , \mathfrak{m}_i(c)) + \gamma^2 \, d_{\mathcal{C}}^2 \, (c \, , c_o) \tag{4.6}$$

Where γ is a parameter that controls the importance of the a priori value.

The interest of a cost function like this is that, in most cases of interest, the corresponding quasisolution depends continuously on the results, in other words is stable under small modifications of data. When the exact solutions were lacking or not stable, one says that the problem has been regularized. There are standard methods to regularize ill-posed problems. One can find for instance a very good study of them in the treatise by A. Tichonov, V. Arsénine quoted in § 1.

However, many problems still are open. They are concerned on one hand in the application of standard methods, or their adaptation, to particularly difficult problems, on the other hand in the difficulties of fixing questions of physical interest, for instance justifying the regularization on physical grounds, or classifying equivalent solutions.

All these questions are met in applied inverse problems, which are certainly the largest stock for examples of ill-posed problems. The first kind of problems has been studied very much in connection with the radiative transfer equation, the Fredholm equations which appear in various sounding, the problems of deconvolution, of analytic continuations, of Laplace or Mellin transforms inversions. In the present book, the lectures by Atkinson, Bertero et al. , Roger, deal with them.

The second kind of problem has been particularly studied in linear inverse problems. We gave in § 3 some references of reviews on this subject. Let us say first that the algebraic methods yield a very good description of the "generalized inverses" of a mapping $\mathfrak{m}$. Following Lanczos, one can show the importance of the eigenvalue of MM*. One uses to divide them in three sets - according to their effect on the results - the ones which are important, the ones which are zero and which can be eliminated by an algebraic trick, and the ones which are "unimportant" for the results, but which in the inversion are responsible for instabilities. The algebraic way of reducing their effect is essentially equivalent to constructing a quasisolution with the cost function $D_2(\gamma)$, γ^2 being called in this language the Marquaralt parameter and being essentially what is asymptotically added to the "unimportant" eigenvalues for regularization. What is weak in algebraic methods, in the point of view of physicists, is their apparent arbitrariness. For this reason, many of them have introduced or used a stochastic analysis of the equation $\mathfrak{m}\ c = e$, which is viewed as a representation of the equality

$$\mathfrak{m}\ p_c + p_n = p_e \tag{4.7}$$

where p_c is a stochastic process describing the model parameters, p_n is a noise process describing errors, p_e is the data process. In the linear case, the Wiener-Kolmogorov theory can be used to determine the optimum estimate of the model parameters. The result again coincides with the quasisolution that minimizes $D_2(\gamma)$, γ giving a measurement of the noise and, by the way, a physical parameter to classify "equivalent" solutions. The algebraic and the stochastic method were not invented for applied inverse problems but they certainly were very much applied and refined for them. In the present book, the lectures by Courtillot et al. and by D.D. Jackson show their importance. However there also exists a method of analysis which has been completely conceived and studied for linearized inverse problems. It is the famous Backus-Gilbert method, in which a very simple idea yields an excellent way of obtaining both a good quasisolution and an appraisal of the extent of non uniqueness. Since its creation (Backus G. Gilbert F. The Resolving Bower of Gross Earth data Geophys. J. R. Astr. Soc. 16 , 169-205 (1968)), this method has been reviewed by many autors, including me, and it has been exposed several times at our meetings. But since many good reviews are available, we do not give any lecture on it in the present book, and I suggest the reader to refer to the paper quoted above. For more information on this method as well as on the others, he also can see the review papers quoted in § (3.2).

Section 5 - Side applications of inverse problems.

Inverse problems define a method to go from datas to parameters. In certain cases, this method, completed by certain a priori assumptions, works even when an uncomplete set of datas is known. Hence it can be used to interpolate or extrapolate the datas. Eventually, the extrapolation can go up to points which are not reached by measurements, solving, in fact, another inverse problem. We give here one lecture on this side application, by Courtillot et al. Needless to say, the weak point in this kind of processing is in the a priori assumptions which are necessary. But when one compare results with experiments, one should admit that it is of real interest.

A new field of applications of inverse problems has been open in the last ten years : the exact solution of non linear evolution equations. The principle of the "inverse method" is not complicated. Suppose we know exact solutions of an inverse problem like those of § 2.3 - 2.4 . Thus the direct problem is solving the equation

$$(\mathcal{D} + V(\underset{\sim}{x})) \, \psi(\underset{\sim}{x}) = 0 \qquad (5.1)$$

completed by conditionsthat ψ must satisfy on a surface Σ . On this same surface (which can be at ∞), one makes certain measurements, which yield an information s, and one knows how to construct V from s .

Now if V is transformed by some operator $\mathfrak{T}$ depending on one parameter t ("time evolution") in such a way that the structure of the problem is preserved (for instance the fact that V should vanish at infinity, etc), and if the corresponding "data" $s(t)$ evolve according to a simple formula, one can easily calculate $V(t)$ by solving the inverse problem $s(t) \rightarrow V(t)$. So as to get this situation, one uses to write conditions for the evolution of ψ that are sufficient to guarantee :

(a) that there exists $V(t)$ such that $\mathcal{D} + V(o) \rightarrow \mathcal{D} + V(t)$ and satisfying the other conditions of (5.1).

(b) that the evolution of $s(t)$ can be written down explicitly.

Since $\mathcal{D}$ usually is a differential operator, it is not surprising that these conditions imply that $V(\underset{\sim}{x},t)$ obeys a non linear partial integrodifferential equation. The surprising point is that one can easily write down conditions that correspond to the transformations of V that are implied by well known partial differential equations in mathematical physics. I have given elsewere a unified treatment of the classical inverse problems in view of these applications (P.C. Sabatier Inverse Scattering Problems for Nonlinear Applications in Proceedings of the Advanced Study Institute on Nonlinear Equations in physics and mathematics, A.O. Barut Ed. Reidel 1978). In the present book, two lectures are concerned with this subject. I. Miodek tries to give an introduction for non-specialists. Calogero and Degasperis give us a deep review of their very powerful wronskian method. On the way, these lectures give good reference lists of comprehensive treatments of the subject. At least one excellent reference however is missing, which we quote here :

M.Y. Ablowitz Lectures on the Inverse Scattering Transform. Studies in Applied Mathematics 58 , 17-94 (1978).

The side applications of inverse problems to solving non linear equations has given an extraordinary "Bain de Jouvence" to theoretical studies of inverse problems. It has justified the interest of scientists in constructing new methods of solution for them (see for example the lectures by Cornille and by Karlsson). It has also shown how these problems are related to so various branches of mathematics. Inverse methods are wide generalizations of using integral transforms to solve differential equations. their relations with the theory of integral equations are obvious, as well as with harmonic analysis. But there is more. As a recent example, Sturm Liouville inverse problems, which could be considered a couple of years ago as a closed subject, were recently shown to be related to the study of periodic solutions of Korteweg de Vries equations, and, through it, to very abstract mathematics like algebraic geometry.

Conclusion

At the time we write this book, theoretical and applied inverse problems are together a justification for a hundred of papers a year, going from very naive to very sophisticated mathematics, everyday experimental work to very abstract speculations. A few years ago, we were convainced that an unified treatise on the theory and applications of inverse problems was possible. Now we realize that it is no longer possible . The aim of this book is only to show the reader a number of directions for research. We hope he will see clearly how the subject is open, and why it is fascinating.

Acknowledgements

Let me first thank the "Centre National de la Recherche Scientifique", who supported this scientific activity and all our meetings through the organization called

"Recherche Coopérative sur Programme n° 264 : Etude Interdisciplinaire des Problèmes Inverses". The value of our meetings is, of course, due to all the scientists who joined with us. The list of our speakers is given in the tables at the end of the book. Those who attended are not listed. However I wish to thank all of them and I know that they will all join with me to recognize the work of our meetings secretary, Mrs Albernhe, who was also responsible for the Bibliography and all practical details.

PART I

FIFTEEN REVIEW LECTURES
ON APPLIED INVERSE PROBLEMS

RAY THEORETICAL INVERSE METHODS IN GEOPHYSICS

B.L.N. KENNETT

Department of Applied Mathematics and Theoretical Physics,
University of Cambridge, Silver Street,
Cambridge CB3 9EW

Summary:

Inverse methods based on ray theory have received considerable development in geophysical applications. Methods now exist for the direct inversion of exact data and the generation of extremal bounds on a solution once the errors in the data are taken into account. Linearised inverse methods may also be used and allow a treatment of the resolution attainable from the observed data.

INTRODUCTION

The study of wave propagation by ray tracing in geophysics has been extensively used in both underwater acoustics and seismology and also in ionospheric research. Inversion techniques based on ray methods have been particularly developed in seismology in an effort to determine the elastic wave speed distribution within the Earth from observations of the arrival times of waves generated by earthquakes and explosions.

In an elastic body, like the Earth, both compressional (P) waves and transverse (S) waves can be propagated. In the very high frequency geometrical ray theory limit these waves can be assumed to travel independently and interconversions only occur at the major discontinuities of structure e.g. the core-mantle boundary. Most work on seismic inversion has concentrated on the P-wave distribution since these waves arrive first (since they have higher wave speed) and are thus most easily distinguished. This allows the arrival times of the P-wave to be determined quite accurately and once the origin time for the source in known, the transit time from source to receiver (the 'travel time'), can be determined. For an earthquake source the determination of the source position and time origin requires the solution of a further inverse problem based on previous knowledge of travel times.

The first approach to travel time inversion for exact seismic data was due to Herglotz (1907) and Wiechert (1910) which was applied to Mohorovičić (1910) in his classic study of the 1909 Kulpatal earthquake, which led to the seismic definition of the Earth's crust. Further development of the technique were made by Slichter (1932) but it was left to Gerver & Markusevitch (1966) to achieve the extension of this approach to the case where velocity inversions occur, i.e. the velocity distribution is not monotonic with depth.

Subsequently inverse methods, which can take account of the errors inherent in seismic data, have been devised to look at the class of acceptable velocity distributions which are compatible with the data. The envelope of possible velocity models can be determined by the methods of McMechan & Wiggins (1972) and Bessonova et al (1974, 1976). A more direct approach which has been adopted by Johnson & Gilbert (1972) who have used linear inverse methods to examine the resolution available from the available data.

Constructive solutions to ray theory inverse problems are available for spherically symmetric or horizontally layered velocity distributions, i.e. where the velocity depends on a single coordinate. Comparatively little progress has been made on ray inverse problems where the velocity depends on more than one spatial coordinate. Belonosova & Alekseev (1967) have however established the existence of an inverse solution for complete exact data when the velocity depends on only two coordinates in a flat geometry.

For a spherically symmetric velocity model the rays will lie in a plane and thus within a circular section. The ray problem can then be conveniently studied in a two dimensional flat geometry of a half space $(z > 0)$ with x as a horizontal coordinate by making a conformal transformation. This 'Earth flattening transformation' takes the form

$$\begin{aligned} x &= R\Delta , \\ z &= R \ln (R/r), \\ \alpha(z) &= \alpha(r)\,(R/r), \end{aligned} \tag{1}$$

where R is the radius of the Earth, Δ the angular distance and α the elastic wavespeed. To simplify the mathematical discussions we will restrict our discussion to a horizontally stratified medium.

RAYS IN HORIZONTALLY STRATIFIED MEDIA

The generalisation of Snell's law for a continuously varying seismic velocity distribution is that the ray parameter

$$p = \sin \vartheta(z) / \alpha(z),$$

is constant along the ray path. Here $\vartheta(z)$ is the local inclination to the vertical and $\alpha(z)$ the local velocity. The ray path will be symmetric about the point where the ray is travelling horizontally (fig. 1), i.e. at the depth $Z(p)$ such that

$$p = 1 / \alpha(Z(p)) .$$

We shall term this level the 'turning point' depth.

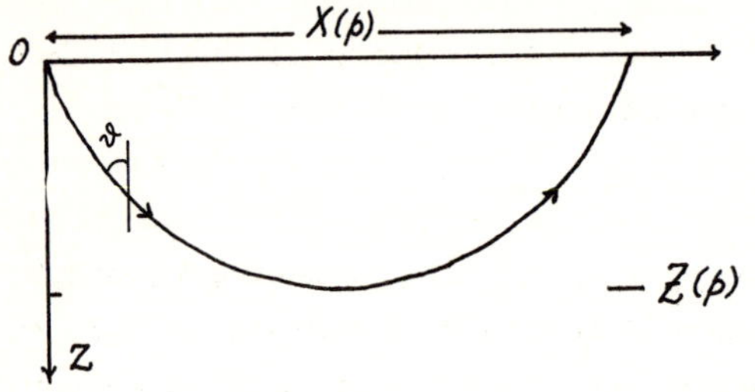

FIG 1

In a small interval ds along a ray, the time increment is related to the depth increment by

$$dt = ds/\alpha = dz / (\alpha \sec\vartheta),$$

and the horizontal increment

$$dx = ds \sin\vartheta = dz \tan\vartheta .$$

Thus in the passage from a surface source to a surface receiver we may parametrise the range $X(p)$ and the travel time $Z(p)$ in terms of the ray parameter p. Thus the range

$$X(p) \;=\; 2p \int_0^{Z(p)} \alpha(z)\,[1 - p^2\alpha^2(z)]^{-1/2}\,dz, \quad (2)$$

and the travel time

$$T(p) \;=\; 2 \int_0^{Z(p)} \alpha^{-1}(z)\,[1 - p^2\alpha^2(z)]^{-1/2}\,dz.$$

Rather than work in terms of velocities $\alpha(z)$ it is convenient to introduce the slowness distribution

$$a(z) \;=\; 1/\alpha(z)$$

and then

$$X(p) \;=\; 2p \int_0^{Z(p)} [a^2(z) - p^2]^{-1/2}\,dz$$

$$T(p) \;=\; 2 \int_0^{Z(p)} a^2(z)\,[a^2(z) - p^2]^{-1/2}\,dz. \quad (3)$$

The ray parameter p also has the significance of the slope of the travel time curve $T(X)$

$$p \;=\; dT/dX.$$

We may also consider the intercept time for the local tangent to the travel time curve

$$\tau(p) \;=\; T(p) - p\,X(p) \;=\; 2\int_0^{Z(p)} [a^2(z) - p^2]^{1/2}\,dz \quad (4)$$

The derivative of $\tau(p)$, in the case of a velocity distribution which increases with depth, is simply related to the range

$$d\tau/dp \;=\; -\,X(p)$$

Whilst the travel time curve $T(X)$ is not necessarily single valued, the $\tau(p)$ relation has the advantage of being almost everywhere single valued and monotonically decreasing with increasing. For example, we consider a reasonably strong velocity gradient

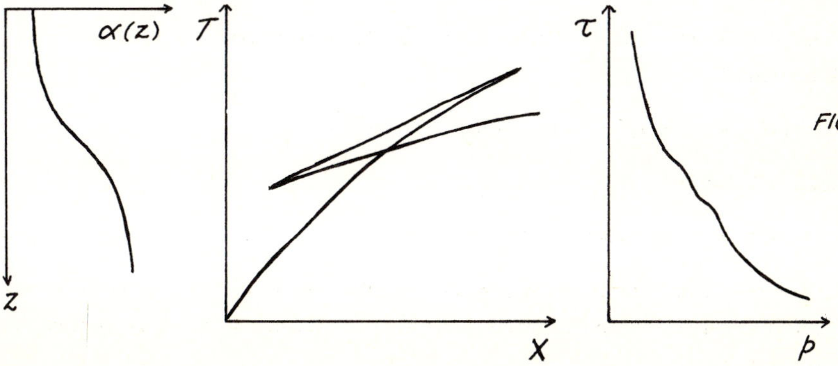

FIG 2

The effect of the gradient is to generate a triplication in the travel curve due to clustering of turning points in the region of the gradient. The $\tau(p)$ relation on the other hand is single valued and the details of the triplication are reflected by the slight structure in the curve.

In the case of a velocity inversion with increasing depth (a 'low-velocity zone') geometrical ray theory leads to the prediction of a shadow zone since no rays can have their turning points in the low velocity material. Diffraction effects will, of course, lead to the presence of energy on the real seismograms in the shadow zone.

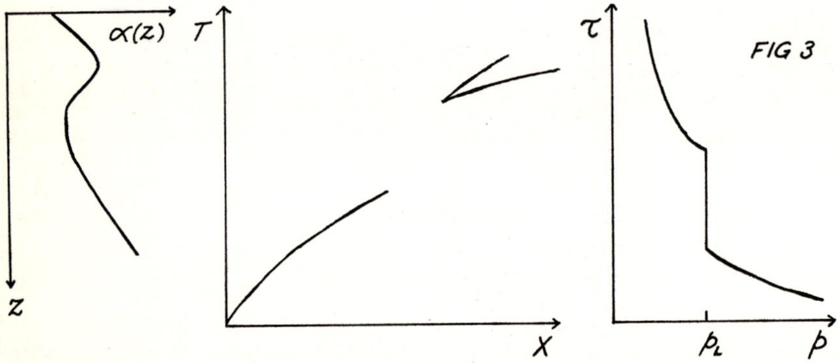

FIG 3

The $\tau(p)$ relation in this case shows a discontinuity at the ray parameter corresponding to the slowness at the lid of the low velocity zone.

The interaction of rays with both a velocity inversion and a velocity gradient are well illustrated in fig. 4.

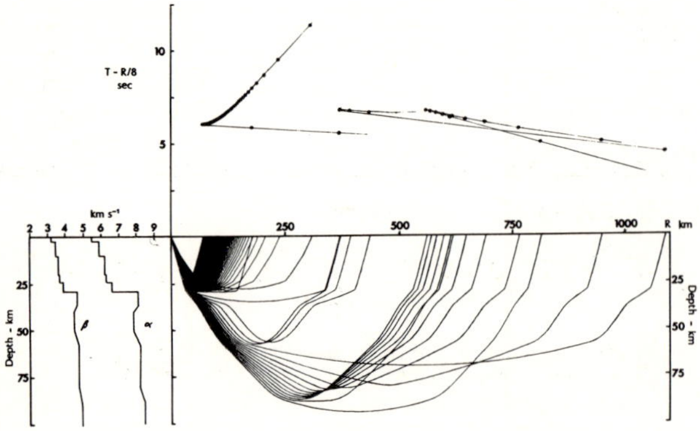

FIG 4

INVERSION OF EXACT DATA

i) In the absence of low velocity zones:

(Herglotz (1907), Wiechert (1910)).

We assume that the range $X(p)$ is known for all ray parameters $p_D < p < a_0$ where a_0 is the surface slowness; then we may determine the velocity distribution down to the depth $Z(p_D)$.

We use the known integral

$$\pi/2 = \int_\alpha^\beta d\xi \; \xi \, [\beta^2 - \xi^2]^{-1/2} [\xi^2 - \alpha^2]^{-1/2}$$

and for a particular ray parameter q we set

$$Z(q) = \int_0^{Z(q)} dz = \frac{2}{\pi} \int_0^{Z(q)} dz \int_q^{a(z)} dp \; p \, [a^2(z) - p^2]^{-1/2} [p^2 - q^2]^{-1/2} \quad (5)$$

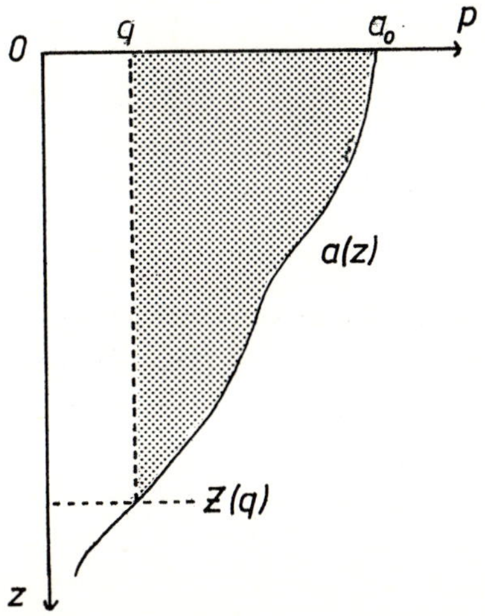

FIG 5

The order of integration may be changed since the function is integrable over the domain (denoted by shading in fig. 5) and thus

$$\begin{aligned} Z(q) &= \frac{1}{\pi} \int_q^{a_0} dp \, [p^2 - q^2]^{-1/2} \int_0^{Z(p)} dz \; 2p \, [a^2(z) - p^2]^{-1/2} \\ &= \frac{1}{\pi} \int_q^{a_0} dp \; X(p) \, [p^2 - q^2]^{-1/2} = \mathcal{Z}(q) \end{aligned} \quad (6)$$

This expression for the turning point depth is due to Herglotz and Wiechart, though derived here by a rather different approach. From the depth $Z(q)$ we may recover the velocity from

$$q = 1/\alpha(Z(q))$$

ii) In the presence of low velocity zones;

(Gerver & Markusevitch 1966).

In this case we need to define the parameters of the regions of velocity inversions in those regions where

$$a(z) > \inf\{ a(y) : 0<y<z \} = b(z)$$

For the kth low velocity zone we define $\underline{z}_k$ as the top of the zone and $\bar{z}_k$ as the lower limit of the zone. Within this zone

$$a(z) > p_k = a(\underline{z}_k)$$

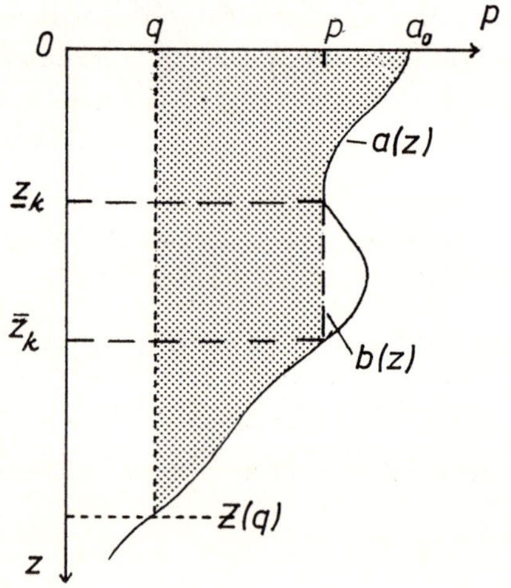

FIG 6

We may once again express the turning point depth $Z(q)$ in terms of the double integral (5) but now we may only change the order of integration over the domain between $b(z)$ and the ray parameter q . (fig.6). Thus we need to take account of additional contributions from all velocity inversions such that $p_k > q$, and so we have

$$Z(q) = \mathcal{Z}(q) + \frac{2}{\pi}\sum_k \int_{\underline{z}_k}^{\bar{z}_k} dz \int_{p_k}^{a(z)} dp\, p\, [a^2(z)-p^2]^{-1/2}[p^2-q^2]^{-1/2} \quad (7)$$

and thus

$$Z(q) = \frac{1}{\pi}\int_p^{a_0} dp\, X(p)\, [p^2-q^2]^{-1/2}$$

$$+ \frac{2}{\pi}\sum_k \int_{\underline{z}_k}^{\bar{z}_k} dz\, \tan^{-1}\{[a^2(z)-p^2]^{1/2}/[p_k^2-q^2]^{1/2}\} \quad (8)$$

The contribution from the low velocity zones is always positive so that in general

$$\mathcal{Z}(q) > \hat{\mathcal{Z}}(q).$$

However below the first low velocity zone the turning point depth for a given ray parameter depends on the nature of the velocity distribution within the low velocity zones as well as the range information $X(p)$. We thus have a fundamental nonuniqueness in the inversion of the travel time curves even for exact data.

No rays have their turning points in the regions of velocity inversion and thus without using further information, e.g. the amplitudes on the seismograms, we are unable to use the available data to control the velocity distribution in this region. This leaves us with a considerable degree of arbitrariness in the choice of $a(z)$ in these low velocity zones and indeed any equally measurable slowness distributions within the zones are equivalent.

THE TAU METHOD (Bessonova et al 1976)

The function $\tau(p)$ was introduced earlier and many recent inverse techniques have been developed to exploit the convenience of the single valued behaviour with ray parameter.

i) In the absence of low velocity zones:

We start from the Herglotz-Wiechert form for the turning point depth (6)

$$\mathcal{Z}(q) = \frac{1}{\pi}\int_q^{a_0} dp\, X(p)\, [p^2-q^2]^{-\frac{1}{2}}$$

and use $d\tau/dp = -X(p)$ to obtain

$$\mathcal{Z}(q) = \frac{1}{\pi}\int_{a_0}^{q} dp\, \tau'(p)\, [p^2-q^2]^{-\frac{1}{2}}$$

Now changing the variable of integration to τ we obtain

$$\mathcal{Z}(q) = \frac{1}{\pi}\int_0^{\tau(q)} d\tau\, [\nu^2(\tau)-q^2]^{-\frac{1}{2}} \qquad (9)$$

where we have written $\nu(\tau)$ for the inverse function to $\tau(p)$. We are therefore able to recover the velocity distribution directly from the $\tau(p)$ relation for exact data.

ii) In the presence of low velocity zones:

As we have already seen, the velocity distribution within a low velocity zone is inaccessible to travel time analysis. We will therefore make the specific assumption that the slowness $a(z)$ decreases with depth within the velocity zone (fig. 7), i.e. that there is a discontinuity in velocity at the top of a zone and that $\alpha(z)$ then increases with depth. Corresponding to these portions of the velocity distribution we introduce $\mathcal{Z}_k^*(p)$ the inverse function to $a(z)$ in the kth low velocity zone.

We define $\underline{z}_k$ and $\bar{z}_k$ as the depth of the top and bottom of the kth

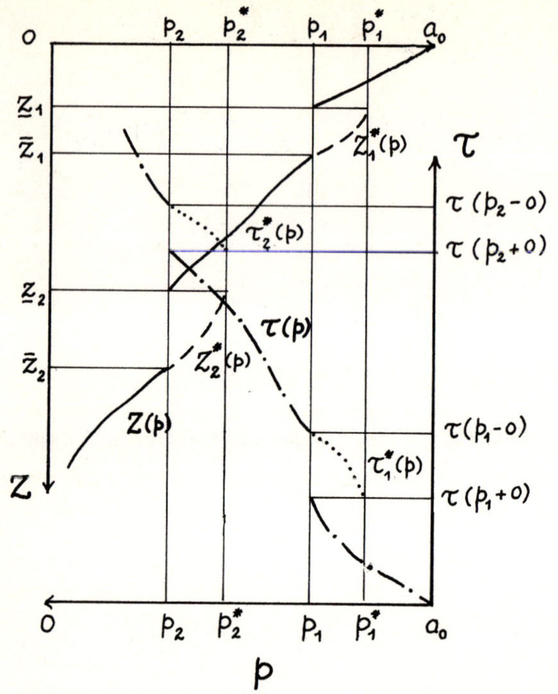

FIG 7

low velocity zone and define the two ray parameters

$$p_k = a(\underline{z}_k - 0),$$

$$p_k^* = a(\underline{z}_k + 0).$$

The equation (5) for the turning point depth can be rewritten in the form

$$\begin{aligned} Z(q) &= \frac{2}{\pi}\int_0^{Z(q)} dz \int_q^{a(z)} dp\; p\,[a^2(z)-p^2]^{-\frac{1}{2}}\,[p^2-q^2]^{-\frac{1}{2}} \\ &= \frac{1}{\pi}\int_q^{a_0} dp\,[p^2-q^2]^{-\frac{1}{2}} \int_0^{Z(p)} dz\; 2p\,[a^2(z)-p^2]^{-\frac{1}{2}} \\ &+ \frac{1}{\pi}\sum_k \int_{p_k}^{p_k^*} dp\,[p^2-q^2]^{-\frac{1}{2}} \int_{\underline{z}_k}^{Z_k^*(p)} dz\; 2p\,[a^2(z)-p^2]^{-\frac{1}{2}}, \end{aligned} \qquad (10)$$

where once again we have changed the order of integration and the summation is to be taken over all low velocity zones such that $p_k > q$.

Now the range as a function of ray parameter is given by

$$X(p) = 2\int_0^{Z(p)} dz\; p\,[a^2(z)-p^2]^{-\frac{1}{2}}$$

and by analogy we introduce

$$X_k^*(p) = 2 \int_{\underline{z}_k}^{z_k^*(p)} dz \; p \, [a^2(z) - p^2]^{-1/2}$$

and thus

$$\mathcal{Z}(q) = \frac{1}{\pi} \int_q^{a_0} dp \; X(p) \, [p^2 - q^2]^{-1/2} + \frac{1}{\pi} \sum_k \int_{p_k}^{p_k^*} dp \; X_k^*(p) \, [p^2 - q^2]^{-1/2} \tag{11}$$

By making a suitable extension of the definition of $\tau(p)$ to correspond to the intervals of low velocity zones we are able to cast this equation into the form (9) which it possesses in the absence of velocity inversions.

We introduce the continuous function

$$\hat{\tau}(p) = \tau(p) - \sum_{p_k > p} \sigma_k \; .$$

where σ_k is the discontinuity in $\tau(p)$ across the kth low velocity zone, and then

$$d\hat{\tau}/dp = - X(p)$$

so that we may write

$$\int_q^{a_0} dp \; X(p) \, [p^2 - q^2]^{-1/2} = \int_0^{\hat{\tau}(q)} d\hat{\tau} \; [v^2(\hat{\tau}) - q^2]^{-1/2} \tag{12}$$

where $v(\hat{\tau})$ is the inverse function to $\hat{\tau}(p)$

For each low velocity zone we introduce the extension $\tau_k^*(p)$, denoted by the dashed segment in fig. 7, and defined as

$$\tau_k^*(p) = \tau(p_k + 0) + \int_{\underline{z}_k}^{z_k^*(p)} dz \; [a^2(z) - p^2]^{1/2}$$

so that $\tau_k^*(p)$ is continuous and almost everywhere satisfies

$$d\tau_k^*/dp = - X_k^*(p)$$

Thus we have

$$\int_{p_k}^{p_k^*} dp \; X_k^*(p) \, [p^2 - q^2]^{-1/2} = \int_{\tau(p_k + 0)}^{\tau(p_k - 0)} d\tau_k^* \; [v^2(\tau_k^*) - q^2]^{-1/2} \tag{13}$$

where now $v(\tau_k^*)$ is the inverse function to $\tau_k^*(p)$. We note that equation (12) may alternatively be written as

$$\int_q^{a_0} dp \; X(p) \, [p^2 - q^2]^{-1/2} = \sum_k \int_{\tau(p_{k-1} - 0)}^{\tau(p_k + 0)} d\hat{\tau} \; [v^2(\hat{\tau}) - q^2]^{-1/2} \tag{14}$$

with $p_0 = a_0$. Inserting the expression (13) and (14) into equation (11) we recover a representation of the turning point depth $\mathcal{Z}(q)$ in terms of τ analogous to equation (9), i.e.

$$\mathcal{Z}(q) = \frac{1}{\pi} \int_0^{\tau(q)} d\tau \; [v^2(\tau) - q^2]^{-1/2} \tag{15}$$

where $\nu(\tau)$ is a single valued function determined for all τ : within the intervals $(\tau(p_{k-1}-0), \tau(p_k+0))$ $\nu(\tau)$ is inverse to $\tau(p)$, and within the intervals $(\tau(p_k+0), \tau(p_k-0))$ $\nu(\tau)$ is inverse to $\tau_k^*(p)$

Thus for exact data we may invert $\tau(p)$ information with specific assumptions as to the form of any low velocity zones.

INVERSION OF INEXACT DATA (Bessonova et al 1974, 1976)

The formulation we have previously considered is appropriate for exact data, but once we take into account the errors in the travel time observations the best we can hope to do is to place bounds on the $\tau(p)$ relation

$$\underline{\tau}(p) \leq \tau(p) \leq \bar{\tau}(p)$$

and from this we can construct bounds on $Z(p)$

$$\underline{Z}(p) \leq Z(p) \leq \bar{Z}(p)$$

and thus place bounds on the velocity distribution.

The original method proposed by Bessonova et al (1974), which has been reviewed by Kennett (1976), is to use the Gerver & Markusevitch form of the inversion formula (8). With the assumption of a lower limit to the velocity in any velocity zone, upper and lower bounds on $Z(p)$ are constructed by averaging (8) over small intervals in ray parameter.

An improved method has recently been proposed by Bessonova et al (1976), but this appears to be more difficult to implement computationally.

For each low velocity zone we place a lower limit α_k on the minimum velocity within the zone, and set

$$r_k = 1/\alpha_k$$

for a surface velocity $\alpha_0 = 1/a_0$. From the observations we also have to define bounds on the ray parameter at the lid of each low velocity zone

$$\underline{p}_k \leq p_k \leq \bar{p}_k$$

and on the discontinuity in $\tau(p)$ at p_k

$$\underline{\sigma}_k \leq \sigma_k \leq \bar{\sigma}_k$$

and we may construct an upper bound on the width of the low velocity zone $(\bar{h}_k)$. This enables us to bound the extensions $\tau_k^*(p)$ for each low velocity zone by

$$\tau_k^*(p) \leq \bar{\tau}_k^*(p) = \bar{\tau}(\bar{p}_k+0) + \bar{\sigma}_k [r_k^2-p^2]^{\frac{1}{2}} [r_k^2-\bar{p}_k^2]^{-\frac{1}{2}}$$

$$\tau_k^*(p) \geq \underline{\tau}_k^*(p) = \underline{\tau}(\underline{p}_k-0) - \underline{\sigma}_k + [\underline{\sigma}_k^2 - \bar{h}_k^2(p^2-\underline{p}_k^2)]^{\frac{1}{2}} \quad (16)$$

and thus to construct bounds on the inverse function

$$\underline{\nu}(\tau) \leq \nu(\tau) \leq \bar{\nu}(\tau)$$

over the whole range (fig. 8). The solid portions of the bounds are those determined directly from the travel time observations, the dashed portions those obtained using (16) for the low velocity zones.

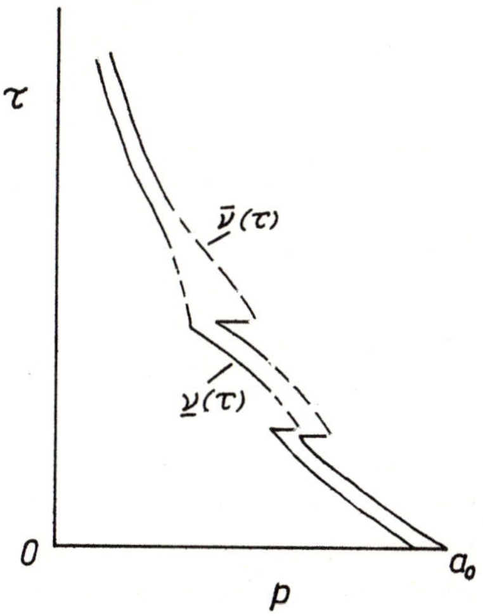

FIG 8

For a ray parameter q_0 we define

$$Z^\beta(p) = \frac{1}{\pi}\int_0^{\underline{\tau}(p)} [\underline{\nu}^2(\tau) - p^2]^{-\frac{1}{2}}\, d\tau,$$

$$Z^\gamma(p) = \frac{1}{\pi}\int_0^{\bar{\tau}(p)} [\bar{\nu}^2(\tau) - p^2]^{-\frac{1}{2}}\, d\tau, \qquad q_0 \leq p \leq a_0$$

and (17)

$$Z^\beta(p) = y \geq Z^\beta(q_0),$$

$$Z^\gamma(p) = Z^\gamma(q_0), \qquad 0 < p < q_0$$

We also introduce

$$\tau_y^\beta(p) = 2\int_0^{Z^\beta(p)} [a_\beta^2(z) - p^2]^{\frac{1}{2}}\, dz,$$

$$\tau^\gamma(p) = 2\int_0^{Z^\gamma(p)} [a_\gamma^2(z) - p^2]^{\frac{1}{2}}\, dz, \qquad 0 \leq p \leq a_0 \tag{18}$$

where $a_\beta(z)$, $a_\gamma(z)$ are the inverse functions to $Z^\beta(p)$, $Z^\gamma(p)$.

Within any low velocity zones we set

$$a_\beta(z) = \underline{a}_k = [\underline{p}_k^2 + \underline{\sigma}_k^2/\bar{h}_k^2]^{1/2} \quad \text{for } Z^\beta(\underline{p}_k+0) < z < Z^\beta(\underline{p}_k-0)$$

$$a_\gamma(z) = 1/\alpha_k \quad \text{for } Z^\gamma(\bar{p}_k+0) < z < Z^\gamma(\bar{p}_k-0)$$

and also

$$a_\beta(z) = q_0 \quad \text{for } Z^\beta(q_0) < z < y$$

We may note that $Z^\beta(p)$, $Z^\gamma(p)$ are constructed to only make use of the lower and upper bounds on τ , respectively, and by our method of construction

$$\begin{aligned} \tau_y^\beta(p) &= \underline{\tau}(p) \\ \tau^\gamma(p) &= \bar{\tau}(p) \end{aligned} , \quad q_0 \leq p \leq a_0 \tag{19}$$

and

$$\begin{aligned} \tau_y^\beta(p) &> \underline{\tau}(p) \\ \tau^\gamma(p) &< \bar{\tau}(p) \end{aligned} , \quad \text{in an interval } p < q_0$$

By using the extrapolated values of $\tau_y^\beta(p)$, $\tau^\gamma(p)$ for $p < q_0$ we can construct bounds on the turning point behaviour since

$$\begin{aligned} Z^\beta(p) &\geq Z(p) \\ Z^\gamma(p) &\leq Z(p) \end{aligned} , \quad \text{in an interval } p < q_0 \tag{20}$$

Thus we find the ray parameter q_1 for which $\tau_y^\beta(p)$ meets $\bar{\tau}(p)$ (fig. 9) and then

$$Z^\beta(q_1) \geq Z(q_0) \tag{21}$$

Similarly we find the ray parameter q_2 for which $\tau^\gamma(p)$ meets $\underline{\tau}(p)$ and then

$$Z^\gamma(q_0) \leq Z(q_2) \tag{22}$$

This enables us to set up an algorithm to determine the bounds on $Z(p)$

$$\underline{Z}(p) \leq Z(p) \leq \bar{Z}(p)$$

We step along the p axis from the surface slowness a_0 to p_D , i.e. the smallest ray parameter for which estimates of bounds on $\tau(p)$ are available. Then for each ray parameter q we find

$$M_k = \min_{p \in (p_D, q)} \left\{ \frac{\bar{\tau}(p) - \tau_y^\beta(p)}{[q^2 - p^2]^{1/2}} \right\}, \quad y = Z^\beta(q) \tag{23}$$

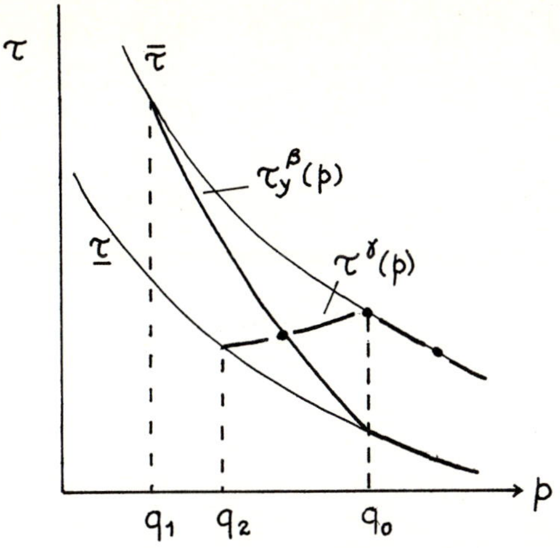

FIG 9

and

$$q^* = \max_{p \in (p_0, q)} \{ p : \underline{\tau}(p) \geq \tau^\gamma(p) \}$$

and from these values we may construct the estimators

$$\underline{Z}(q^*) = Z^\gamma(q)$$

and

$$\bar{Z}(q) = Z^\beta(q) + M_k$$

This allows us to draw the bounds on the turning point depth, which will be dependent on the choice for the lowest velocity within the low velocity zones. These bounds then define limits on the depth range at which a particular velocity can occur and thus bound the velocity distribution itself.

As an example we consider data from a long range refraction profile in France which was considered in the study of Kennett (1976). Fig. 10 shows the bounds on the $\tau(p)$ distribution for velocities greater than 7.8kms^{-1}, there is an indication of a low velocity zone for a p value of 0.123skm^{-1} (8.13kms^{-1}). The corresponding velocity bounds are also shown, the solid curves correspond to the assumption of no low velocity zone; whilst the dashed line is the revised shallower bound when the lowest velocity within the low velocity zone is constrained to 7.5kms^{-1}.

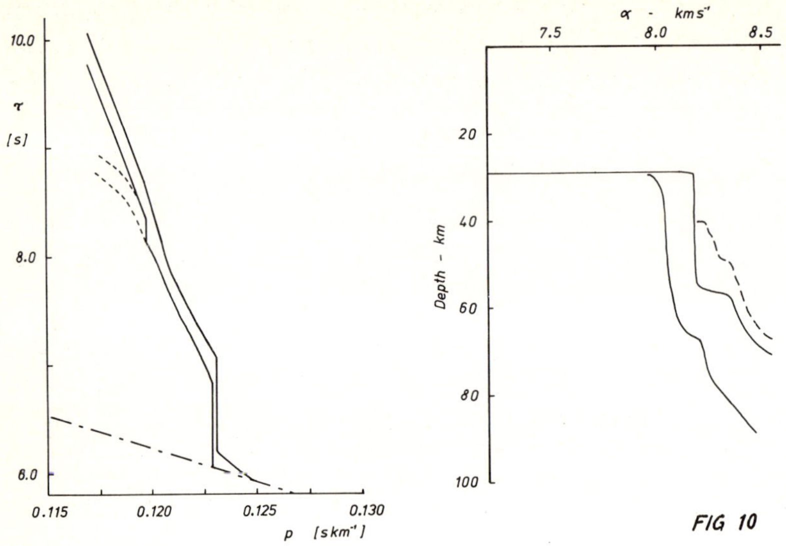

FIG 10

LINEARISED INVERSION

The convenience of the $\tau(p)$ representation for travel time data led Johnson & Gilbert (1972) to set up a linearised inversion scheme to recover the velocity profile. This approach has also been followed by Kennett (1976). We consider, the $\tau(p)$ function appropriate to surface source and receiver

$$\tau(p) = 2\int_0^{z_p} [a^2(z) - p^2]^{\frac{1}{2}}\, dz, \tag{24}$$

where z_p is equal to the depth of reflector for reflected waves and to the turning point $Z(p)$ for refracted waves. Although equation (24) is non-linear , a small perturbation δa will lead to a small change $\delta\tau$ in τ

$$\delta\tau(p) = 2\int_0^{z_p} \delta a(z)\, a(z)\, [a^2(z) - p^2]^{-\frac{1}{2}}\, dz. \tag{25}$$

If we consider a set of ray parameter values p_i , the corresponding perturbations $\delta\tau_i(p_i)$ will be linear functionals, to this level of approximation, so we may write

$$\delta\tau_i = \int_0^{z_M} G_i(z)\, m(z)\, dz, \tag{26}$$

where we have introduced the relative perturbation

$$m(z) = \delta a(z)/a(z) = -\delta\alpha(z)/\alpha(z).$$

The kernel G_i is given by

$$G_i(z) = 2a^2[a^2 - p^2]^{-1/2}, \quad 0 \le z < z_p,$$
$$= 0 \quad , \quad z \geq z_p .$$

We choose z_M to be the greatest depth to be considered in any velocity model.

From the observed travel time data we construct N 'observed' $\tilde{\tau}_i(p_i)$ with variances σ_i^2, and using the values $\tau_i(p_i)$ calculated from some test model we construct

$$\gamma_i = \tilde{\tau}_i - \tau_i , \qquad i = 1, 2, \ldots, N.$$

We then try to find a perturbation $m(z)$ to bring the model into agreement with the observations, i.e. we attempt to find $m(z)$ from the local linear approximation

$$\gamma_i - \sigma_i \leq \int_0^{z_M} G_i(z)\, m(z)\, dz \leq \gamma_i + \sigma_i , \qquad i = 1, 2, \ldots, N. \tag{27}$$

We attempt to satisfy these equations in a least squares sense, and need to remove the singularity in G_i by integrating by parts

$$\int_0^{z_M} G_i(z)\, m(z)\, dz = J_i(0)\, m(0) + \int_0^{z_M} J_i(z)\, m'(z)\, dz, \tag{28}$$

with

$$J_i(z) = \int_z^{z_p} G_i(s)\, ds ,$$

which corresponds to the travel time from z to the bottoming point z_p.

At each stage we seek the 'flattest' perturbation following Backus & Gilbert (1969), i.e. we seek the $m(z)$ which minimises

$$\frac{1}{2} \int_0^{z_M} [m'(z)]^2\, dz + \frac{1}{2} [m(0)]^2 \tag{29}$$

subject to the constraints imposed by the data (27). We introduce N Lagrange multipliers ν_j such that

$$m'(z) = \sum_{j=1}^{N} \nu_j J_j(z) , \qquad m(0) = \sum_{j=1}^{N} \nu_j J_j(0) , \tag{30}$$

and then we require

$$\gamma_i - \sigma_i \leq \sum_{j=1}^{N} A_{ij} \nu_j \leq \gamma_i + \sigma_i , \qquad i = 1, 2, \ldots N \tag{31}$$

with

$$A_{ij} = \int_0^{z_M} J_i(s)\, J_j(s)\, ds + J_i(0)\, J_j(0) .$$

Following Gilbert (1971) we construct linear combinations of the original γ_i by means of a matrix transformation T which ranks these linear combinations Γ_l in order of increasing relative standard error ξ_l. This transformation simultaneously diagonalises the covariance matrix of the original data E and the kernel matrix A. After transformation the constraints (31) become

$$\Gamma_l - \xi_l \leq N_l \leq \Gamma_l + \xi_l \tag{32}$$

where

$$\Gamma_l = T_{lj} \gamma_j ,$$

and the new vector $\mathcal{N}$ is related to the Lagrange multipliers by

$$\nu_j = T_{lj} N_l .$$

The minimisation criterion now becomes that we seek the minimum of

$$\frac{1}{2} \int_0^{z_m} [m'(z)]^2 dz + \frac{1}{2} [m(0)]^2 = \sum_l N_l^2 \tag{33}$$

subject to the new constraints (32). Thus we choose the N_l to be as small as possible subject to (32). To avoid excessive contamination of details of the model perturbation $m(z)$ by errors, we regularize the solution by rejecting Γ_l with relative standard errors above a certain threshold, i.e. the corresponding N_l are set equal to zero. The choice of threshold is somewhat arbitrary but can reasonably be set where the relative error is greater in magnitude than Γ_l .

Once the N_l have been determined we may determine the Lagrange multipliers ν_j and from this the model perturbation $m(z)$. This enables us to find a new model which then has to be tested against the original data and if the agreement with the $\tilde{\tau}_i$ is not satisfactory we adopt the new model as the test model and iterate the procedure described above until a 'solution' is found.

At each stage in the construction scheme we assume local linearity but the final model which we construct to be compatible with the observations is not necessarily linearly related to the starting model.

The iterative technique described above normally converges in three or four iterations from any reasonable starting model, e.g. one which lies within the bounds obtained from the tau inversion described before. However for situations with well developed low velocity zones, some problems with convergence can arise and it is convenient to constrain the linearised inverse by introducing an upper limit on the velocity at any depth obtained from the extremal tau analysis and an arbitrary, but non-zero, lower limit (Kennett 1976). These constraints may be conveniently applied by modifying intermediate models so that they be on or between the specified bounds.

MODEL RESOLUTION

At the present time we are only able to look at the intrinsic resolution within a model using the linear approximation. Thus we need to assume that the final model constructed by the scheme we have just described is linearly close to the 'real earth'. For this nonlinear problem the resolving power calculations may well be in error (Wiggins et al 1973), however, as noted by Kennett (1976), a comparison of the predicted resolution for a number of models can be very informative.

Backus & Gilbert (1968, 1970) have demonstrated that at best we can only obtain linear averages of the 'true model'. Thus, if we wish to obtain a good estimate of the 'true velocity' at z_0 from a final model $\alpha_0(z)$ we may construct

$$\langle \alpha(z_0) \rangle = \int_0^{z_M} A(z,z_0)\,\alpha_0(z)\,dz\,, \tag{34}$$

where $A(z,z_0)$ is to be built up from the original data kernels $G_i(z)$

$$A(z,z_0) = \sum_i \eta_i(z_0)\,G_i(z)\,. \tag{35}$$

We want $A(z,z_0)$ to be concentrated in the neighbourhood of z_0, if we are indeed to produce as good an estimator, and we measure the 'spread' of A by the quantity

$$s(z_0) = 12\int_0^{z_M} [H(z+z_0-z_M) - \int_{z_M}^{z} A(s,z_0)\,ds]^2\,dz \tag{36}$$

As a measure of the error made in constructing the linear average we use

$$\epsilon(z_0) = \sum_i \sum_j \eta_i(z_0)\,\eta_j(z_0)\,E_{ij}\,, \tag{37}$$

where E_{ij} is the covariance matrix of the observed $\tilde{\tau}_i(p_i)$ data. In terms of the $\eta_i(z_0)$ the spread takes the form

$$s/12 = \sum_i \sum_j \eta_i \eta_j S_{ij} - 2\sum_j \eta_j R_j + k\,, \tag{38}$$

with

$$S_{ij} = \int_0^{z_M} J_i(z)\,J_j(z)\,dz\,, \qquad R_j = \int_0^{z_0} J_j(z)\,dz$$

To obtain the best compromise between minimum spread and minimum error we seek to minimise a linear combination of the two with respect to the coefficients $\eta_j(z_0)$

$$K(z_0) = \mu\, s(z_0) + \epsilon(z_0) \tag{39}$$

which leads to the following equations for $\eta_j(z_0)$: $(\mu' = 12\mu)$

$$(\mu' S_{ij} + E_{ij})\,\eta_j = \mu' R_j \tag{40}$$

we may note that for fixed μ' the operator $(\mu' S_{ij} + E_{ij})$ is independent of z_0. If therefore we again simultaneously diagonalise S_{ij} and E_{ij} we can solve the set of equations (40) for all μ' and z_0 for the cost of a single diagonalisation. This substantially reduces the labour of calculating the complete 'spread-error' curve for each value of z_0.

The complete specification of the model resolution would require the presentation of spread-error curves for all values of depth. However for may purposes it is adequate to fix a compromise value of μ (e.g. the 'knee' of the spread-error curve at the closest approach to the origin) and to then consider the nature of the $A(z,z_0)$ functions with depth.

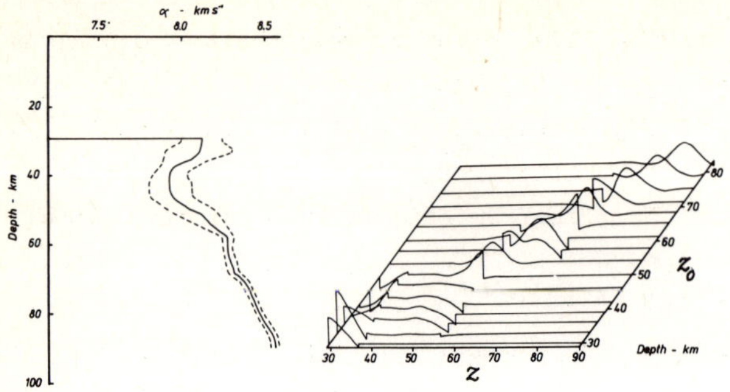

FIG 11

An example of a linearised inversion from τ data, taken from the study of Kennett (1976) is shown in fig. 11. The dotted bounds bordering the velocity model indicate the error estimate obtained from the resolution analysis. The projective plot shows the resolving kernels $A(z, z_0)$ plotted for a sequence of values of z_0 as a function of depth z. The resolution is found to be closely related to the density of ray turning points in the structure, for example within the pronounced low velocity zone no resolution is achieved as would be expected in a technique based on geometrical ray theory.

REFERENCES

Backus, G.E. & Gilbert, F., (1968) The resolving power of gross Earth data, Geophys. J. R. astr. Soc., 16, 169-205.

Backus, G.E. & Gilbert, F., (1969) Constructing P-velocity models to fit restricted sets of travel-time data, Bull. Seis. Soc. Am., 59, 1407-

Backus, G.E. & Gilbert, F., (1970) Uniqueness in the inversion of inaccurate gross Earth data, Phil. Trans. Roy. Soc., A266, 123-192.

Belonosova, A.V. & Alekseev, A.S., (1967) On a formulation of the inverse kinematic seismic problem for a two dimensional continuous inhomogeneous medium, in "Certain Methods and Algorithms for the Interpretation of Geophysical Data", Nauka, Moscow.

Bessonova, E.N., Fishman, V.M., Ryaboyi, V.Z. & Sitnikova, G.A., (1974) The tau method for the inversion of travel-times - I. Deep seismic sounding data, Geophys. J. R. astr. Soc., 36, 377-398.

Bessonova, E.N., Fishman, V.M., Johnson, L.R., Shnirman, M.G. & Sitnikova, G.A., (1976) The tau method for the inversion of travel-times - II. Earthquake data, Geophys. J. R. astr. Soc., 46, 87-108.

Gerver, M.L. & Markushevich, V., (1966) Determination of a seismic wave velocity from the travel-time curve, Geophys. J. R. astr. Soc., 11, 165-173.

Herglotz, G., (1907) Über das Benndorfsche Problem der Fortpflanzungsgeshwindigkeit der Erdbebenstrahlen, Phys. Z., 8, 145-147.

Johnson, L.E. & Gilbert, G., (1972) Inversion and inference for teleseismic ray data, in Methods in Computational Physics 12, ed. B.A. Bolt, 231-266.

Kennett, B.L.N., (1976) A comparison of travel time inversions, Geophys. J.R. astr. Soc., 44, 517-536.

McMechan, G.A. & Wiggins, R.A., (1972) Depth limits in body wave inversions, Geophys. J. R. astr. Soc., 28, 459

Mohorovicic, A., (1910) Das Beben vom 8.10.1909, Jahrb. Meteorol. Obs. Zagreb für 1909, Band 9, Teil 4, Abschn. 1.

Slichter, L.B., (1932) The theory of the interpretation of seismic travel-time curves in horizontal structures, Physics 3, 273-295.

Wiechert, E., (1910) Bestimmung der weges der Erdbebenwellen im Erdinnern, I, Theoretisches, Phys. Z., 11, 294-304.

Wiggins, R.A., McMechan, G.A. & Toksoz, M.N., (1973) Range of Earth structure non-uniqueness implied by body wave observations, Revs. Geophys. Sp. Phys., 11, 87-113.

INVERSE METHODS APPLIED TO CONTINUATION PROBLEMS IN GEOPHYSICS

by V. Courtillot, J. Ducruix and J. L. Le Mouël

Equipe de Géomagnétisme Interne

Laboratoire d'Etudes Géophysiques des Structures Profondes

(CNRS - LA 195)

Institut de Physique du Globe

2 place Jussieu 75230 Paris Cedex 05

November 1977

ABSTRACT

This paper focuses on the application of inverse methods to continuation problems in geophysics. It is divided in three sections. The first introductory section lists a number of continuation problems which are of interest to geophysicists and gives a brief review of earlier work on the subject. The second section is devoted to the solution we have developed in three papers from 1973 to 1975, which we have further generalized and termed a global inverse method. This method is equally applicable to two- and three-dimensional problems. A particular solution of the problem is found as a linear combination of continuation kernels expressed at the observation points (they generate a vector space E_n); this solution is acceptable i.e. consistent with the data. The general solution of the problem requires that solutions from the null space of the kernels be added to the particular solution belonging to E_n. The properties of the Gram matrix built from the continuation kernels are investigated. One important result provided by the global inverse method is that an analytic expression of the elements of this matrix is easily obtained when the concept of images is introduced. We also show how an approximate Green's function for the irregular surface over which the data points are distributed can be found. In the third section we compare the global inverse method with generalized inverse matrix theory. When the latter theory is generalized to the case when one dimension becomes infinite the two formalisms are equivalent. We show how the concepts of resolution and information density are related in the two methods and investigate the properties of the base of eigenfunctions of the problem. Expanding the solution over this base leads to the concept of ranking and winnowing (Gilbert, 1971) and allows the computation of various covariance matrices for model components. Also, in the case of continuation problems, the relationship between the two sets of eigenvectors which appear in the theory of generalized inverse matrices is made clear and is quite enlightening.

Acknowledgements : an early version of this paper was given at the seminar on inverse problems organized in late 1976 by professor P.C. Sabatier (CNRS RCP 264) and was published in french in the Ph.D. thesis of Courtillot (april 1977,pp. 6 to 25).

I - INTRODUCTION

Geophysicists are often faced with the problem of continuation of scalar or vectorial functions which have been measured at a number of points, along given lines or given surfaces of the three dimensional space. Most of the time, the problem can be reduced either to the continuation of a scalar harmonic potential ψ ($\nabla^2 \psi = 0$) or of the solution of the Helmholtz equation ($\nabla^2 \psi + k^2 \psi = 0$). We will now give a number of examples in which these problems are encountered (geomagnetism,gravimetry, heat flow, electromagnetism) ; this list is by no means complete.

I - 1 Computation of the geomagnetic potential

The main part of the geomagnetic field can be expressed, outside the Earth, as the gradient of a harmonic potential ($\vec{B} = -\vec{\nabla}\psi$, $\nabla^2 \psi = 0$). This field (its components) are measured either at the surface of the globe ($r = a$), or aboard artificial satellites with a low perigee. The question of continuation is different in both cases.

In the case where measurements are performed over the surface of the Earth, the problem is a classical and elementary one <u>if</u> one doesn't take into account the discrete distribution of data points : the sphere $r = a$ is a coordinate surface of the system of geocentric spherical coordinates. If, on the other hand, one measures $\vec{B}$ along low-altitude satellite trajectories, the problem is less classical ; moreover, one measures only the intensity B of the field $\vec{B}$ (we will come back to this later on in this paragraph) . One could then think of applying techniques related to the inverse problem, such as the ones we will introduce in this paper. Up to now, in practical cases, things have not been done in this way : let P_i ($i = 1, 2, \ldots, n$) be the data points. $\psi(P)$ is written in the form of a spherical harmonic expansion limited to order and degree m :

$$(1) \qquad \psi(P) = \sum_{i=1}^{m} \sum_{j=-i}^{i} f_i^j \left(\frac{a}{r}\right)^{i+1} Y_i^j(\theta, \phi)$$

where the Y_i^j are surface harmonics ; the coefficients f_i^j are computed so as to minimize the quadratic distance :

$$(2) \qquad \delta^2 = \sum_{i=1}^{n} |\vec{B}(P_i) + \vec{\nabla}\psi(P_i)|^2 ,$$

where the $\vec{B}(P_i)$ are the measured values.

This a brute force method ; it should not really be called an inverse method, like all adjustement (or fitting) methods that rely on least squares. However, it gives good results, which can easily be understood : the number of observations is enormous (hundred or thousands) compared to the number of unknown coefficients in the expansion, which is smaller than, say, 200 . Very similar problems are of course encountered in the study of the gravity potential (see for example Gaposchkin and Lambeck, 1970 ; Gaposchkin, 1974) .

Remark : aboard satellites one only measures the intensity B of $\vec{B}$, using nuclear or atomic resonance magnetometers ; the " distance " which one seeks to minimize is then :

$$(3) \qquad \delta^2 = \sum_{i=1}^{n} (B(P_i) - |\vec{\nabla}\psi(P_i)|)^2 ;$$

the f_i^j of (1) are computed through iterative minimization of δ^2 . This is a continuation problem which was studied by Backus (1968 , 1970 , 1974) ; Backus shows, among other results, that a knowledge of B over the sphere $r = a$ does not determine a unique continuation of $\vec{B}$ in outer space but that, under certain conditions which are always met in practice, a knowledge of B in an open part of outer space leads to a unique continuation of $\vec{B}$ in all outer space (except for a sign ambiguity). Backus notes that this result is not as trivial as might seem, being in particular false in two-dimensional space. All proofs imply that intensity measurements of B be error-free.

I - 2 Continuation of magnetic or gravity anomalies :

The problem of continuation of local or regional magnetic (or gravity) anomalies is a common concern of both theoretical and applied geophysicists. For example, a magnetometer measures a component E of the abnormal field created by magnetized rocks in the crust ; this magnetometer travels along a curve (C) which is kept as close as possible from the ocean bottom. This situation is encountered in practice in the case of "deep-tow" surveys (see Parker and Klitgord, 1972 and, more recently, Miller, 1977) . The problem is to determine E in the half space which has (C) as a lower bound, assuming a two dimensional geometry for the sources. A corresponding example in the three dimensional case is that of satellite data acquired at varying altitudes (Bhattacharrya, 1977) . We will come back in detail in paragraph II to the solution of this problem, which can be formulated as an inverse problem.

I - 3 Topographical correction of heat flow measurements :

One measures the vertical component of heat flow $-K\partial T/\partial z$ (K being the thermal conductivity, T the temperature, Oz the upward vertical) over part of the surface of the Earth. One of the many corrections which must be applied to the raw data, in order to free them from surface effects and allow one to compare them, is the topographic correction. This correction attempts to account for the distorsions which topography imposes on isothermal surfaces (see for example Courtillot and Francheteau, 1976) . The computation of this correction amounts to the solution of a Dirichlet problem, but with discrete data. It is of course analogous to the previous one (I - 2) and has been solved by Ducruix et al (1974 b) .

I - 4 Electromagnetic induction

The heat flow problem mentionned above is a steady state one. Now, if we consider a harmonic time-variation, the Laplace equation should be replaced by the Helmholtz equation. However, this last equation is more frequently encountered in electromagnetic induction problems, which can in some instances be stated as follows ; find a function ψ such that :

$$(4)\qquad \begin{cases} \nabla^2\psi + k_1^2\psi = 0 & P \in D_1 \\ \nabla^2\psi + k_2^2\psi = 0 & P \in D_2 = \complement D_1 \\ \psi \text{ and } \partial\psi / \partial n \text{ continuous across } (S) \text{, the boundary between } D_1 \text{ and } D_2 . \end{cases}$$

This is not strictly a continuation problem but it can be solved with the same formalism we will develop in this paper (see Le Mouël et al, 1975 and Menvielle, 1977) .

I - 5 "Non - inverse" solutions :

Contrary to what might be inferred from their apparent simplicity, problems I - 2 and I - 3 have been the subject of numerous, often careful and complex, studies. Such is the case for example for the heat flow topographic correction. Jobert (1960), for instance, has given an exact analytical solution in the case when (C) is a circular cubic with double point, a curve which can lead to reasonable topographic models.

The non inverse solutions to the various problems which we have summarized are many and we will not attempt to describe them all. One should of course mention purely numerical methods, or mesh methods (such as finite differences) ; these are not always convenient in the case where measurements are performed on very irregularly distributed points, which is often the case in geophysics. One must then use an interpolation procedure which leads to a new set of transformed data, this time on a regular grid. Let us remark in passing that the methods we introduce in this paper also provide a means of interpolation.

Much work has been devoted to these continuation problems in the Soviet Union ; we will briefly mention those papers which explicitely refer to geophysical problems. Strakhov and Devitsyn (1965) reduce the solution of the problem to that of a Fredholm integral equation of the first kind under the restrictive condition that a plane surface be found which both lie under the irregular surface over which measurements are performed and be above the sources of the potential. In other words, there cannot be any data point lower than the top of the sources. Polonskii (1966) succeeds in keeping the same formalism without this severe restriction : the only requirement is to find a surface which lie between the measurement surface and sources ; however, no practical method of computation is proposed. In order to do without (or almost) any hypothesis concerning the relative position of measurement points and sources, Tsirul'skiy (1968) expands the measured anomaly in a MacLaurin series as a function of altitude. He then truncates the expansion, computes its Fourier transform and reduces the problem to the solution of a Fredholm integral equation of the second kind, which is then solved by iteration (all this requires a few additional assumptions which the author doesn't think to be too much restrictive). One can mention a paper by Solov'ev (1967) whose presentation in terms of operators leads to an application of the method of successive approximations ; however, the method does not seem to have led to any practical development. All these methods stem from the Cauchy integral formula (they apply to the two-dimensional problem) . Strakhov has devoted a number of papers to that subject ; let us mention just one (Strakhov, 1972) in which he starts from a generalized Cauchy formula :

$$(5) \qquad f(s) = \frac{1}{2i\pi} \int_{\Gamma} \frac{f(\sigma)\ \phi'(\sigma)}{\phi(\sigma) - \phi(s)}\ d\sigma$$

$\phi(\sigma)$ is analytic and is chosen in such a way as to permit a transformation of the generalized Cauchy formula into an other analytic expression (a series or an integral) which itself can then be transformed in order to allow the numerical techniques of harmonic analysis to be used. These various transformations amount to conformal transformations, the choice of which is guided by intuition.

I - 6 "Inverse" solutions

Having set aside all methods which reduce the problem to finding the solution of an integral equation (such as those proposed by Strakhov), the number of papers devoted to methods which are strictly speaking inverse ones is very limited. A study of the Izvestia (Fisika Zemli series, english translation) from 1965 to 1975, of the Bulletin Signalétique du CNRS from 1969 to 1976 and a recent bibliography on inverse problems (RCP 264, 1976), only yields two papers that doubtlessly fall under our classification. One is by Savinskii (1967) and the other one by Parker and Klitgord (1972). This last paper is well known and in some respects resembles the paper published by Strakhov in the same year. To this list, we may add our contribution (Courtillot et al, 1973 ; Ducruix et al, 1974 ; Le Mouël et al, 1975). The present paper contains both a summary and a generalization of the method developed in these earlier publications.

We discovered the paper by Savinskii (1967) in the course of the bibliographical study which was intended for the present review. In its principle, Savinskii's method bears similarities with our own : Savinskii looks for a solution of the problem under the form of a linear combination of kernels of the integral continuation equation expressed at the n data points. He then builds the Gram matrix of these kernels, which is analougous to our (g_{ij}) matrices (see paragraph II - 2) , and even proposes to use only a sub-matrix of the Gram matrix in the final computation. This, of course, reminds one of the "ranking and winnowing" technique proposed by Gilbert (1971) . Let us mention the fact that Savinskii only studied the problem of downward continuation of a potential field measured on a (horizontal) coordinate line ; he did not, to our present knowledge, fully use the potential strength of his first ideas.

Let us now recall the method proposed in 1972 by Parker and Klitgord. They solve the problem stated in paragraph I - 2 (continuation from an uneven profile) ; the aim is to do without the hypothesis that no data point be lower than the highest source point, since this hypothesis is never satisfied in deep tow surveys. Their method is based on the Schwarz-Cristoffel transformation. The irregular data profile (C) is approximated by a polygonal line with vertices z_j (in the complex plane $z = x + iy$; see figure 2) . The authors next look for the conformal transformation $w = g(z)$ that transform this polygonal line into the real axis in the w plane (vertices z_j have images w_j) . The inverse transform $z = f(w)$ can be written :

$$f(w) = \int_{w_0}^{w} \prod_{j=0}^{n} (w' - w_j)^{-\theta_j/\pi} \, dw' \tag{6}$$

Angles θ_j between segments of the original profile (C) are shown in figure 2 . The w_j are computed in order that transformation $z = f(w)$ map the real axis ($\mathrm{Im}\, w = 0$) into profile (C) , with $z_j = f(w_j)$. This is done by a rapidly converging iterative technique. Once $f(w)$ has been computed, the inverse transformation g is computed from $(dg / dz)(df / dw) = 1$, where :

$$dg / dz = \prod_{j=0}^{n} (g - w_j)^{\theta_j/\pi} \tag{7}$$

Finally, in order to continue the measured values to horizontal profile (L) - see figure 2 - , the solution of differential equation (7) is computed at points z of (L) ; one thus obtains a transformed profile (L') in the w-plane. There only remains to continue a harmonic function from axis Im(w) = 0 to profile (L') and this is a classical problem.

Let us notice that the method cannot be extended to the three-dimensional problem. In what follows, we describe a method which we have developed in the last years (Courtillot et al, 1973 ; Ducruix et al, 1974 ; Le Mouël et al, 1975) which is applicable to three-dimensional problems. However, the presentation given here is more general than in earlier papers and we try to point out clearly the relationship between our method and a generalized version of the theory of generalized inverse matrices, along the lines of our 1975 paper.

II - A GLOBAL INVERSE METHOD

In what follows, we focus on the continuation of a scalar function which is a solution of the Laplace or the Helmholtz equation. Both equations govern a large number of phenomena which are of interest to geophysicists since the wave equation, the heat flow equation, the induction equation can be reduced to a Helmholtz equation in the case of harmonic time-variations. The gravity and static geomagnetic fields are of course governed by the Laplace equation.

Let P_i ($i = 1, 2, \dots, n$) be a set of points where a function $\psi(P)$ which is a solution of the Laplace or Helmholtz equation has been measured. We want to compute a continuation of ψ in a domain containing points P_i. We will not attempt to be more precise in defining this domain : this is easily done in each particular case.

II - 1 The inverse problem approach :

Let (S) be a coordinate surface $w = w_o$ in a curvilinear coordinate system (u, v, w) in which the Helmholtz equation is separable ; (S) is chosen so as to be "close" from points P_i. Here again there is no need to be more precise ; a choice is easily made in each particular case. There are eleven separable systems for the three dimensional Helmholtz equation but, for obvious geophysical reasons, we will be more interested in the cartesian, the spherical and the ellipsoïdal systems.

We know the Green's function $\mathcal{F}$ for surface (S) and differential operator ∇^2 (or $\nabla^2 + k^2$), both in the case of the (internal or external) Dirichlet or Neumann problem. For example, in the case of the external Dirichlet problem (figure 3) :

$$(8) \qquad \psi(P) = -\frac{1}{4\pi}\int_S \frac{\partial \mathcal{F}}{\partial n_M}(P,M)\,\psi(M)\,dS_M$$

$$= \frac{1}{4\pi}\int_S G(P,M)\,\psi(M)\,dS_M$$

$$= (G,\psi) \qquad P \text{ being outside } (S)$$

with $G(P,M) = -\frac{\partial \mathcal{F}}{\partial n_M}(P,M)$, $\vec{n}_M$ being the outward normal to (S) at M . The third equality defines the scalar product of two functions on (S) . In order to simplify equations, we will assume that all functions are real, but the formalism is equally valid in the case of complex functions (imaginary k^2) .

Let $\psi(M) = \psi(u,v,w_o)$ be the function of two coordinates (u,v) which is generated by values of ψ for points over (S) - in the following, we will always denote points on (S) by the letter M. In general, (S) does not contain any source (point source, linear or surface source) of ψ nor does it cross regions containing volume sources of ψ ; in short, we will assume that (S) does not include any singularity of ψ, and that $\psi(M)$ belongs to the set of functions which are continuous (and everywhere bounded) on (S) : $\psi \in C_S$.

We will first compute a continuation $\psi(M)$ for points of (S) from measured values $\psi(P_i) = \gamma_i$. We have n equations :

(9) $(\psi, G(P_i, M)) = (\psi, G_i) = \gamma_i$

$i = 1, 2, \ldots, n$

When no two points P_i coincide, the $G(P_i, M)$ are linearly independant functions of M, at least mathematicaly, if not always in numerical practice. They generate a vectorial subspace E_n of C_S (if no P_i is on (S)).

We will compute a continuation $\psi_n(M)$ belonging to E_n (an acceptable model in the words of Backus and Gilbert) such that $(\psi_n, G_i) = \gamma_i$. Let (g_{ij}) be the matrix made up of scalar products of the G_i ; it is also the Gram matrix of these vectors. Also, (g_{ij}) is the twice covariant metric tensor associated to the vector base (G_i), or again the matrix of the first fundamental quadratic for relative to base (G_i). It is a symmetrical (or hermitian in the complex case) definite positive matrix :

(10) $(G_i, G_j) = g_{ij}$

Let (g^{ij}) be the corresponding twice contravariant tensor : $g_{ij}\, g^{jk} = \delta_i^k$. The solution ψ_n can be written :

(11) $\psi_n(M) = g^{ij}\, \gamma_j\, G_i(M) = g^{ij}\, \gamma_j\, G(P_i, M)$

We can present this reasoning in another way, directly inspired from the Backus and Gilbert formalism (Backus and Gilbert, 1967, 1968, 1970). Recall that these authors look for what they call a smoothed model, here a smoothed function $\psi_B(M)$, which yields a smooth (averaged) value of the true model ψ in the neighbourhood of each point M_o. In order to do this, they compute a unimodular kernel $A(M, M_o)$ which be as close as possible from a Dirac distribution. Their choice of A leads to a model which is not acceptable. It is in fact simple to obtain a kernel which leads to an acceptable smoothed model.

Thus, let δ_S be the Dirac distribution over (S) :

(12) $\forall \psi \in C_S \ (\delta_S(M, M_o), \psi(M)) = \psi(M_o)$

Let us compute the projection of δ_S in E_n :

(13) $\delta_n(M, M_o) = g^{ij} G_i(M) G_j(M_o)$

δ_n is the Dirac distribution for all functions of E_n and we can use it as the kernel A (it is the Dirichlet kernel) :

(14) $$\psi_B(M_o) = (\delta_N, \psi) = g^{ij} (G_i, \psi) G_j(M_o)$$
$$= g^{ij} \gamma_i G_j(M_o)$$

This is indeed the same result as that given by equation(11); δ_n is a projection operator which projects functions of C_S in E_n. We will come back in II - 5 to the null space (or again kernel) Ker (δ_n) of this operator. We will also show in III-3 that $\delta_n(M, M_o)$ is the resolution "matrix" of the generalized inverse matrix theory (Lanczos, 1961 ; Bjerhammar, 1973) .

II - 2 Computation of the (g_{ij}) matrix. Continuation formula.

Let us recall the explicit formulation of equation (10) :

(15) $$g_{ij} = \frac{1}{4\pi} \int_S G(P_i, M) \, G(P_j, M) \, dS_M$$

In a number of cases this integral is readily evaluated, which is an essential feature of our methód. Such is in particular the case for the Laplace equation in cartesian, spherical or ellipsoïdal coordinates. In that case the function $G(P, Q) = -\frac{\partial \mathcal{T}}{\partial n_Q}(P, Q) = -\frac{1}{h_3(u,v,w)} \left.\frac{\partial \mathcal{T}}{\partial w}\right|_{(u,v,w)=(u_Q,v_Q,w_Q)}$ (h_3 being the third scale factor) is not an harmonic function of Q anymore. But a function H (P , Q) can be found which is harmonic in Q and such that $H(P, Q) \equiv G(P, Q)$ when $Q = M$

is on (S) . Furthermore, P_i has an image $\overline{P_i}$ with respect to (S) , such that $H(P_i, M) = H(\overline{P_i}, M)$ (figure 3) . As a function of Q, $H(\overline{P_i}, Q)$ is a solution of the Laplace equation with a singularity at $Q = \overline{P_i}$, which is now inside (S) . It can be continued using equations (8) and (15) :

$$(16) \qquad g_{ij} = \frac{1}{4\pi} \int_S G(P_i, M)\, H(\overline{P_j}, M)\, dS_M$$

$$= H(\overline{P_j}, P_i)$$

Indeed, $H(\overline{P_j}, M)$ can be continued outside (S) - in the present case of an external Dirichlet problem - which is not the case for $H(P_j, M)$.

Moreover, we obtain at once not only the continuation of ψ over (S) (i.e. the surface function $\psi_n(M)$) , but also the continuation of ψ in a whole neighbourhood of (S) and points P_i :

$$(17) \qquad \overline{\psi}(Q) = g^{ij}\, \gamma_j\, H(\overline{P_i}, Q)$$

This function coincides with ψ_n when $Q = M$ is on (S) and is a solution of the Laplace equation ; it is acceptable, which means that $\overline{\psi}(P_i) = \gamma_i$. Equation (17) completely solves the problem. In the appendix we give a few examples of choices of function H and computation of the g_{ij} .

(17) is valid, in the case we are interested in here, in the whole space outside (S) (figure 3) . $\overline{\psi}(Q)$ also exists inside (S) as long as $Q \neq \overline{P_i}$ and even there may give a good approximation of the true function ψ if sources of ψ are located "below" (with respect to coordinate w) image points $\overline{P_i}$. An obvious example is the downward continuation of gravity or magnetic anomalies below the measurement surface. Furthermore, ψ as given in (17) depends on the choice of the reference surface (S) , that is on w_o . It may be interesting to explore the family of functions $\overline{\psi}$ obtained by varying w_o . Indeed, a choice of w_o is equi-

valent to the choice of a physical filter applied to ψ . The farther away (S) is from (Σ) , the stronger the corresponding smoothing of ψ .

Le Mouël et al (1975) have shown how the problem can be solved when (S) and (Σ) intersect.

Remark : Determination of an approximate Green's function for surface (Σ):

Suppose we want to compute the Green's function for the external Dirichlet problem (figure 3) for surface (Σ) , on which points P_i lie. Equation (17) implies that function :

$$\mathcal{H}(P_i, Q) = g^{ij} H(\overline{P_j}, Q) \tag{18}$$

tends towards the normal derivative of the Green's function for the external Dirichlet problem and surface (Σ) as the density of points P_i increases (with a multiplicative factor $-1/4\pi$) . We will not attempt here to be precise concerning the nature of this convergence.

In the same way we have introduced the concept of an approximate Green's function for surface (Σ) , we will soon encounter the corresponding concept of approximate elementary functions for the Dirichlet problem for surface (Σ) (paragraph III-3) .

II - 3 The null space and the general solution on (S) :

The general acceptable solution is readily built from the particular solution $\psi_n(M)$ which has minimum norm :

$$\psi_S(M) = \psi_n(M) + (\delta_S - \delta_n, f) \tag{19}$$

where δ_S is the Dirac distribution on (S) , δ_n the function $g^{ij} G_i(M) G_j(M_o)$ (equation 13) and $f(M_o)$ any continuous (arbitrary) function of C_S . One can easily check that :

$$(\psi_S, G_i) = \gamma_i \tag{20}$$

Equation (19) generalizes to the continuous case the formula by Bjerhammar (1973) which gives the general solution of a system of linear equations using his generalized inverse matrix formalism (see also Courtillot et al, 1974; Le Mouël et al, 1975, and paragraph III of this paper). Thus, at least in theory, we can have an idea of what the null space Ker $\{G_i\}$ of kernels (G_i) looks like by computing $(f - f_n)$, where f_n is the projection of f in E_n using projector δ_n, and this for a number of functions f belonging to C_S. One can add to the minimal norm solution $\psi_n(M)$ any function $(f - f_n)$ from the null space in order to meet some particular condition when required. Parker and Huestis (1974) use a similar procedure when they add to a particular solution of the problem what they call an annihilator in order to obtain a solution with zero mean. We will not discuss the problem further here.

III - COMPARISON OF OUR FORMALISM WITH THE GENERALIZED INVERSE MATRIX FORMALISM

III - 1 Generalized eigenvector analysis and the Lanczos inverse :

Let us first recall the formalism of generalized eigenvector analysis as proposed by Lanczos (1961) and exposed by Jackson (1972) with only a slight change in notation. Suppose m unknowns ψ_i are related to a set of n data γ_i through n linear relations :

(21) $$\gamma_i = G_{ij} \psi_j$$

or (22) $$\Gamma = G \psi \qquad \Gamma \in \mathbb{R}^n \text{ and } \psi \in \mathbb{R}^m$$

Two sets of eigenvectors $\underline{u}_i$ and $\underline{v}_j$ may be found for the $n \times m$ matrix G of rank p :

$$\left\{\begin{array}{ll} (23) & G\,\underline{v}_i = \lambda_i\,\underline{u}_i \\ (24) & G^T\,\underline{u}_i = \lambda_i\,\underline{v}_i \end{array}\right.$$

leading to :

$$\left\{\begin{array}{ll} (25) & G^T G\,\underline{v}_i = \lambda_i^2\,\underline{v}_i \\ (26) & G G^T\,\underline{u}_i = \lambda_i^2\,\underline{u}_i \end{array}\right.$$

no summation $i = 1, \dots, p$

G can be factored into the product :

$$(27) \quad G = U \Lambda V^T$$

where U is an $n \times p$ matrix whose columns are the eigenvectors $\underline{u}_i$ $(i - 1, \dots, p)$, V an $m \times p$ matrix whose columns are the eigenvectors $\underline{v}_i$ $(i = 1, \dots, p)$ and Λ the diagonal matrix of eigenvalues λ_i $(i = 1, \dots, p)$. By the orthonormality of the eigenvectors :

$$(28) \qquad U^T U = I_p \quad \text{and} \quad V^T V = I_p$$

(27) and (28) imply that :

$$(29) \qquad G G^T = U \Lambda^2 U^T \quad \text{and} \quad G^T G = V \Lambda^2 V^T$$

one then introduces the inverse :

$$(30) \qquad K = V \Lambda^{-1} U^T$$

It is indeed a generalized inverse according to Bjerhammar's (1973) definition that $GKG = G$. The solution (the model) is given by :

$$(31) \qquad \hat{\psi} = K\Gamma = (KG)\psi \qquad ;$$

$R = KG = VV^T$ is called the resolution matrix ; the " theoretical " data corresponding to the model are :

$$(32) \qquad \hat{\Gamma} = G\hat{\psi} = (GK)\Gamma$$

$S = GK = UU^T$ is called the information density matrix. R and S are in general different from, respectively, I_m and I_n. We will be interested here in the strictly underdetermined case where $p = n < m$. In that case $S = I_n$ and :

$$(33) \qquad K = G^T (G G^T)^{-1} \quad ,$$

since $G G^T$ can now be inverted (cf 27 , 29 and 30) . This inverse is termed G_{IO}^{-1} by Bjerhammar.

III - 2 Generalization of the formalism of § III - 1

Equation (9) can be formally recast in the matrix notation (22) if $\psi(M)$ is considered as a vector belonging to a manifold (an abstract vector space) with infinite non denumerable dimensions. M can be viewed formally as a continuous index replacing the discrete index j of (21) . The implicit finite summation over the repeated index j should be replaced by the scalar product (8) . This approach is that advocated by Le Moüel et al (1975) and first used by Courtillot et al (1973) . In a recent review paper on inverse theory, Parker (1977) also follows that approach.

Let us write down the matrix dimensions as indices for clarity. The elements of matrix $G_{n\infty} G^T_{\infty n}$ of equation (29) are simply the g_{ij} of equation (10) and those of "matrix" $G^T_{\infty n} G_{n\infty}$ are given by $\sum_{i=1}^{n} G_i(M)\, G_i(M')$, with continuous "indices" M and M' (M and M' being two points of S) .

Equation (11) which yields our solution of the problem (our model) may now be formally written in matrix form :

$$(34) \qquad \psi_{\infty 1} = G^T_{\infty n} (G G^T)^{-1}_{nn} \, \Gamma_{n1}$$

This is identical to (31) with the special inverse (33) ; elements of $(G G^T)^{-1}$ are our g^{ij} and indeed $(G G^T)$ of (29) can be "naturally" inverted.

The eigenvectors $\underline{u}_i$ which belong to $\mathbb{R}^n$ have components u_i^k (these are the elements of matrix U^T). But the eigenvectors $\underline{v}_j$ now are functions $v_i(M)$ of C_S . Matrix equation (23) can be written explicitly with our conventions :

(35) $$\frac{1}{4\pi}\int_S G(P_k, M)\, v_i(M)\, dS_M = \lambda_i u_i^k$$

The scalar product on the left hand side of this equation defines the covariant components of the function $v_j(M)$ in the $\{\underline{G}_k\}$ base (see § II - 1). Matrix equation (24) can be written explicit :

(36) $$\sum_{k=1}^{n} G(P_k, M)\, u_i^k = \lambda_i v_i(M)$$

The quantities (u_i^k / λ_i) thus appear to be the contravariant components of $v_i(M)$ in the $\{\underline{G}_k\}$ base. We can say that the covariant components of $\underline{v}_i$ in the $\{\underline{G}_k\}$ base are equal to their contravariant components, within the constant multiplicative factor λ_i^2. They are naturally related through the metric tensor g_{ij} (see equation (10)) :

(37) $$g_{rs}(u_i^s / \lambda_i) = (\lambda_i u_i^r) \qquad \text{(no summation on } i\text{)}$$

or : $$g_{rs} u_i^s = \lambda_i^2 u_i^r$$

Equation (37) is nothing but equation (26) ; this illustrates how the formal identification between our formalism and the extension of generalized inverse matrix theory works. Some further consequences of this identification will now be investigated.

III - 3 Resolution and information density

According to (31) and (33) the resolution matrix is :

(38) $$R = VV^T = G^T(GG^T)^{-1}G$$

With our formal identification, this can be written explicitely :

(39) $$R(M, M') = \sum_{i=1}^{n} v_i(M)\, v_i(M')$$

$$= G(P_i, M)\, g^{ij}\, G(P_j, M')$$

This can also be demonstrated directly using (36) and (26). It shows that the resolution "matrix" is identical to the Dirac distribution δ_n for functions of E_n which we defined in equation (13).

Thus, although no "deltaness" criterion was required in order to build resolution "matrix" (or resolving kernel) R(M, M'), this kernel happens to be the most resolving, the most delta-like one and the geometrical reason for this has been made clear. As we have seen in the strictly underdetermined case we have considered so far, matrix U is orthogonal and $U^T U = U U^T = I_n$. Thus, the information density matrix reduces to the identity matrix and the model is always acceptable, which was shown in paragraph II - 1 ($\hat{\Gamma} = \Gamma$).

We have already seen from the second equation (28) that the $\underline{v}_i$, considered as functions of C_S, form an orthogonal set over surface (S) according to the definition (8) of the scalar product (see also Parker, 1977):

$$(40) \qquad \frac{1}{4\pi} \int_S v_i(M)\, v_j(M)\, dS_M = \delta_{ij}$$

Moreover the $v_i(M)$ can be continued outside (S) in the same way the $G_i(M)$ were continued in paragraph II - 2 (equation 16). These continued functions $v_i(P)$ happen to be also orthogonal over the set of observation points P_k (over the "surface" (Σ) of figure 3 as the number of points is increased to infinity - see the remark at the end of § II - 2). Indeed, according to (16) and (36):

$$(41) \qquad \sum_{k=1}^{n} v_i(P_k)\, v_j(P_k) = \sum_k \left(\frac{u_i^r}{\lambda_i} H(\overline{P}_r, P_k)\right) \left(\frac{u_j^s}{\lambda_j} H(\overline{P}_s, P_k)\right)$$

$$= \sum_k \left(\frac{u_i^r}{\lambda_i}\right) g_{rk} \left(\frac{u_j^s}{\lambda_j}\right) g_{sk}$$

which transforms, according to (37), to:

$$(42) \qquad \sum_{k=1}^{n} v_i(P_k)\, v_j(P_k) = \lambda_i \lambda_j \sum_k u_i^k u_j^k$$

$$= \lambda_i^2\, \delta_{ij}$$

since $U^T U = I_n$. These important properties of the $\underline{v}_i$, which may be considered as elementary functions of the problem, make them a particularly appropriate base to work in .

III - 4 Treatment of inaccurate data :

Let the set of data points P_i be given. In order to compute the continuation as given by equation (17), in principle one only needs to invert matrix $G G^T = (g_{ij})$. In general, $G G^T$ is a well conditionned matrix since, for a given value of index i , the g_{ij} , considered as a function of index j , look more or less like bell shaped curves which are shifted one with respect to another. The matrix is often easily inverted, without any need for particular precautions (e.g. Le Mouël et al, 1975) . However, if some data points are too close together or too far away from surface (S) , or again if the number n of data points is too high, matrix $G G^T$ may become numerically singular. In any case, there are many instances in which it may not be wise to proceed to a direct inversion.

We will assume that errors bearing on measurements γ_i performed at points P_i are independant, with variance σ^2 (more general cases in which σ^2 depends on i / or in which errors are correlated are easily solved) : the covariance matrix of the γ_i is then $B_\Gamma = \sigma^2 I_n$.

Let us now expand ψ in the base of orthonormal functions $v_i (M)$:

$$(43) \qquad \psi (M) = \gamma'^i v_i (M) \qquad \text{or} \quad \psi = V \Gamma'$$

The column vector Γ' of contravariant components γ'^i of ψ in the $\{\underline{v}_i\}$ base is related to the column vector Γ of covariant components γ_i in the $\{ \underline{G}_i \}$ base through :

$$(44) \qquad \Gamma' = (\Lambda^{-1} U^T) \Gamma$$

This is readily established with (30) , (31) and (43) . Thus the covariance matrix for Γ' is :

$$(45) \qquad B_{\Gamma'} = \Lambda^{-1} U^T B_\Gamma U \Lambda^{-1} = \sigma^2 \Lambda^{-2}$$

The orthogonality of the $\underline{v}_i$ leads to the property that the errors on the γ'^i are uncorrelated. The smaller the eigenvalues λ_i (associated to eigen-vector $\underline{v}_i$) , the larger the error on component γ'^i of ψ in the orthogonal base $\{\underline{v}_i\}$. Thus, in the expansion of ψ in the $\{\underline{v}_i\}$ base, in practical computations, only terms involving the q " largest " eigenvalues should be kept. Let us give a numerical example in the case of the downward continuation (for the Laplace equation) of data which have been measured on a horizontal line, one grid spacing apart. For n = 50 data points, the ratio of the largest to the smallest eigenvalue (which is called the condition number of matrix $G G^T$) is 20 for a continuation of half a grid spacing (h = 1/2) ; it is 100 when h = 1 and 10^4 when h = 1.5 .

This problem had already been mentionned in the geophysical literature by Savinskii (1967) . Gilbert (1971) attracted the attention of western geophysicists to this point in his often quoted paper on " ranking and winnowing " . The study was taken over by Wiggins (1972) , Jackson (1972) , Jordan (1973) and Parker (1977), among others.

So, the final model is a truncated one :

$$(46) \qquad \psi_q(M) = \sum_{\alpha=1}^{q} \gamma'^\alpha v_\alpha(M) \qquad\qquad q < n$$

and is not acceptable any more. But it is the model from space E_q (generated by the $\underline{v}_\alpha$, $\alpha \leqslant q$) which is closest, according to the L^2 norm, to the acceptable model ψ_n in E_n .

ψ_q can also be expanded in the $\{\underline{G}_i\}$ base (for in practice the $\underline{v}_\alpha$ are not necessarily computed) :

$$(47) \qquad \psi_q(M) = \sum_{i=1}^{n} \gamma''^i G_i(M)$$

Let us compute the components γ''^i.

From (27) , using again indices to denote matrix dimensions, we obtain :

$$(48) \qquad V_{\infty n} = G^T_{\infty n} \, U_{nn} \, \Lambda^{-1}_{nn}$$

(46) involves the matrix $V_{\infty q}$ built from the q first vectors $\underline{v}_\alpha$; using corresponding restrictions of U and Λ :

$$(49) \qquad V_{\infty q} = G^T_{\infty n} \, U_{nq} \, \Lambda^{-1}_{qq}$$

Let Γ'_{q1} be the column vector built from the q useful components of Γ'_{n1} and Γ''_{n1} be the column vector with components γ''^i. From the matrix forms of equations (46) and (47), using (49), we have :

$$(50) \qquad \Gamma''_{n1} = U_{nq} \, \Lambda^{-1}_{qq} \, \Gamma'_{q1}$$

The matrix of covariances of this vector is :

$$(51) \qquad B_{\Gamma''} = \sigma^2 \, U_{nq} \, \Lambda^{-4}_{qq} \, U^T_{qn}$$

since the restriction of (45) to Γ'_{q1} is $B_{\Gamma'_{q1}} = \sigma^2 \, \Lambda^{-2}_{qq}$.

We can also compute the norm of " error " $\psi_n - \psi_q$ as a function of observations γ_i :

$$(52) \qquad \| \psi_n - \psi_q \|^2 = \sum_{i=q+1}^{n} (\gamma'^i)^2 = \Gamma'^T_{1\,n-q} \, \Gamma'_{n-q\,1}$$

where :

$$(53) \qquad \Gamma'_{n-q\,1} = \Lambda^{-1}_{n-q\,n-q} \, U^T_{n-q\,n} \, \Gamma_{n1}$$

with obvious notations for block submatrices ($\Lambda_{n-q\,n-q}$ for example is the block of discarded eigenvalues λ_i , i = q+1 , ..., n). Thus :

$$(54) \qquad \| \psi_n - \psi_q \|^2 = \Gamma^T_{1n} (U_{n\,n-q} \, \Lambda^{-2}_{n-q\,n-q} \, U^T_{n-q\,n}) \, \Gamma_{n1}$$

which should be compared with :

$$(55) \qquad \| \psi_n \|^2 = \Gamma^T_{1n} (U_{nn} \, \Lambda^{-2}_{nn} \, U^T_{nn}) \, \Gamma_{n1}$$

Equation (54) gives the smallest possible distance (for the 2-norm) between models of E_q and the unique acceptable model in E_n. It is this distance which is minimized and not the squared two-norm misfit to the data $(\bar{\Gamma}-\Gamma)^T(\bar{\Gamma}-\Gamma)$ as is done in classical least squares methods (here $\bar{\Gamma}$ is the vector of theoretical data corresponding to the truncated model ψ_q of equation 46).

In the case the truncated model $\psi_q(M)$ is used in place of the acceptable model $\psi_n(M)$, on can define new resolution and information density matrices according to the lines of Jackson (1972). The new resolution "matrix" replacing (39) is not any more identical to δ_n and some resolution is lost. On the other hand, the information density matrix is not the trivial I_n and the model is not acceptable anymore. The misfit to the data is :

$$(56) \qquad \bar{\Gamma}_{n1} - \Gamma_{n1} = U_{nq}\,\Lambda_{qq}\,\Gamma'_{q1} - U_{nn}\,\Lambda_{nn}\,\Gamma'_{n1}$$

$$= -\,U_{n\,n-q}\,\Lambda_{n-q\,n-q}\,\Gamma'_{n-q\,1}$$

and

$$(57) \qquad \|\bar{\Gamma}-\Gamma\|^2 = \Gamma'^{T}_{1\,n-q}\,\Lambda^2_{n-q\,n-q}\,\Gamma'_{n-q\,1}$$

This equation was derived (with slightly different notations) by Parker (1977) who remarked that components γ'^i associated with the smallest (discarded) eigenvalues λ_i "contribute relatively little to the misfit compared with their contributions to the solution or its uncertainty" (compare (45), (51), (54) and (55) which involve Λ^{-2} or Λ^{-4} to (57) which involves Λ^2). The concept of data importance introduced by J.B. Minster et al (1974) in relation with the amount of information associated with each datum can also be used (this concept was further developed by J.F. Minster et al (1977). Of course the ranking and winnowing at the same time improve the statistical reliability of model components γ''^i and decreases resolution, with a classical trade off between the two (see for example Parker's (1977) review).

III - 5 Conclusion

In this paper, we have attempted to give a brief review of the work concerning the continuation of potential fields measured at a number of points distributed in a (possibly) irregular fashion over an irregular surface. We have recalled a method for solving the problem when it is formulated as an inverse problem (Courtillot et al, 1973 ; Ducruix et al, 1974). The method is equally applicable to two and three-dimensional space and to any coordinate system of geophysical interest (in which the Laplace and Helmholtz equations are separable) as shown by Le Mouël et al (1975) . We show that our formalism is strictly equivalent to that of generalized inverse matrices (e.g. Jackson, 1972) when the latter theory is itself formally extended to the case of models belonging to an infinite-dimensional vector space (Le Mouël et al, 1975 ; Parker, 1977) . We proceed to show which concepts correspond to resolution and information density in our formalism. In theoretical cases an acceptable model can be found which provides the best possible resolution. It is interesting to investigate also some properties of the base $\{ \underline{v}_i \}$ of eigenfunctions of the problem. Expanding the model over this base naturally leads to the idea of ranking and winnowing (Gilbert, 1971 ; Parker, 1977) , and allows the computation of various covariance matrices for model components. Also, in this particular continuation problem, the relationship between eigenvectors $\underline{u}_i$ and $\underline{v}_i$ of the theory of generalized eigenvectors analysis is made clear and is quite enlightening (§ III - 2) .

APPENDIX

In the appendix, we give a few examples of computation of the g_{ij} . These are the elements of the Gram matrix of vectors G_i are given by equation (10) and (15) (see II - 1 for the meaning of (S) , M , P , Q) .

a) Cartesian coordinates : we chose here (S) as the plane $z = 0$ the sources being in the lower half space $z < 0$. Then :

$$G(P_i, M) = \frac{1}{2\pi} \frac{z_i}{((x_i - x)^2 + (y_i - y)^2 + (z_i)^2)^{3/2}}$$

$$H(P_i, Q) = \frac{1}{2\pi} \frac{z - z_i}{((x_i - x)^2 + (y_i - y)^2 + (z_i - z)^2)^{3/2}}$$

Let $\overline{P_i}\,(x_i, y_i, -z_i)$ be the image of $P_i\,(x_i, y_i, z_i)$:

$$H(\overline{P_i}, Q) = \frac{1}{2\pi} \frac{z + z_i}{((x_i - x)^2 + (y_i - y)^2 + (z_i + z)^2)^{3/2}}$$

We have $H(\overline{P_i}, M) = G(P_i, M)$ and H is harmonic with respect to Q (and with respect to $\overline{P_i}$ in the present case) in the upper half space. Thus :

$$g_{ij} = H(\overline{P_j}, P_i) = \frac{1}{2\pi} \frac{z_i + z_j}{((x_i - x_j)^2 + (y_i - y_j)^2 + (z_i + z_j)^2)^{3/2}}$$

The element g_{ij} is given by the same expression as $G(P_i, M)$ using the following correspondance :

$$(x, y, 0) \longmapsto (x_j, y_j, -z_j)$$

b) Spherical coordinates : here (S) is the sphere $\rho = a$ and the sources are inside this sphere. We have :

$$G(P_i, M) = \frac{1}{4\pi a} \frac{\rho_i^2 - a^2}{(\rho_i^2 - 2\rho_i a \cos\gamma + a^2)^{3/2}}$$

with $P_i\,(\rho_i, \theta_i, \phi_i)$, $M(a, \theta, \phi)$ and $\cos\gamma = \cos(OP_i, OM)$

We chose :

$$H(P_i, Q) = \frac{1}{4\pi a}\left(\frac{\rho_i}{a}\right)\frac{\rho^2 - \rho_i^2}{(\rho_i^2 - 2\rho_i\rho\cos\gamma + \rho^2)^{3/2}}$$

where the coordinates of Q are (ρ, θ, ϕ). Let $\overline{P_i}(a^2/\rho_i, \theta_i, \phi_i)$ be the image of the point P_i then :

$$H(\overline{P_i}, Q) = \frac{1}{4\pi a}\left(\frac{a}{\rho_i}\right)\frac{\rho^2 - (a^2/\rho_i)^2}{((a^2/\rho_i)^2 - 2\rho(a^2/\rho_i)\cos\gamma + \rho^2)^{3/2}}$$

$$= \frac{1}{4\pi a}\frac{(\rho_i\rho/a)^2 - a^2}{((\rho_i\rho/a)^2 - 2\rho_i\rho\cos\gamma + a^2)^{3/2}}$$

$H(\overline{P_i}, M) = G(P_i, M)$ is satisfied and H is harmonic with respect to Q (but not harmonic with respect to $\overline{P_i}$ in this case) outside (S). Finally :

$$g_{ij} = H(\overline{P_j}, P_i) = \frac{1}{4\pi a}\frac{(\rho_i\rho_j/a)^2 - a^2}{((\rho_i\rho_j/a)^2 - 2\rho_i\rho_j\cos\gamma + a^2)^{3/2}}$$

Remark : contrary to the cartesian case, the fonction $H(P, Q)$ is different from the fonction $\frac{\partial}{\partial r_Q}\Gamma(P, Q)$ which can be considered as the natural continuation of $G(P, M)$.

Actually, obtaining the g_{ij} is simple only in the case of the Laplace equation. In the case of the Helmholtz equation the following procedure can be used. For example in order to solve the interior Dirichlet problem :

$$\begin{cases} \nabla^2\psi = \lambda\psi & \text{inside (D)} \\ \psi(M) = f(M) & \text{f being a given function on (C)} = \partial D \end{cases}$$

we write :

$$\psi = \psi_o + \lambda\,\psi_1 + \ldots + \lambda^n\,\psi_n ,$$

with :

$$\begin{cases} \nabla^2 \psi_o = 0 \quad \text{inside (D)} & ; \ \psi_o(M) = f(M) \ , \quad M \in (C) \\ \nabla^2 \psi_1 = \psi_o & ; \ \psi_1(M) = 0 \ , \quad M \in (C) \\ \cdot \\ \cdot \\ \cdot \\ \nabla^2 \psi_n = \psi_{n-1} & ; \ \psi_n(M) = 0 \ , \quad M \in (C) \end{cases}$$

ψ_o is computed using the global inverse method of section II. The result happens to have such a form that a particular solution of the Poisson equation $\nabla^2 \psi_1 = \psi_o$ can be found easily (Menvielle, 1977). Then the iteration method leading to ψ is straightforward : at each step there is an obvious particular solution of equation $\nabla^2 \psi_n = \psi_{n-1}$; the general solution is obtained by adding to the particular solution an harmonic function provided by our global inverse method.

REFERENCES

Backus, G.E., Determination of the external geomagnetic field from intensity measurements, Geophys.Res.Lett., 1, 21, 1974.

Backus, G.E., Non-uniqueness of the external geomagnetic field determined by surface intensity measurements, J.Geophys.Res., 75, 6339-6341, 1970.

Backus, G.E., Applications of a non-linear boundary-value problem for Laplace's equation to gravity and geomagnetic intensity surveys, Quart.J.Mech.Appl. Math., 21, 195-221, 1968.

Backus, G.E. and Gilbert, F., Uniqueness in the inversion of inaccurate gross Earth data, Phil.Trans.Roy.astr.Soc., A 266, 123-192, 1970.

Backus, G.E. and Gilbert F., The resolving power of gross Earth data, Geophys.J. Roy.astr.Soc., 16, 169-205, 1968.

Backus, G.E. and Gilbert F., Numerical applications of a formalism for geophysical inverse problems, Geophys.Roy.astr.Soc., 13, 247-276, 1967.

Bhattacharyya, B.K., Reduction and treatment of magnetic anomalies of crustal origin in satellite data, J.Geophys.Res., 82, 3379-3390, 1977.

Bjerhammar, A., Theory of errors and generalized matrix inverses, 420p , Elsevier, Amsterdam, 1973.

Courtillot, V., Sur l'analyse de certaines variations spatiales et temporelles du champ magnétique terrestre, Ph.D.thesis, Université Paris VII, 1977.

Courtillot V. et Francheteau, J., Géothermie ch.40, in Traité de Géophysique interne, vol.2, J. Coulomb et G. Jobert ed., 449-500, Masson, Paris, 1976.

Courtillot V., Ducruix, J. et Le Mouël J.L., Le prolongement d'un champ de potentiel d'un contour quelconque sur un contour horizontal, Ann.Géophys., 29, 361-366, 1973.

Ducruix J., Le Mouël, J.L. and Courtillot, V., Continuation of three dimensional potential fields measured on an uneven surface, Geophys.J.Roy.astr.Soc., 38, 299-314, 1974 a .

Ducruix J., Le Mouël, J.L. and Courtillot, V., Une méthode simple d'évaluation de la correction topographique dans les problèmes de flux de chaleur, C.R.A.S., B 278, 841-843, 1974 b .

Gaposchkin, E.M., Earth's gravity field to the eighteenth degree and geocentric coordinates for 104 stations from satellite and terrestrial data, J.Geophys.Res., 79, 5377-5411, 1974.

Gaposchkin, E.M., and Lambeck, K., Earth's gravity field to the sixteenth degree ans station coordinates from satellite and terrestrial data, J.Geophys.Res., 76, 4855-4883, 1970.

Gilbert, F., Ranking and winnowing gross Earth data for inversion and resolution, Geophys.J.Roy.astr.Soc., 23, 125-128, 1971.

Jackson, D.D., Interpretation of inaccurate, insufficient and inconsistent data, Geophys.J.Roy.astr.Soc., 28, 97-107, 1971.

Jobert, G., Perturbation du flux de chaleur dans la croûte terrestre due au relief, C.R.A.S., 250, 3209-3210, 1960.

Jordan, T.H. Estimation of the radial variation of seismic velocities and density in the Earth, thesis, Cal.Inst.Tech., 199 p., 1973.

Lanczos, C., Linear differential operators, D. Van Nostrand Co., London, 564 pp., 1961.

Le Mouël, J.L., Courtillot, V. and Ducruix, J., A solution of some problems in potential theory, Geophys.J.Roy.astr.Soc., 42, 251-272, 1975.

Menvielle, M., Induction électromagnétique et anomalies de conductivité : exemples des Pyrénées et du Maroc, doctorat de spécialité, Université P. et M. Curie, 1977.

Miller, S.P., The validity of the geological interpretations of marine magnetic anomalies, Geophys.J.Roy.astr.Soc., 50, 1-21, 1977.

Minster, J.B., Jordan, T.H., Molnar, P. and Haines, E., Numerical modeling of instantaneous plate tectonics, Geophys.J.Roy.astr.Soc., 36, 541-576, 1974.

Minster, J.F., Minster, J.B., Treuil, M., and Allègre, C.J., Systematic use of trace elements in igneous processes, Contrib.Mineral.Petrol., 1-29, 1977.

Parker, R.L., The inverse problem of electrical conductivity in the mantle, Geophys.J.Roy.astr.Soc., 22, 121-138, 1970.

Parker, R.L., Understanding inverse theory, Ann.Rev.Earth Planet.Sci., 5, 35-64, 1977.

Parker, R.L., and Huestis, S.P., The inversion of magnetic anomalies in the presence of topography, J.Geophys.Res., 79, 1587-1593, 1974.

Parker, R.L., and Klitgord, K.D., Magnetic upward continuation from an uneven track, Geophysics, 37, 662-668, 1972.

Polonskyi, A.M., The calculation of the anomalies ΔZ and Δg above a curved observational surface, Isv.Earth Phys., 11, 77-82, 1966.

R C P 264 (C N R S), Bibliographie du problème inverse, Laboratoire de Physique mathématique, Université des Sciences et techniques du Languedoc, Montpellier, 1972 and (revision) 1974.

Savinskii, I.D., Solution of an improperly posed problem in the continuation of a potential field to subjacent levels, Isv.Earth.Phys., 6, 72-92, 1967.

Solov'ev, O.A., Analytic continuation of potential fields by iteration, Isv. Earth.Phys., 4, 92-93, 1967.

Strakhov, V.N., Analytic continuation of two-dimensional potential fields into a region in the lower half-plane, Isv.Earth.Phys., 11, 38-55, 1972.

Strakhov, V.N. and Devitsyn, V.M., Reduction of the observed values of potential fields to a single level, Isv.Eart.Phys., 4, 60-72, 1965.

Tsirul'skiy, A.V., The reduction of observed potential fields to a single level, Isv.Earth Phys., 3, 85-89, 1968.

Wiggins, R.A., The general linear inverse problem : implication of surface waves and free oscillations for Earth structure, Rev.Geophys.Space Phys., 10, 251-285, 1972.

F I G U R E   C A P T I O N S

Figure 1 : A schematic example of a deep-tow profile. Observation points P_i are distributed along the irregular profile (C) which is kept as close as possible from the ocean bottom (shaded magnetic sources). The problem is to compute the anomalous magnetic field along the horizontal profile (L) (points M) from the knowledge of n values of the field at points P_i (after Parker and Klitgord,1972).

Figure 2 : The correspondance between the two complex planes of the z and w. The z-plane corresponds to the original observational space (see figure 1). The w-plane is obtained through the conformal mapping g(z) (see text; after Parker and Klitgord, 1972).

Figure 3 : The notations used in the text with the example of the exterior Dirichlet problem for surface (S). The observation points are the P_i lying on the irregular surface (Σ). They have images $\overline{P}_i$ with respect to (S) (coordinate surface $w=w_o$). Variable points are denoted by M when on surface (S) (outward normal $\underline{n}_M$) and by Q otherwise.

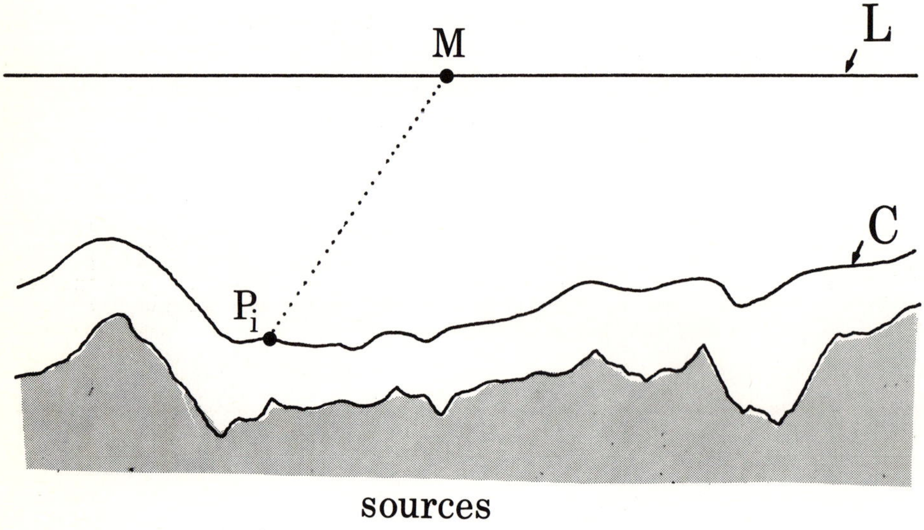

Figure 1 :

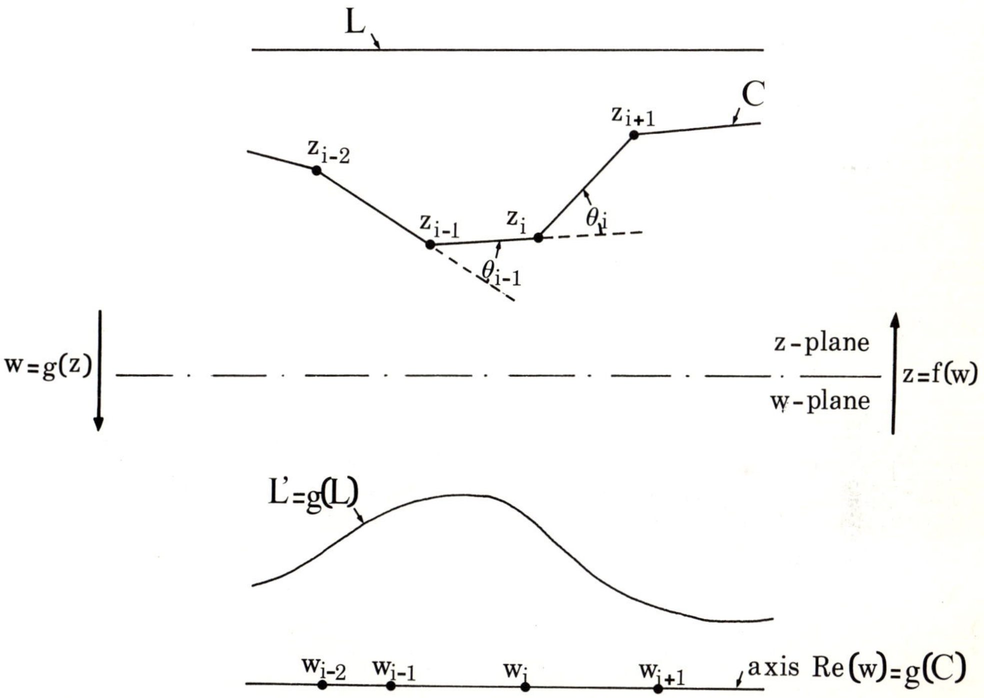
L
C
z_{i+1}
z_{i-2}
z_{i-1}
z_i
θ_i
θ_{i-1}
w=g(z)
z-plane
w-plane
z=f(w)
L'=g(L)
w_{i-2}
w_{i-1}
w_i
w_{i+1}
axis Re(w)=g(C)

Figure 2

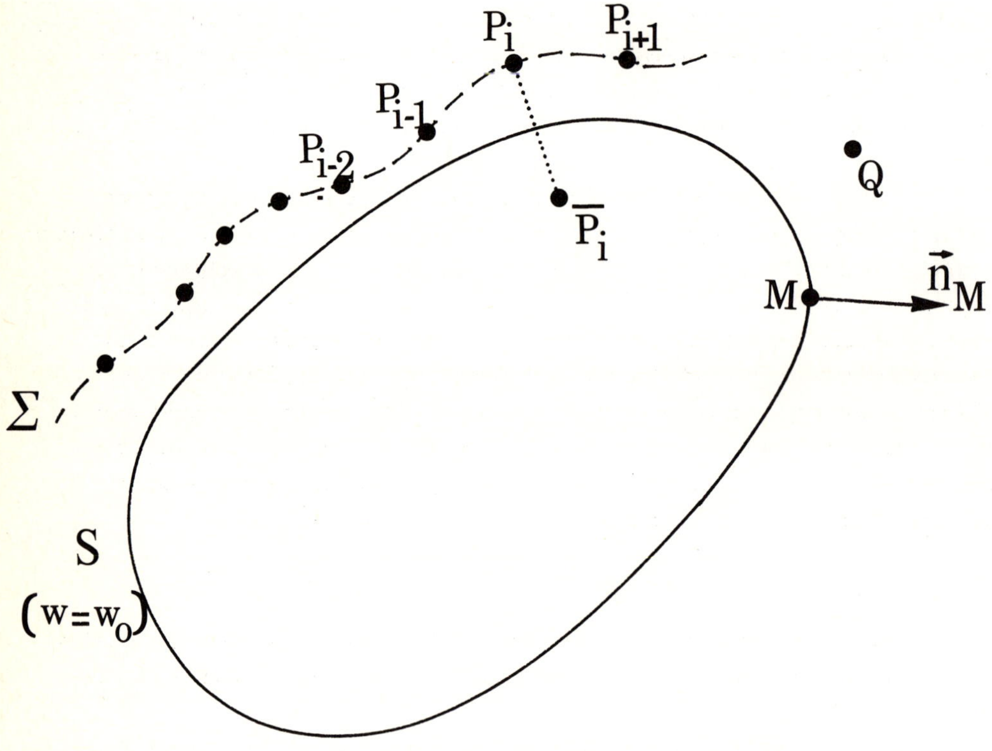

Figure 3 :

LINEAR INVERSE THEORY WITH A PRIORI DATA

David D. Jackson
Department of Geodesy and Geophysics
Cambridge University

Madingley Rise, Madingley Road
Cambridge CB3 0E7, England

I. Abstract

Methods of solution of linear inverse problems are reviewed, with emphasis on the relationship between algebraic and probabilistic approaches. Numerical methods for stabilizing the solution of ill-conditioned equations and interpretive methods for calculating the resolving power of experimental data require implicit assumptions about the solution. In many cases this information may be provided explicitly in the form of a '*a priori* data'. This results in considerable simplification of the numerical procedures required both for estimating the solution and for calculating resolving power.

The method is illustrated with a geophysical example. The earths' upper mantle beneath oceanic regions is characterized by a "lid" of high seismic velocity overlying a low velocity "channel". Gravity and seismic surface wave data are inverted to find the shape of the lid and channel beneath the East Pacific Rise. Without the use of *a priori* data, the problem is horribly underdetermined, and the resolving kernels do not provide much guidance in identifying physically meaningful "average" structures which can be resolved by the data. However, a great deal is known about the earth's upper mantle from a number of completely independent previous experiments. When these data are incorporated, the gravity and seismic data then lead to a very reasonable solution.

II. Introduction

Scientific inference based on experimental data usually proceeds by a series of iterative refinements to a mathematical model. The process follows the four steps listed below.

A. First, the problem is abstracted from the physical world into mathematics; that is, a mathematical formulation is constructed. This has the general form

$$y = f(x_c) + e \qquad (1)$$

where y represents a set of experimental data, f represents a mathematical operation describing the theoretical values of the data, x_c represents a set of unknown parameters on functions belonging to the general set x, and e represents the contribution of all those effects which are not explicitly modelled. The presumption is that the unmodelled phenomena will have a random effect on the data, so that e may be treated as a vector of random variables called "errors". A complete formulation involves a definition of the functions f, a statement of the domain of x, and some statistical statement about the "data errors"

I shall assume for the sake of simplicity that there is a finite number n of real data values, so that y, f, and e may each be written as a real column vector of n elements. I shall also assume that a "model" x consists of a finite number m of unknown real parameters, so that x may be written as a real column vector of m elements. The operation f then maps the m dimensional "model space" into the n dimensions "data space".

I also assume that there is a unique "correct" answer x_c, although there may be an infinite manifold of vectors x which satisfy all available data. This assumption is equivalent to the assumption that the "real earth", in all of its infinite detail, would be transformed by our mathematical abstraction into a unique model vector x_c. The converse is not assumed. Thus we are dealing from the beginning with a "watered down" version of the real problem, and must admit that some information has been lost just passing from the physical world to the mathematical world.

B. Second, some method must be provided for evaluating the operation f(x). This is usually called the "forward" or "direct" problem, and may be represented by the equation

$$f_o = f(x_o) \tag{2}$$

where x_o is some hypothetical model

C. Third, the inverse problem is "solved" in a mathematical sense. This generally involves some operation on the data which may be expressed

$$x_h = h(y) \tag{3}$$

The operation described by h may be the evaluation of a well defined mathematical function, or it may be simply a prescription for some sort of random search. The estimate x_h must then be tested for consistency with the mathematical conditions set forth in the formulation of the problem; specifically x_h must lie in the proper domain, and the residuals

$$r = y - f(x_h) \tag{4}$$

must be reasonably probable under the assumed properties of the errors e.

There are several different usages of the phrase "inverse problem" which are occasionally the source of some confusion:

(a) The <u>exact</u> inverse problem is to find an operator h which exactly inverts the operator f. This operator should satisfy the equations

$$f(h(y)) = y \tag{5}$$

for all admissible data (y). Uniqueness would require that

$$h\ (f(x)) = x \tag{6}$$

for all admissible model vectors x.

(b) The <u>optimum</u> inverse problem is to find an estimate x_m which minimizes some objective function Q(x). The least squares problem which we all know and love is an example for which the objective function is

$$Q = ||y-f(x)||^2 \tag{7}$$

(c) A third type of inverse problem is the <u>complete</u> problem, in which one seeks to find all possible solutions that satisfy the relevant constraints, and one in good enough agreement with the data. Such solutions should satisfy

$$Q(x) \leq Q_e \tag{8}$$

where Q_e is some prescribed threshold value. Because the set of solutions to (8) is generally rather unwieldy, it is often preferable to represent the complete solution by a set of <u>extremal solutions</u>, which minimize or maximize some preassigned functions b(x) subject to equation (8).

D. Finally, the mathematical results must be given a physical interpretation. Allowance must be made for any inadequacies in the original mathematical formulations of the problem. The solutions must be tested against any physical constraints which were not explicitly included in the mathematics. The residuals must be examined for consistency with the assumption of randomness. If a family of possible solutions has been identified, these should be examined for features which are common to all solutions.

In most physical problems the above steps are repeated iteratively. Examination of the answers, and equally important, of the residual between theory and observation, may lead to improvements in the mathematical model or to improved data.

In this discussion I shall focus primarily on discrete, linear inverse problems which may be expressed in the form

$$y = Ax_c + e \tag{9}$$

in which y is a column vector of n observations, x_c is a column vector of m unknown parameters, and A is an n xm matrix of real coefficients. There are relatively few problems for which equation (9) is the most precise formulations which we could imagine. However, there are many problems for which nonlinearities may be conveniently treated as "random" errors in the data, at least for one cycle through the interpretation procedure. After obtaining a solution to the inverse problem, it is then important to check the residuals for systematic behaviour which would imply that nonlinear effects must be explicitly modelled. There is also a large class of problems in which the unknown x may really contain an infinite number of degrees of freedom. For instance, the variation of density with position in the earth should properly be modelled by a piecewise continuous function of three independent variables. For practical purposes it is usually necessary to impose a finite representation. We do this either by expanding the density in a finite set of basis functions with unknown coefficients, or by assuming the earth to have constant density within a finite number of preassigned "shells". In doing so, we assume that any variations within the actual earth which cannot be modelled in this way are of little significance in making a physical interpretation, and that their effect on the data can be treated as random. Clearly a proper parameterization requires a good physical understanding of the object which is being parameterized. A test for the validity of the parameterization is that the ultimate physical conclusions should not depend on the parameterization itself.

Parker (1978) has adopted an alternate approach to the treatment of unknown functions having an infinite number of degrees of freedom. In this approach the "true" answer is assumed to lie in an infinite dimensional space. This true answer is then assumed to be the sum of a "model" vector which can be represented by a finite number of parameters, and a correction function which is orthogonal to the model in the infinite dimensional space. A set of extremal models is then defined in terms of an assumed threshold value of a prescribed norm of the correction function. My personal feeling is that errors caused by incomplete parameterization are best treated in terms of their probable effects on the data, which may then be viewed in the context of other data errors.

Because of the assumed random nature of the data errors e, a test for the acceptability of a given model will ultimately be based on a probabilistic model for the errors. What is needed is a convenient test of hypothetical solutions, so that one could imagine a model, or family of models, having certain physical consequences, and then test it, preferably by inspection. The most direct way to do this would be invert the data as if they were error free, and then to map the probability density function (pdf) for the errors, e, directly into a pdf for estimation errors in the model space. Given any estimation operator h as in equation 3, it will be possible, at least in principle, to construct a pdf for the estimate x_h. This will also be a pdf for the estimation error only in the case that each point in the data space corresponds to a unique point in the model space, and that the estimator h provides the inverse mapping to f. In this case f and h must satisfy equation 6. Operationally, the procedure would work as follows. First, the pdf of e would be mapped into a pdf on $f(x) = y - e$. This pdf need be defined only on the range of f, which is also the domain of h. This would solve the possible problem that random errors may take the data y outside the range of f. Second, the pdf of x would be obtained by an integration over the domain of h.

Under certain circumstances, a linear inverse problem may be eligible for direct mapping of the pdf for data errors into a pdf for estimation errors. Assume that equation (9) holds, and that an estimate x_h is defined by a linear operation on the data, say

$$x_h = Hy \tag{10}$$

From (9) and (10) we may now compute the estimation errors,

$$x_h - x_c = (HA - I)\, x_c + He \tag{11}$$

The estimation errors are of two types: a "bias" or resolving error which depends on the actual values of the true model x_c, and a "random" error which depends on the data errors. The uniqueness of the solution to the exact inverse problem requires that equation (6) is satisfied, which in this case is equivalent to the vanishing of the resolving error. Equation (11) may then be used to integrate the pdf of e to provide a pdf of $x_h - x_c$.

In many cases, the pdf of a vector random variable may be described simply in terms of a vector of means, and a covariance matrix. If it assumed that the data errors have zero mean and covariance matrix C_e, and again that resolving errors vanish, then the estimation errors of equation 11 will have zero mean and covariance matrix

$$C_h = HC_eH^T \tag{12}$$

If the resolving errors do not vanish, then one cannot make a statistical statement about estimation errors without some <u>a priori</u> information on x. As you may already have guessed, this is to be the subject of further discussion below.

The conditions for construction of a pdf for x are rather strict: that there be a unique solution to the exact inverse problem whenever a solution exists. In other cases, it is still possible to make a meaningful hypothesis test in the data space. The reasoning goes like this: suppose I have a candidate model x_h and I propose that it is in fact the true model x_c. In this case the residuals r must be equal to the random errors e. A test of these residuals against the pdf for e will then provide a consistency test of the hypothesis that the candidate model is correct. Such reasoning is aimed primarily at eliminating from consideration those models which are inconsistent with the data.

Assume for a moment that the data errors e are Gaussian with zero mean. The general pdf for mean free Gaussian errors is

$$p(e) = (2\pi^n D)^{-\frac{1}{2}} \quad \exp\left(-\tfrac{1}{2}e^T C_e^{-1} e\right) \tag{13}$$

where n is the number of data, C_e is the covariance matrix for data errors e, and D is the determinant of C_e. The hypothesis test for a given model x_h is made by computing the residuals from equation 4, then substituting these for e in equation 13. The computation is simplified by the fact that the model x_h enters only in the form

$$Q(X_n) = r^T C_e^{-1} r \tag{14}$$

which is essentially the weighted sum of squares of the residuals. Furthermore, if r are presumed to satisfy (13), then $Q(x_h)$ is a chi square variable with n degrees of freedom, so that it is not even necessary to evaluate (13). The quantity Q plays an important role in both the optimum and the extremal inverse problems. The maximum likelihood estimate x_m is that which maximizes (13), or equivalently, minimizes (14). As a criterion for an extremal inversion a logical choice would be that the residuals should agree with (13) at a specified confidence level. A value of Q_e corresponding to this choice can then be obtained from a "chi-squared table".

There are important problems for which the errors cannot be expected to be Gaussian. A somewhat more general statistical model is

$$p(e) = a \exp\left(- (b/k) \sum_{1=1}^{n} \left|\frac{e_i}{s_i}\right|^k\right) \tag{15}$$

where e_i is an element of e, k and s_i are parameters of the distribution, and a and b are normalization factors. This pde is still rather specific in that it is symmetric in e_i, and that as written it is appropriate only for statistically independent data. The argument of the exponential function in equation 15 provides a useful objective function for optimal inversion. Criteria for extremal inversion may likewise be defined in terms of the confidence intervals derived from (15). The confidence intervals are much wider than those corresponding to Gaussian variables, for confidence levels greater than about 75%. For k=1 the relevant optimization criterion is a weighted sum of absolute residuals. Minimization of this objective function requires a non linear estimation procedure, even when the data satisfy a linear equation such as (9). Claerbout and Muir (1975) have studied this case in detail, and present a computational algorithm for calculating x_m. To my knowledge, no one has addressed the problem of constructing a pde for the estimation errors, nor the complete inverse problem. The case k=2 is the Gaussian case, which will be discussed in more detail below. Another interesting case results when k approaches infinity. Then the pde for a single variable becomes rectangular, effectively putting hard inequality constraints on the data. In this case it is only the complete inverse problem which has any meaning, since all models that satisfy the constraints are in equal agreement with the data. The solution of linear equations with inequality constraint is the subject of "linear programming", about which numerous books have been written (e.g. Dantzig, 1971). Again, linear estimation procedures such as (10) are not appropriate, even when the fundamental mathematical formulation is linear.

The introduction of non-Gaussian errors obviously complicates the inverse problem considerably. The situation gets worse for mixed data sets, in which there are some data having k=1, for instance, and others k=2. Some progress has been made in the case of a mixture of k=2 and $k=\infty$. This is essentially the "least squares problem with inequality constraints". The optimal inverse problem has been discussed by Stoer (1971), Lawson and Hanson (1974), Bartels (1975) and Elden (1977). To my knowledge, no one has addressed the complete solution to this problem, nor attempted to derive a pdf for the solution.

There is much to be said for ignoring some of the problems discussed above, at least on the first pass through the interpretation scheme. The difference between Gaussian statistics and some other variety may be negligible for many problems. The primary role of statistics is to determine the weighting to be given to the various data. If the data are relatively complete and mutually consistent, the weighting won't matter much. In many cases, the appropriate scaling factors s_i in equation 14 may be unknown. Errors in estimating the s_i are likely to be more serious than an error in the assumed value of k. For example, a rectangular pdf with s=1 is better approximated by a Gaussian pdf with s=1 than by a rectangular pdf with s=2. Even when the precise form of the pdf for the errors is important, it needn't always be built in precisely. It may be a useful approach to solve an extremal inverse problem assuming Gaussian data, and then to test the resulting extremal models against the proper pdf. These models may no longer be extremal, but they may be sufficient to answer the physical questions which motivated the inversion study in the first place.

III. Overconstrained Problems

In the discussion to follow, I shall assume a linear parametric inverse problem as in equation 9, for which the data errors are Gaussian with zero mean and known covariance, as in equation 13. I shall concentrate on optimal and extremal inversion problems, since the solutions to the exact inverse problem follows in a straightforward way as a special case of the optimal inverse problem.

A natural optimization criterion is provided by (14). If there is a unique solution which minimizes this criterion, it deservedly carries the name "maximum likelihood solution". Formally, the solution is

$$X_m = N^{-1}A^T C_e^{-1} y \tag{16}$$

where

$$N = (A^T C_e^{-1} A) \tag{17}$$

is commonly known as the "normal matrix". The validity of the formal solution clearly requires that N be non-singular, which shall be assumed until further notice. Equation 16 is a linear operation on the data of the form (10) with

$$H = N^{-1}A^T C_e^{-1} y \tag{18}$$

so that (11) may be used to describe the estimation errors. The resolving errors vanish, so that this is one of the special cases for which a pde can be written for the solution space. By equation (12), we see that N^{-1} is the covariance matrix both for the estimate itself, and for the estimation errors.

The extremal problem has been addressed by this author (Jackson, 1977) for the case $b(x) = b^T x$, where b is an arbitrary vector in the solution space. By appropriate choice of b, the function $b^T x$ may represent a single element of x, a weighted average of the elements, the difference between successive elements, etc. The extremal models for a given b are

$$x_e = x_m \pm wC\, b \tag{19}$$

where

$$C = N^{-1} \tag{20}$$

and

$$w = ((Q_e - Q_m) \,/\, (b^r C\, b))^{\frac{1}{2}} \tag{21}$$

Here x_m is the maximum likelihood solution, Q_m is the minimum value of Q, and Q_e is the prescribed threshold value. It is natural to choose Q_e=n, the expected value of Q for the "true" solution. This is approximately equivalent to requiring that the residuals agree with (13) at the 70% confidence level. The expected value of Q_m is n-m, where m is the number of unknown parameters.

Because we have a pdf for x, we may just as well peform an extremal inversion in model space using an objective function of the form $Q = (x-x_m)^T C^{-1}(x-x_m)$. Here, it is natural to take as the threshold value $Q_e = m$, the number of unknown parameters, which is the expected value of Q for the correct model. This will lead to the above result if Q_m = n-m, its expected value.

It is commonly assumed that incrementing any of the elements of the maximum likelihood solution by its standard deviation will lead to an extremal model. However, the expected value of Q for such a model is only n-m+1, whereas the expected value for the correct answer is n. Thus the standard deviation underestimates the range of models which will satisfy the data. The correct model can be expected to lie $\pm$ m standard deviations from the maximum likelihood model just as the correct data y-e can be expected to lie $\pm$ n standard deviations from the observed data. This is a result of the fact that the confidence interval for a chi square variable grows in approximate proportion to the number of degrees of freedom.

IV. Regularization of ill conditioned problems

What happens if the normal matrix is singular? Two cases can be considered, although the distinction between them is rather fuzzy. The first case is that in which the singularity is relatively minor, and can be address by "conditioning" or regularizing the normal matrix. The second case is that in which the singularity results from a massive deficiency of data relative to the number of parameters which must be determined. In this case, special methods may be formulated to identify certain pieces of information which can be pulled out of the data. These methods will be discussed in the following section.

Singularity of the "normal matrix" is equivalent to the vanishing of any or all of its eigenvalues. "Poor conditioning", or "instability", is due to some eigenvalues having very small values. Consider an eigenvector v and associated eigenvalue t^2 of the normal matrix, such that

$$Nv = t^2 v \tag{22}$$

The eigenvalue can be denoted t^2 because the normal matrix must be non-negative definite. Now consider a model of the form

$$x = x_m + av \tag{23}$$

where x_m is a maximum likelihood solution, and a is an arbitrary scalar . The residual criterion then becomes

$$Q(x) = Q_m + a^2 t^2 \tag{24}$$

where Q_m is the minimized residual criterion. Clearly when t^2 is small, the residual criterion is rather insensitive to the size of the perturbation, and when $t^2 = o$, the maximum likelihood solution becomes nonunique. For machine computations, there is some non zero threshold below which an eigenvalue is effectively zero, so that the distinction between true singularity and near singularity is blurred.

There are two basic methods in common usage for dealing with near singularity of the normal matrix. In one method, the solution is estimated by a linear operation on the data as in (10) where H is a generalized inverse of the matrix A. Generalized inverses are considered in detail in Lanczos (1964) and Rao and Mitra (1971). An essential feature of these inverses is that

$$Hv = o \tag{25}$$

for all those solutions of (22) for which t^2 is less than some threshold. This has the effect of stabilizing the estimation procedure in the sense that C_h of equation 12 will have smaller values. However, the resolving errors will no longer vanish for this estimator, so that C_h is only the covariance of the estimate about its mean, but is <u>not</u> the covariance of estimation errors. Equation 25 also guarantees that

$$v^T x_h = o \tag{26}$$

that is, the estimate is orthogonal to those eigenvectors corresponding to small eigenvalues. Equation 26 may be viewed as the inclusion of constraints, equal in number to the small eigenvalues, along with the experimental data. In fact, identical results would be obtained by including (26) as a set of data equations, given extremely large weight.

In another method, the estimate is computed just as in (16), except that (17) is replaced by

$$N = (A^T C_e^{-1} A + S) \tag{27}$$

where S is chosen to guarantee that (27) is non singular. The effect of the modification is to increase the size of the eigenvalues so that the estimation procedure is stable. Resolving errors will not be zero for this estimation, so that C_h will not be the covariance matrix for estimation errors. It will be demonstrated below that using (27) is equivalent to the addition of supplemental data, and that under certain circumstances the inverse of N as defined in (27) may actually be viewed as the covariance matrix for estimation errors.

Suppose that there are two independent data sets which are both to be explained by the same model. Then

$$y_1 = A_1 x_c + e_1 \quad \text{and} \quad y_2 = A_2 x_c + e_2 \tag{28}$$

For a given x, residuals for the two data sets can be defined as in equation 4, and if the two data sets are Gaussian with covariance matrices C_1 and C_2, then the maximum likelihood solution will be that which minimizes

$$Q = r_1^T C_1 r_1 + r_2^T C_2 r_2 \ . \tag{29}$$

The solution is

$$x_m = (A_1^T C_1^{-1} A_1 + A_2^T C_2^{-1} A_2)^{-1} (A_1^T C_1^{-1} y_1 + A_2^T C_2^{-1} y_2) . \tag{30}$$

Comparison of (30) with (18) and (27) will show that they are the same in the special case

$$A = A_1 \qquad S = A_2^T C_2^{-1} A_2 \qquad y_2 = 0 \tag{31}$$

that is to say, the result of the regularization method of (27) is identical to that of adding additional data of the form

$$A_2 x = 0 \tag{32}$$

with covariance matrix C_2. This may be reasonable if the unknown vector x represents a correction to a model inferred from a previous data set. In this case the data y_1 must also be the residual between the actual observations and the theoretical predictions of the *a priori* model. This perturbation technique is commonly used for non linear inverse problems, but it is not commonly recognized that it may be appropriate for linear problems as well if (16) and (27) are to be used together. Alternatively, the *a priori* data may be added directly to the data set as in equation 30. In this case, the effects of the *a priori* model need not be subtracted out as described above.

Care must be taken when using (27) that the matrix S be chosen to reflect a defensible statement about the correct solution x_c. In many cases S is chosen according to some criterion involving only the condition number of the normal matrix, rather than something that is known about the solution. It should be remembered that any choice of S is the equivalent of adding data, and the solution will be meaningful only if the data used are also meaningful. In many cases, crude bounds may be put on the solution either from physical constraint or from previous observations. Consider for example the inverse problem of determining density variations in the earth's crust from gravity observations. Densities of igneous rock samples from the continental crust average about 2.5 Mg/m^3 with a standard deviation of about 0.5 Mg/m^3, based on direct observation. More precise information will be relevant for specific problems. These data may be added to the gravity data set with quite reasonable results.

V. Resolving power

Backus and Gilbert (1967, 1968, 1970) have addressed the problem of massive nonuniqueness by dealing from the start with "averages" of the parameters. This approach is especially sensible when the unknown parameters represent consecutive samples of a continuous function such as, for example, density within the earth. The aim is to form a set of averages from the data themselves, using a linear operation of the form

$$x_a = H_a \, y \tag{33}$$

Because the data contain information about the correct solution by virtue of equation (9), the linear operation (33) provides an "averaged" version of the correct model,

$$x_a = Rx_c + e_a \tag{34}$$

where

$$R = H_a \, A \tag{35}$$

and e_a is a random error about which a statistical statement is to be made.

The matrix R provides a window through which the correct solution may be viewed with a random error e_a. Each row of the matrix R depends only on the corresponding row of the matrix H_a, and may be interpreted as a "resolving kernel" for the corresponding element of x_c. Backus and Gilbert have given a number of methods for choosing H_a to optimize the shape of the kernels. The usefulness of the averaged model x_a depends on the shape of the resolving kernels. If they are all sharply peaked at the corresponding element of x_c, then of course the average x_a may be interpreted as an estimate of x_c. If the kernels are more or less boxcar shaped, then the averages may still have a useful interpretation, provided that x_c is known to be relatively well behaved in the appropriate interval. The meaning of "well behaved" may be examined by considering the difference between the "averaged" estimate x_a and the correct solution x_c:

$$x_a - x_c = (H_a A-I)\, x_c + He \quad (36)$$

From (36) it is evident that if the value of x_c is large and $R = H_a A-I$, then the error in approximating x_c by its average x_a will also be large. A consequence of (36) is that the Backus and Gilbert technique may be used much more effectively on small perturbations to an a priori model than on the the total model vector itself, unless of course the a priori model has values near zero. Let us presume that here the unknown vector does in fact represent a correction to some a priori model, and that we have reason to believe that the elements of the correction vector are small. In fact, lets go even further: let us treat the elements of the correction vector as random variables, which have mean zero and covariance matrix denoted by C_x. The matrix C_x then expresses the degree of confidence which we have in the a priori model. We may then consider the difference between the averaged and correct model to be a random variable, which will have mean zero and covariance matrix

$$C = (H_a A-I)\, C_x\, (H_a A-I)^T + HC_eH^T \quad (37)$$

presuming that the data which led to the a priori model are independent from the experimental.

It is perhaps dangerous to push too far the interpretation of the unknown parameters as random variables. In many cases the information which we have about the parameters cannot be described in terms of means and covariances of some probability density function. On the other hand, (37) tells us that we must have some a priori information about the solution in order to give an interpretation to an averaged solution. A solution about which absolutely nothing is known in advance can be represented well as a vector of gaussian random variables with variances approaching infinity. Common sense tells us that an average over such a solution is meaningless, even if the random errors contributed by the data are small. Equation (37) tells us the same story.

When the solution can be treated as a vector of random variables with a known mean and covariance matrix, then (37) may be used to optimize the tradeoff between resolving errors (the first term) and data errors (the second term). One may choose H_a so as to minimize the variances of the total averaging errors given by (37). The result is

$$H_a = C_x A^T\, (AC_x A^T + C_e)^{-1} \quad (38)$$

which is algebraically equivalent to

$$H_a = (A^T C_e^{-1} A + C_e^{-1})\; A^T C_e^{-1} \quad (39)$$

Jordan and Minster (1972) derived an expression equivalent to (38) using a related but slightly different optimization criterion. Equation 39 will be readily recognized as the equivalent of (30), with $A_2 = I$ and $y_2 = o$. Thus the "averaged" solution x_a obtained using this method is just exactly the estimated solution x_m which would be obtained by including as data the equations

$$x = o \tag{40}$$

with covariance matrix C_x, and using the maximum likelihood method. The covariance matrix for estimation errors is rather simple for the operation defined by (38) or (39); from equation (37) we find that

$$C = (A^T C_e^{-1} A + C_x^{-1})^{-1}$$

the inverse of the augmented normal matrix.

My conclusion from all this is that there is no way to find optimal solutions to undetermined or poorly conditioned inverse problems without the implicit use of a priori data. Given that situation, it seems to me that the most straightforward approach is simply to include the a priori data explicitly, weighted using a reasonable uncertainty estimate, as in equation (30). In practice the two data sets y_1 and y_2, may be "welded" together and treated as a single data set, with similar treatment for A_1 and A_2, to yield identical results to (37). If all data, including a priori data, may be treated as gaussian, then the inverse normal matrix will be the covariance matrix for estimation errors, and will include the effects of both resolving errors and random data errors of equation 11. There will be no need to carry out elaborate optimization procedures to find resolving kernels. One may simply select a resolving kernel at will as one chose the vector b in equation 19. In fact, the role of b in (19) is identical to that of the resolving kernel. Because of the non singularity of the augmented normal matrix, one can easily compute the variance of the "average" $b^T x$, as

$$\text{var}\,(b^T x) = b^T C\, b \tag{42}$$

where C is the inverse normal matrix. One may also use the "most-squares" procedure (Jackson, 1976) to find the extremal values of $b^T x$ consistent with the data.

VI. Geophysical Example

According to the theory of plate tectonics, the earth's crust and upper mantle are capped with a relatively rigid "lithosphere", which overlies an anelastic " asthenosphere". The asthenosphere in turn rests on top of a more rigid "mesosphere". These regions seem to correspond with the seismologically observed high velocity "lid", low velocity "channel", and high velocity "subchannel". Although the correspondence is not known to be exact, it is worthwhile to assume the correspondence, map the structure of the lid, channel, and subchannel using seismic waves, and examine the consequences implied for plate tectonics.

The lid is known from seismological studies to be very thin under the oceanic ridges, where new lithosphere is being formed out of asthenosphere rock, while the existing lithosphere is spreading away from the ridges. This situation implies the vertical migration of asthenosphere rock immediately beneath the ridge, but the depth of this flow and the thickness of the asthenosphere itself are the subjects of heated debate. Other questions of interest involve the density and seismic velocities within the lid and channel. Major candidates for the composition of the lid are "pyrolite", a mixture of pyroxene, olivine, and basalt having a density of about 3.3 Mg/m^3, and eclogite, a high pressure form of basalt having a density of about 3.5 Mg/m^3. The average density contrast between the lid and channel is also of interest, because the bouyancy provided by a low density asthemosphere could provide a significant driving force on the lithospheric plates. Previous investigators (Forsyth, 1973, Schlue and Knopoff, 1976) have suggested anisotropy in shear velocity in the channel, which could be evidence for oriented inclusions of partially melted rock. The work described in this section is lifted from the Ph.D thesis of Burkhard (1977), and addresses the above questions.

The upper mantle was parameterized by conceptually dividing it into three crustal layers, a lid, a channel, and a subchannel. These layers were subdivided into "age regions" with vertical boundaries roughly parallel to the East Pacific Ridge. Eight age regions, numbered consecutively, were selected with average ages of 150, 122.5, 97.5, 72.5, 50, 30, 15, and 5 Myear. The basic parameters in the problem are the average seismic velocities, density and thickness of each layer within each age region. However, a number of constraints were imposed:

(1) The seismic compressional wave velocity was assumed known everywhere.

(2) The properties of the uppermost crustal layer, and the properties of the subchannel and all lower layers, were assumed known.

(3) The shear velocity was assumed to be isotropic in the crust and below the channel. In the lid and channel, separate velocities were allowed for vertically and horizontally polarized shear waves.

(4) The thicknesses of all layers in age region 2, 4, 5, and 7 were constrained to be linear interpolations of the thickness vs age curve determined by the remaining age regions.

(5) The density and velocities of each layer were assumed to be independent of age.

Assumptions (1) to (3) were based on calculations which showed that the geophysical data used here are relatively insensitive to variations within the a priori uncertainty of the parameters in question. Assumption number (4) was imposed to assure that the thickness of the layers be smooth functions of geologic age. Assumption (5) was originally based on the presumption that the identity of the various layers is the result of geochemical variations or phase changes which will affect the density and velocities much more than the variations of temperature and pressure within the layers. However, this presumption is by no means certain, and is not necessary for the interpretation of the models presented here; the reported values should be treated as averages only. If density, for example, varies continuously with depth in a given age region, it may be possible to describe the variation quite well by specifying the depths to the top and bottom of several layers with a prescribed average density, especially if average density increases with depth.

The data used include free air gravity anomaly values, and seismic surface wave phase velocities for paths which in general cross several age regions. The gravity values are primarily sensitive to the integrated mass in a vertical column and are not very sensitive to the distribution of mass within the vertical column. The data show that regions as large as the age zones used here are essentially isostatically compensated; that is, the integrated mass per unit area within a vertical column is the same in all age regions. The surface waves include Love waves, which involve horizontally polarized shear motion only, and Rayleigh waves, which involve compressional and vertically polarized shear motion. The anisotropy reported above is deduced from the fact that anisotropic models have not been able to fit both Love and Rayleigh wave velocities for these same paths. Both Love and Rayleigh wave velocities are sensitive to the density contrast between adjacent layers as it effects the impedance contrast. Rayleigh waves are also mildly sensitive to absolute density.

The data set also included the assumed a priori values of the unknown parameters. The starting model is given in table 1a. These values were determined by Burkhard's assessment of the results of published investigations using data that are for the most part different in type from Burkhard's own data. The a priori uncertainties assumed here are generous enough to account for the fact that some of the surface wave data have been used twice (once by Schlue and Knopoff in determining the a priori model, and once again by Burkhard). The velocities V_{SV} and V_{SH} are assumed equal in the a priori model, but the uncertainty allows for substantial anisotropy. The assumed density in the lid is 3.4 ± 0.2 Mg/m^3, which would allow either the pyrolite or eclogite hypotheses for composition of the lithosphere. There is no density contrast between lid and channel in the a priori model, but the assumed uncertainties would allow a rather substantial increase or decrease in density at this boundary. The shape of the crust, lid, and channel assumed in the a priori model is shown in figure 1a.

The data were inverted using equations 33 and 39, to obtain an "averaged" solution. Resolving kernels were calculated using equation 35, in the hope that they would suggest some interpretable features of the model which could be well determined. This did not prove to be the case, as the averaging kernels are too complicated to warrant a physical interpretation. The complications are compounded by the fact that the parameters are of several different types with different units. However, the averaged solution itself is rather reasonable, and fit the data fairly well (the residual criterion Q as defined in (14) was 754, which is just less than the number of data, 785). Furthermore, the covariance matrix for estimation errors, given by (42) indicates that the estimation errors, including both resolving errors and random data errors, are in many cases quite small.

The final model is shown in table 1b and in figure 1b. The average density in the lid is 3.29 ± 0.015 Mg/m^3, providing strong confirmation of the pyrolite hypothesis. The average density in the channel is 3.33 ± 0.011 Mg/m^3. This average value does not prevent a slight density reversal just below the base of the lid, but it does rule out a large density reversal which could provide significant plate driving forces. The anisotropy in shear velocity in the channel is confirmed, but anisotropy within the lid is not needed to explain the data. The low velocity channel is shown to become thinner with age at both the top and the bottom. The anisotropy and shape of the channel are both consistent with the hypothesis that the channel is the result of partial melting. The shape of the channel would then be determined by the intersections of the temperature-depth curve with the melting "solidus" curve for rocks of pyrolite composition. The temperature at a given depth is known to be highest beneath the ridge and to decrease with age away from the ridge. This would explain the thinning of the channel at the top and the bottom. The anisotrophy can be explained by the effect of small, flat, horizontal inclusions of partial melt which have been preferentially oriented by horizontal shear flow in the asthenosphere.

In interpreting the results of this inversion, we must of course take care to allow for the assumptions made in reducing this problem from a physical to a mathematical problem. The strongest assumption is certainly number (5), that the velocities and density do not vary within a given layer. This may be particularly important for interpreting the shape of the bottom of the channel. The surface wave velocities for the periods used here are known to be very insensitive to anything at that depth. Therefore the shape of the channel beneath the ridge can be seen as a result of the gravity data, which require isostacy, and the calculated positive density contrast between lid and channel. This implies a positive mass anomaly at the top of channel, which in our model can only be compensated by depressing the boundary between channel and subchannel. One might ask if the compensation could be achieved by a lower density within the channel instead. Some "back of envelope" calculations indicate that it probably could. A density difference in the channel of about 2% between the oldest and youngest age regions would preserve isostatic balance if the bottom of the channel were a constant 180 km deep independent of age. A smaller density difference would be required if the bottom of the channel were deeper, but at least a 1% density variation is required for any reasonable depth. This would require an average temperature difference between old and new channel of between 500 and 250°C, depending on the assumed bottom depth. Such temperature differences are substantial but not inconsistent with heat flow or other relevant observations. However, this model would require a horizontal temperature gradient at the bottom of the channel as well. Because any reasonable physical explanation of the bottom of the channel would require that its location be dependent on temperature and pressure, we are led again to a model in which the bottom of the channel is deepest beneath the ridge, and becomes shallower as the channel cools with age.

In summary, the results of the inversion are certainly quite good in the numerical sense that the data have reduced the a priori uncertainties quite substantially and the final model fits the data well. The physical conclusions seem clear and reasonable, but the constraints placed on the solution during the formulations of the inverse problem make a rigorous test of alternate hypotheses rather difficult. The results presented here should be viewed as merely one cycle through the inference cycle proposed in the introduction. Based on the results of the first cycle, there seems to be reason for hope that these same data might be capable of resolving density or velocity variations within the lid and channel. Had the numerical results been inconclusive even with the assumptions imposed here, there would clearly be no point in pursuing the inversion without significantly better data.

Table 1

Physical Properties of models

A. *A priori* model

	Density	V_{SH}	V_{SV}	V_p
Water	1.03	0.0	0.0	1.52
Crust 1	2.00	1.00	1.00	1.65
Crust 2	2.65 ± .05	3.00	3.00	5.15
Crust 3	2.90 ± .04	3.90	3.90	6.80
Lid	3.40 ± .20	4.60 ± 0.20	4.60 ± 0.20	8.10
Channel	3.40 ± .20	4.10 ± 0.20	4.10 ± 0.20	7.60
Subchannel	3.50	4.55	4.55	8.20
Mantle 3	3.96	5.40	5.40	9.80
Mantle 4	4.21	5.90	5.90	10.00
Mantle 5	4.95	6.38	6.38	11.48

B. Final Model

	Density	V_{SH}	V_{SV}
Crust 2	2.56 ± .05		
Crust 3	2.84 ± .04		
Lid	3.29 ± .015	4.58 ± .021	4.58 ± .021
Channel	3.33 ± .011	4.24 ± .028	4.08 ± .031

V_{SH} = Velocity of horizontally polarized shear waves
V_{SV} = Velocity of vertically
V_p = Velocity of compressional waves
Units: density in Mg/m^3, velocity in km/sec.
± values indicate standard deviation. Where no standard deviation given, parameter was constrained at *a priori* value. Blanks in Final Model correspond to values so constrained.

References

Backus, G.E., and Gilbert, F.J., 1967. Numerical applications of a formalism for geophysical inverse problems, Geophys. J. Roy. Astr. Soc. 13, 247-267.

Backus, G.E., and Gilbert, F., 1968. The resolving power of gross earth data, Geophys. J. Roy. Astr. Soc., 16, 169-205.

Backus, G., and Gilbert, F., 1970. Uniqueness in the inversion of gross earth data, Phil. Trans. Roy. Soc. London, 266, 123-192.

Bartels, R., 1975. Constrained least squares, quadratic programming, complementary pivot programming, and duality, Johns Hopkins Univ., Technical Report No. 218.

Burkhard, Norman R., 1977. Ph.D. Thesis, Univ. Calif. at Los Angeles.

Claerbout, J., and Muir, F., 1973. Robust modeling with erratic data, Geophysics, 38, 826-844.

Dantzig, G., 1963. Linear programming and its extensions, Princeton Univ. Press.

Elden, Lars, 1977. Algorithms for least squares problems with banded inequality constraints, Linkoping University, Report LiTH-MAT-R-1977-20.

Forsyth, D.W., 1973. Anisotropy and structural evolution of the oceanic upper mantle. Ph.D. thesis, Mass. Inst. Tech., Cambridge.

Jackson, David D., 1976. Most-squares inversion. J. Geophys. Res., 81, 1027-1030.

Jordan, T.H., and Minster, J.B., 1972. Application of a stochastic inverse to the geophysical inverse problem, in The Mathematics of Profile Inversion, ed. L. Colin, Marcel Dekker Inc., New York.

Lawson, Charles L., and Hanson, Richard J., Solving least squares problems, Prentice Hall, 1974.

Parker, R.L., 1977. Linear inference and underparameterized models. Rev. Geophys. Space, Phys. 15, 446-456.

Sabatier, P.C., 1977. On geophysical inverse problems and constraints, J. Geophys., 43, 115-137.

Schlue, J.W., and Knopoff, L., 1976. Shear wave anisotropy in the upper mantle of the Pacific Ocean. Geophys. Res. Lett., 3, 359-362.

Stoer, J., 1971. On the numerical solution of constrained least square problems, SIAM J. Numer. Anal. 8, 382-411.

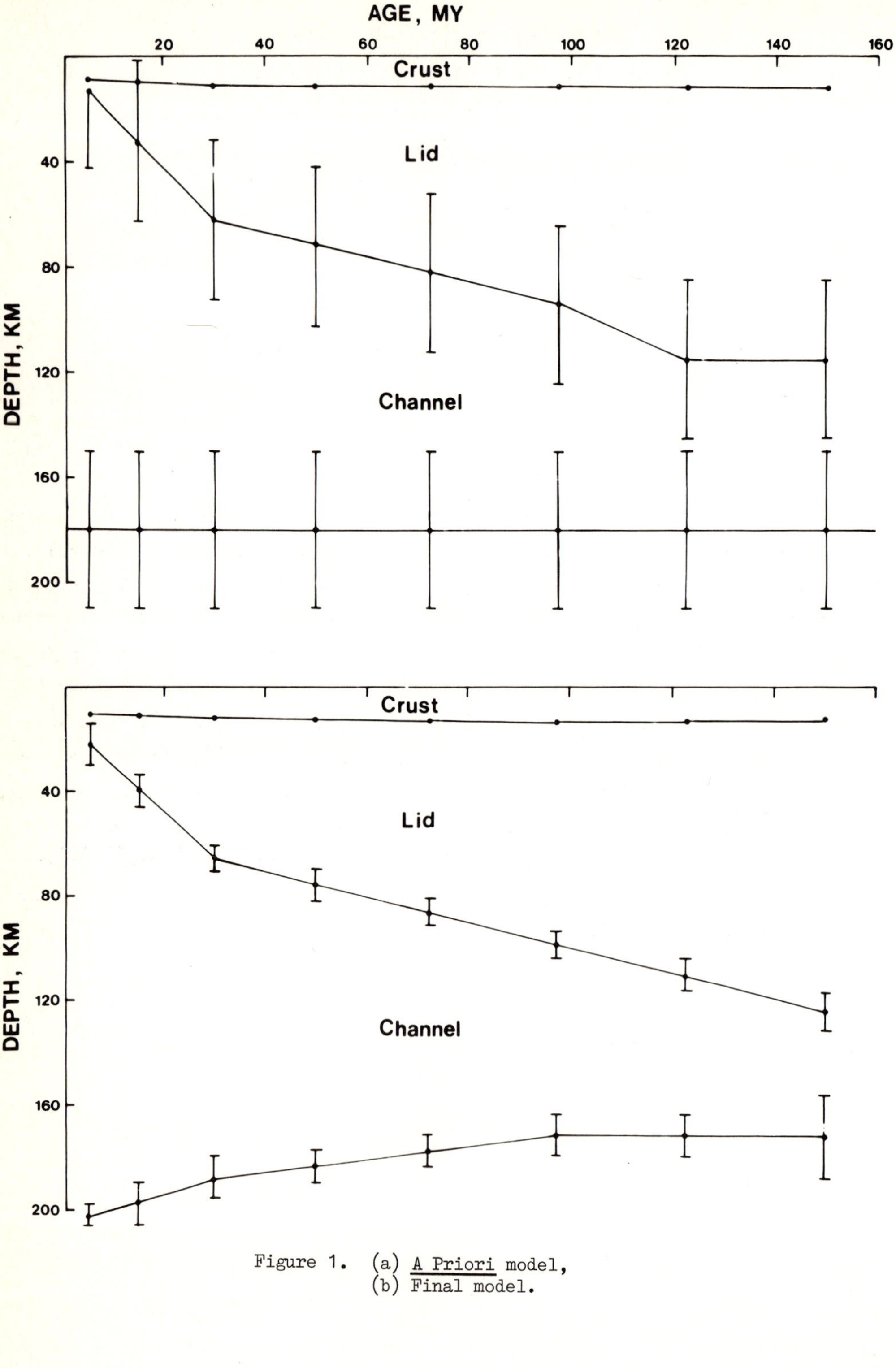

Figure 1. (a) A Priori model,
(b) Final model.

APPLICATION OF LINEAR PROGRAMMING TO

THE INVERSE GRAVITY OR MAGNETIC PROBLEM

BASIC NUMERICAL TECHNIQUES

M. CUER

Département de Physique Mathématiques

Université des Sciences et Techniques du Languedoc

34060 MONTPELLIER CEDEX, FRANCE

Summary : Linear programming using the Bartels-Golub decomposition of the basis and the "steepest edge" strategies for column pivoting is relevant to linear inverse problems.

1. Introduction

The inverse gravity problem is to obtain informations on subsurface mass distribution from the measurement of the earth gravity field. A similar problem lies in extracting informations on the subsurface magnetization from the geomagnetic field. In most cases, the measurement is an anomaly that is to say the field excess of default in relation to a standard model. The subsurface mass distribution is described in term of density contrast.

The approach taken in this paper is based on the following idea proposed by Sabatier (1977a-1977b) and used by Safon et all (1977) and Bourmatte (1977). The data are the anomalies $\gamma_i \pm e_i$ $1 \leq i \leq mp$ (e_i are errors measurement) at stations i $1 \leq i \leq mp$. The studied domain of the earth is divided into a great number of parallelipipeds ω_j $1 \leq j \leq np$. The value of the unknown function in ω_j is ρ_j and from petrological or geological considerations some bounds ρ_j^{min} and ρ_j^{max} can be assigned to ρ_j . Thus the unknown vector $\rho = \{\rho_j\}_{1 \leq j \leq np}$ must satisfy :

$$\gamma - e \leq G\rho \leq \gamma + e \tag{1}$$

$$\rho^{min} \leq \rho \leq \rho^{max}$$

where $\gamma = \{\gamma_i\}_{1 \leq i \leq mp}$;

$e = \{e_i\}_{1 \leq i \leq mp}$;

$\rho^{min} = \{\rho_j^{min}\}_{1 \leq j \leq np}$;

$\rho^{max} = \{\rho_j^{max}\}_{1 \leq j \leq np}$;

and $G = \{g_{ij}\}_{\substack{1\le i\le mp \\ 1\le j\le np}}$ is an $mp\times np$ matrix, the element g_{ij} being the effect at station i of ω_j when $\rho_j = 1$.

Introducing the vector $x = \{x_j\}_{1\le j\le n}$ with $n = np+mp$, $x_j = \rho_j - \rho_j^{min}$ if $1\le j\le np$ and $x_{np+i} = \gamma_i + e_i - (G\rho)_i$ if $1\le i\le mp$, the system (1) takes the most used form

$$(2)\qquad \begin{aligned} &Ax = b \\ &0 \le x \le x^{max} \end{aligned}$$

where A is the $mp\times n$ matrix $[G, I_{mp}]$, I_{mp} being the identity matrix in $\mathbb{R}^{mp}$;

$$b = \gamma + e - G\rho^{min}$$

and $x^{max} = \{x_j^{max}\}_{1\le j\le n}$ with $x_j^{max} = \rho_j^{max} - \rho_j^{min}$ if $1\le j\le np$ and $x_{np+i}^{max} = 2e_i$ if $1\le i\le mp$.

A well known mathematical result, the proof of which can be found in the books of Stoer and Witzgall (1970) or Rockafellar (1970) says that the set of solutions of (2) is a <u>convex polyhedron</u> and that, if this set is bounded, the general solution of (2) is described by the formula

$$(3)\qquad x = \sum_{k=1}^{K} \lambda_k\, x^{(k)}$$

where

λ_k $1\le k\le K$ are positive numbers such that $\sum_{k=1}^{K} \lambda_k = 1$;

and

$x^{(k)}$ $1\le k\le K$ are particular solutions of (2) called extreme points of the polyhedron or <u>basic feasible solutions</u> of (2).

These particular solutions are computed in the following way :

i) from the matrix A with columns $A_1,\dots,A_n$, we extract an $mp\times mp$ invertible matrix $A_J = [A_{j_1},\dots,A_{j_{mp}}]$ defined by an <u>index set</u> $J = \{j_1,\dots,j_{mp}\} \subset \{1,2,\dots,n\}$; such a matrix is called a <u>basis</u> ;

ii) from $\{1,2,\dots,n\}\setminus J$ we extract an <u>index set</u> J_m (that can be empty) and solve the linear system $A_J y = b - \sum_{j\in J_m} x_j^{max} A_j$;

iii) the basic solution defined by J and J_m is then $x = \{x_j\}_{1\le j\le n}$ with $x_{j_i} = y_i$ if $j_i \in J$, $x_j = x_j^{max}$ if $j \in J_m$ and $x_j = 0$ otherwise.

A basic feasible solution is a basic solution which satisfies $0 \le x \le x^{max}$. Since the index sets J and J_m can be choosen in $2^{n-mp}\, C_n^{mp}$ ways, the set of solutions of the inverse problem associated with the measurements $\gamma_i \pm e_i$ $1 \le i \le mp$ and the parallelipipeds ω_j $1 \le j \le np$ is represented by $K \le 2^{np}\, C_{np+mp}^{mp}$ particular solutions.

2. The problems which can be solved

A realistic interpretation of a local three dimensional gravity anomaly requires the size problem $mp \simeq 100\text{-}300$, $np \simeq 400\text{-}800$ and in order to have some insight about the bias introduced by the choice of the measurements and the parallelipipeds the computation must be executed with different sets of data. The bound $2^{np}\, C^{mp}_{np+mp}$ shows that the computation of all basic feasible solutions is impracticable. But the following problems can be solved :

p1 : compute a basic feasible solution of $Ax = b \quad 0 \leq x \leq x^{max}$ or show that the problem has no solution ;

p2 : if the problem has solutions and c^T is an n-components row vector, compute a solution of $Ax = b \quad 0 \leq x \leq x^{max}$ which minimizes the number $c^T x$ (if c is replaced by $-c$, the same algorithm can be used to compute the maximum of $c^T x$); the number $c^T x$ can represent the total mass or some moment of the mass distribution ;

p3 : given two vectors $c_1 \in \mathbb{R}^n$, $c_2 \in \mathbb{R}^n$ draw the functions :

$$p \in \Big[\min_{\substack{Ax=b\\ 0\leq x\leq x^{max}}} c_1^T x \quad , \quad \max_{\substack{Ax=b\\ 0\leq x\leq x^{max}}} c_1^T x\Big] \longrightarrow \min_{\substack{Ax=b\\ 0\leq x\leq x^{max}}} \text{and} \max c_2^T x \; ;$$

if $c_1^T x$ represents the total mass and $c_2^T x$ the first moment of the mass distribution these curves give bounds on the center of mass of the body which produces the anomaly ;

p41 : given a vector $x^o \in \mathbb{R}^n$ and an index set $N'' \subset \{1,2,\ldots,n\}$, compute a solution of $Ax = b \quad 0 \leq x \leq x^{max}$ which minimizes the number $\max_{j \in N''} |x_j - x_j^o|$;

it is thus possible to construct some realistic models ;

p42 : given the solution of p41 and a vector $\delta x^o \in \mathbb{R}^n$ find solutions of p41 when x^o is replaced by $x^o + p\,\delta x^o$ for some values of the parameter p ;

p52 : solve the problem p51 when $\max_{j \in N''} |x_j - x_j^o - p\delta x_j^o|$ is replaced by $\sum_{j \in N''} |x_j - x_j^o - p\,\delta\, x_j^o|$.

Before showing that all these problems can be solved by linear programming techniques let us recall that given a matrix $A \in \mathbb{R}^{m,n}$ $(n > m)$ and vectors $c \in \mathbb{R}^n$, $b \in \mathbb{R}^m$, a standard linear programming problem is to find the minimum of the number $c^T x$ when the vector $x \in \mathbb{R}^n$ is subject to the constraints $Ax = b \; x \geq 0$. A good theoritical account of these techniques can be found in the books of Dantzig (1963) and Golstein and Youdine (1973). Usually the simplex algorithm of Dantzig is used to solve the standard linear programming problem. When a basic feasible solution x of $Ax = b \; x \geq 0$

is known, the basis being $A_J = [A_{j_1},\dots,A_{j_m}]$ with $J = \{j_1,\dots,j_m\}$ and the solution being defined by $x_{j_i} = (A_J^{-1}\, b)_i$ if $j_i \in J$ and $x_j = 0$ otherwise, this algorithm consists of the following steps (we suppose that the m rows of A are linearly independant) :

i) if $\gamma \in \mathbb{R}^m$ is defined by $\gamma_i = c_{j_i}$, compute the simplex multipliers vector $v = A_J^{-T}\gamma$ where A_J^{-T} is the inverse of the transpose of the basis ;

ii) compute the reduced costs $u_j = c_j - A_j^T v$; then :

if $u_j \geq 0$ for $1 \leq j \leq n$ the number $c^T x$ is minimum (because if the solution x is replaced by $x + \delta x$ with $A\delta x = 0$ and $x + \delta x \geq 0$, the number $c^T x$ is replaced by $c^T x + c^T \delta x = c^T x + \sum_{j \notin J} u_j\, \delta x_j \geq c^T x)$;

either jin being an index such that $u_{jin} < 0$ (so that $jin \notin J$) compute the m-components vector $y = A_J^{-1}\, A_{jin}$;

iii) then if $y_i \leq 0$ for $1 \leq i \leq m$ the number $c^T x$ is not bounded (because if x is replaced by $x + \theta\delta x$ with $\delta x_{jin} = 1$ $\delta x_{j_i} = -y_i$ if $j_i \in J$ and $\delta x_j = 0$ if $j \notin J \cup \{jin\}$ the constraints are satisfied for $\theta \geq 0$ but $c^T(x + \theta\delta x) = c^T x + \theta u_{jin} \underset{\theta \to +\infty}{\to} -\infty$) ;

either defined the index $jout = j_{ip} \in J$ by $\frac{x_{j_{ip}}}{y_{ip}} = \min \{\frac{x_{j_i}}{y_i}$ for $y_i > 0\ j_i \in J\}$

iv) replace the index jout by jin in J, sothat J becomes $J \cup \{jin\} \setminus \{jout\}$ and return to i) after having computed the new basic feasible solution.

Since the number of basic feasible solutions is finite, the algorithm converges in a finite number of steps if θ satisfies always $\theta > 0$ (because $c^T x$ becomes $c^T x + \theta u_{jin} < c^T x$ after iv)). If θ becomes null the cycling of the algorithm is possible but this phenomenon is very exceptional and can be saved by a slight improvement of the algorithm.

After having computed a basic solution of $Ax = b$, to find a basic feasible solution one can minimize the linear form defined by $c_j = 0$ if $x_j \geq 0$ and $c_j = -1$ if $x_j < 0$ at each step.

Now, since the constraints $Ax = b$, $0 \leq x \leq x^{max}$ can be written :

$$(4) \qquad \begin{aligned} &Ax = 0 \\ &x + x' = x^{max} \\ &x \geq 0\ x' \geq 0\ (x' \in \mathbb{R}^n) \end{aligned}$$

we conclude that the simplex algorithm allows to solve the problems p1 and p2. The problem p3 is a parametric linear programming problem which can be solved using techniques described in the book of Golstein and Youdine (1973). The problem p41 can be written :

find min α when

$$Ax = b$$

$$x + x' = x^{max} \quad (x' \in \mathbb{R}^n)$$

$$x_j + x''_j - \alpha = x^o_j \text{ for } j \in N''$$

$$x_j - x'''_j + \alpha = x^o_j \text{ for } j \in N''$$

with $x \geq 0$ $x' \geq 0$ $x''_j \geq 0$ and $x'''_j \geq 0$ for $j \in N''$ and $\alpha > 0$ and can be solved by the simplex algorithm.

The equivalence of (5) and p41 follows from the fact that given a solution of $Ax = b$, $0 \leq x \leq x^{max}$ the least possible α in (5) is $\alpha = \max_{j \in N''} |x_j - x^o_j|$ because

$$x_j + x''_j - x = \alpha^o_j \text{ and } x''_j \geq 0 \Rightarrow x_j - x^o_j = \alpha - x''_j \leq \alpha \text{ and}$$

$$x_j - x'''_j + \alpha = x^o_j \text{ and } x'''_j \geq 0 \Rightarrow -\alpha \leq -\alpha + x'''_j = x_j - x^o_j \quad .$$

The problem p51 can be written :

$$\text{find min } \sum_{j \in N''} x''_j + x'''_j \text{ when}$$

$$Ax = b$$

$$x + x' = x^{max} \quad (x' \in R^n)$$

$$x''_j - x'''_j = x^o_j - x_j \text{ for } j \in N''$$

$$x \geq 0 \quad x' \geq 0 \quad x''_j \geq 0 \quad \text{for } j \notin N''$$

(The variables x''_j and x'''_j cannot be simultaneously in basis so that if x''_j is "in basis" $x'''_j = 0$ and $x''_j + x'''_j = x''_j = x^o_j - x_j \geq 0$ and if $x'''_j$is "in basis" $x''_j = 0$ and $x''_j + x'''_j = x'''_j = x_j - x^o_j \geq 0$).

Lastly the problems p42 and p52 are solved by parametric linear programming techniques.

It must be noticed that the simplex algorithm is never applied to problems such as p1,p2,p41,p42 in the raw form that it has been described. A preliminary study is always necessary. In the case of the constraints (4) it is easily shown that a basic solution of (4) can be described in term of index sets $J = \{j_1, \ldots, j_{mp}\} \subset N = \{1,2,\ldots,n\}$ and $J_m \subset N \setminus J$ such that $A_J = [A_{j_1}, \ldots, A_{j_{mp}}]$ is an invertible $mp \times mp$ matrix and $x_j = x^{max}_j$ if $j \in J_m$. Similar but more complicated results hold for problems p41 and p51. In all cases the order of the matrix to invert at each step is mp, the number of measurements.

3. Available numerical techniques

The first numerical problem is the recurrent computation of the different inverse basis A_J^{-1}. The most known technique is as follows.
If the index set J becomes $J' = J \cup \{jin\} \setminus \{j_{ip}\}$ we have
$A_{J'} = A_J + (A_{jin} - A_{j_{ip}})\, e_{ip}^T$ where e_{ip} is the mp dimensional vector whose components are null except the ip^{th} which is 1. Using the Sherman Morrison formula (see Householder (1964) p. 123) one finds that :

$$(7) \qquad A_{J'}^{-1} = \left(I_{mp} + \frac{e_{ip} - y}{y_{ip}}\, e_{ip}^T\right) A_J^{-1}, \quad \text{where } I_{mp} \text{ is the identity matrix in } \mathbb{R}^{mp}$$

and $y = A_J^{-1} A_{jin}$
Likewise the reduced costs u_j become $u_j' = u_j - u_{jin} A_j^T v_p$ with $v_p = A_{J'}^{-T} e_{ip}$. But since the ip^{th} diagonal element of

$I_{mp} + \dfrac{e_{ip} - y}{y_{ip}}\, e_{ip}^T$ is $\dfrac{1}{y_{ip}}$, the formula (7) is instable. In order to ensure numerical stability it is necessary to make some complete reinversions of the basis from time to time using a stable Gaussian elimination. A great number of complete reinversions increases the CPU time.
At present we use FORTRAN subroutines which are adjustments of the ALGOL program of Bartels Stoer and Zenger (1971). These subroutines exploit the stable Bartel Golub decomposition (Bartels (1971)) where the basis A_J is handled with matrices R and L such that :

$$(8) \qquad LA_J = R$$

R upper triangular

If J becomes $J' = J \cup \{jin\} \setminus \{j_{ip}\}$, by moving columns of R of order > ip forward by one place, R becomes the upper Hessenberg matrix :

(9) $\overset{\infty}{R} =$

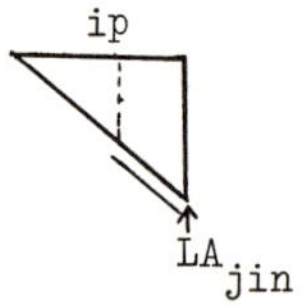

Some eliminations on rows of order $\geq$ ip, perhaps including interchanges between adjacent rows, suffice to restore the upper triangular form in a numerical stable way. Thus the number of necessary complete reinversions is reduced.
The second numerical problem is the choice of the index jin. At present we use the method of Dantzig :

(10) $$u_{jin} = \min_{j \notin J} u_j .$$

But is has long been known (Kuhn and Quandt (1963)) that significantly less iterations are necessary if the "steepest edge" is always taken, that is to say if

(11) $$\frac{u_{jin}}{\|\delta x^{(jin)}\|_2} = \min_{j \notin J} \frac{u_j}{\|\delta x^{(j)}\|_2}$$

where $\|\delta x^{(j)}\|_2$ is the euclidian norm of the vector $\delta x^{(j)} \in \mathbb{R}^n$ such that

$\delta x_j^{(j)} = 1 \quad \delta x_{j_i}^{(j)} = -y_i$ with $y = A_J^{-1} A_j$ if $j_i \in J$ and $\delta x_k^{(j)} = 0$ otherwise.

For large problem explicit computation of all norms $\|\delta x^{(j)}\|_2$ at each step is too expensive. Fortunately Goldfard and Reid (Reid (1975)) have shown that the Sherman Morrison formula yieds to the updating formula for $\eta_j = \|\delta x^{(j)}\|_2^2$:

$$\eta'_{j_{ip}} = \frac{\eta_{jin}}{y_{ip}^2} \quad \text{where} \quad y = A_J^{-1} A_{jin}$$

$$\eta'_j = \eta'_j - 2(A_j^T y_p)(A_j^T v_p) + (A_j^T v_p)^2 \eta_{jin} \quad \text{with} \quad y_p = A_J^{-T} y$$

$$\text{and} \quad v_p = A_{J'}^{-T} e_{ip} .$$

4. Conclusion

Some tests have shown that the computation of the numbers η_j is practicable. On an IBM 360/65 we have find the following CPU time to solve a gravity problem with 100 measurements and 400 parallelipipeds :

9mn 31.34s to obtain a first feasible solution

20mn 53.62s to solve a problem p41.

Numerical experiment leads to the conclusion that when careful implementation is made the linear programming techniques can be used in linear inverse problems if the number of measurements is less than 200-300 and the number of unknowns less than 400-800.

REFERENCES

Bartels, R.H., Stoer, J., Zenger C.H. (1971) : A realisation of the simplex method based on triangular decomposition in Handbook for automatic computation vol 2 : J.H. Wilkinson and C. Reinsh editors, Berlin, Springer Verlag p 152-190.

Bartels, R.H. (1971) : A stabilisation of the simplex method Numer Math 46, 414-434.
Bourmatte (1977) : Thèse de 3ième cycle. CGG Montpellier.
Dantzig, G.B (1963) : Linear programming and extensions. Princeton University Press.
Golstein, E., Youdine, D. (1973) : Problèmes particuliers de la programmation linéaire. Mir. Moscou.
Householder, A.S (1964) : The theory of matrices in numerical analysis Blaisdell Publishing Compagny New York p 123.
Kuhn, H.W., Quandt, R.E (1963) : An experimental study of the simplex method in Proceeding of Symposia an Applied mathematics vol XV. Amer. Math. Soc. Providence. p 107-124.
Reid, J.K (1975) Sparce in-core linear programming in Lecture Notes in Mathematics vol 506 p 176-189 Springer-Verlag.
Rodkafellar, R.T (1970) : Convex analysis. Princeton University Press.
Sabatier PC (1977a) : Positivity constraints in linear inverse problem : I General theory : Geophys. J. Roy. Astr. Soc. vol 48 p 415-422.
Sabatier PC (1977b) : Positivity constraints in linear inverse problems : II Applications. Geophys, J. Roy. Astr. Soc. Vol 48 p 443-469.
Safon C, Vasseur G, Cuer M : (1977) Some applications of linear programming to the inverse gravity problem. Geophysics, Vol 42 N°6 p 1215-1229.
Stoer, J. Witzgall, C. (1970) : Convexity and optimization in finite dimensions I Springer Verlag.

Analytic Extrapolations and Inverse Problems

D. Atkinson

University of Groningen, The Netherlands.

Abstract

The stabilized analytic extrapolation techniques of Ciulli and co-workers are explained. Certain applications in the field of high-energy physics are used as illustrative examples.

I want to discuss some techniques that have much in common with ill-posed problems, namely certain algorithms for making the analytic continuation of a function, that is measured experimentally in one region, to another region of interest. One might characterize the problem, and its difficulty, by two light-hearted theorems:

Divine Theorem: If one knows an analytic function on a segment of a line, inside its domain of analyticity, one knows it throughout this domain of analyticity.

Diabolical Theorem: If the function is not known exactly on the line segment, but only within an error corridor of width 2ε, then the uncertainty in the continued function is such that its value at any given point, in the domain of analyticity, can be any number whatsoever, and this for any ε, no matter how small. (See appendix A, Ciulli 75, for a central European proof of this theorem).

The devilishness of the latter theorem arises from the fact that we cannot exclude the possibility that the function has insanely rapid oscillations within the error corridor, and these oscillations can explode exponentially as one continues away from the experimental region.

Since our experimental colleagues seem to be incapable of giving us error-free measurements, we must take the diabolical theorem very seriously: we need to study stabilizing conditions, which serve to exorcize the wildest oscillations, and I will sketch some recent ideas of S. Ciulli and co-workers. Some of the details can be found in the Physics Report article, Ciulli 75, while other ideas have not yet been published. First of all, we must make a distinction between exterior-interior and interior-interior extrapolations. I shall discuss each in turn, sketching possible applications in the field of high-energy physics.

Exterior-Interior Extrapolations

By this description we understand that an analytic function is measured on certain cuts, and that its value is required at some point away from the cuts. An example of this kind of problem is the determination of the pion-nucleon coupling constant from the measured forward pion-nucleon scattering amplitude. This amplitude is a real-analytic function with a cut on the real axis of the variable $s = E^2$, from $E = M + m$ to $E = \infty$, where E is the total energy in the cms system, and where M and m are respectively the masses of the nucleon and the pion. The values of the amplitude, say F(s), are physically accessible, from a study of πN scattering, if s is on the cut; and we wish to make an extrapolation to the point $s = M^2$, where F(s) has a pole, and to extract the residue of this pole, which is the square of the πN coupling constant that we wish to estimate. (In the full problem, there is a second cut, $(M + m)^2 \leqslant u < \infty$, and a second pole, at $u = M^2$, where $u = 2(M^2 + m^2) - s$. However, these u-channel effects can be easily accommodated by using crossing symmetry, and we shall simply neglect them here, in order to keep our illustrative example as simple as possible).

A simple way of obtaining the residue is by means of a dispersion relation, followed by a numerical extrapolation, and this is the way the problem was first solved, Chew 59 and Hamilton 63. However this method, which I will not explain further, has the disadvantage that it is difficult to estimate the effects of the various uncertainties in the data. These uncertainties may be broadly divided into three classes:

A. The statistical experimental errors in the measurements of the cross-sections and polarizations.

B. The inherent ambiguities in the determination of the amplitudes from the cross-sections and polarizations, Bowcock 75. One needs both the real and imaginary parts of the amplitude, and it has been shown that πN phase-shift analysis is subject to non-negligible continuum ambiguities, Atkinson 76.

C. The dispersion relation involves an integral of the amplitude out to infinity, and in practice the experimental data only extend to a finite energy.

The method of Ciulli is designed to make the extrapolation in an optimal manner, and to estimate in an honest way the effect of the above uncertainties. The first step(naturally!) is to make a conformal mapping $s \to \omega$, such that the cut s-plane becomes the unit disk $|\omega| < 1$. The real axis, $-\infty < s < (M + m)^2$, is mapped onto the real line, $-1 < \omega < 1$, and the pole at $s = M^2$ is now at the origin, $\omega = 0$. The upper and lower lips of the cut, $(M + m)^2 < s < \infty$, are mapped respectively onto the upper and lower halves of the unit circle, $|\omega| = 1$. Let us define f by

$$F(s) = \frac{f(\omega)}{\omega} , \qquad (1)$$

so that $f(\omega)$ has no pole at the origin, and we want to evaluate $f(0)$. We suppose that F(s) has been determined from $s = (M + m)^2$ to some highest available energy, which means that $f(\omega)$ is known approximately on part of the unit circle, a domain Γ_1 say, but subject to the uncertainties A and B above. Let us designate by $d(\omega)$ a fit to this data, for $\omega \in \Gamma_1$, with estimated error $\Delta d(\omega)$, arising from sources A and B. Then the function $f(\omega)$ should be analytic for $|\omega| < 1$, and it should satisfy

$$|f(\omega) - d(\omega)| \leqslant \Delta d(\omega) \tag{2}$$

for $\omega \in \Gamma_1$. Unfortunately this is not enough to determine $f(0)$, nor to limit the πN coupling constant in any way! The difficulty is that we do not know $f(\omega)$ on the rest of the unit circle, say $\omega \in \Gamma_2$, since this domain represents energies beyond the reach of the available experiments. By making $f(\omega)$ sufficiently disgusting for $\omega \in \Gamma_2$, one could manage to satisfy (2) but to obtain <u>any</u> value for $f(0)$! This is an example of the diabolical theorem; and what we need is a stabilizing condition on Γ_2 in order to make any prediction at all.

It is necessary to make some theoretical assumption about the nature of $f(\omega)$ in the experimentally inaccessible region, Γ_2. If one has a phenomenological model of the process in question, then one can parametrize the function in the high-energy region, $\omega \in \Gamma_2$, with perhaps some uncertainties as to the correct values of the various parameters. In our paradigm, πN forward scattering, such a model would come from Regge theory, with an important contribution from Pomeron exchange in the t-channel. It is sufficient for our purposes to use the model to construct an upper bound for the modulus of the amplitude, so that we can assert

$$|f(\omega)| \leqslant M(\omega), \tag{3}$$

for $\omega \in \Gamma_2$. For πN scattering, one would simply take the Regge parametrization of $f(\omega)$, calculate its modulus, and then play with the Regge parameters to find the worst, i.e. the greatest acceptable value of $M(\omega)$ for each ω in Γ_2. For technical reasons, it is convenient to make $M(\omega)$ (and also $\Delta d(\omega)$) a continuous, and even a differentiable function, and this can always be managed without difficulty. We can now demonstrate that the problem of making the extrapolation to $\omega = 0$ (or to any other point inside the unit circle), is well-posed, in the sense that the diabolical theorem has been laid to rest.

The next step is the construction of an exterior function, $h(\omega)$, i.e. a function that is real-analytic and zero-free in $|\omega| < 1$. We define

$$h(\omega) = \exp\left[\sum_{n=0}^{\infty} h_n \omega^n\right], \tag{4}$$

where the real coefficients are determined by the conditions

$$\tfrac{1}{2}\sum_{n=0}^{\infty} h_n(\omega^n + \omega^{-n}) = \log \Delta d(\omega)\ , \ \omega \in \Gamma_1, \tag{5A}$$

$$\tfrac{1}{2}\sum_{n=0}^{\infty} h_n(\omega^n + \omega^{-n}) = \log M(\omega)\ , \ \omega \in \Gamma_2. \tag{5B}$$

Since the domains Γ_1 and Γ_2 together make up the unit circle $|\omega| = 1$, eqs. 5 are simply Fourier cosine series, as we see by setting $\omega = e^{i\theta}$. As we noted above, we have the freedom to make $\Delta d(\omega)$ and $M(\omega)$ as continuous as we wish, and it is convenient (although not essential) to assume that these functions have a Hölder-continuous derivative, even at the points at which Γ_1 and Γ_2 join. Then the Fourier series is absolutely convergent; and the Fourier coefficients can be determined by inverting eqs. 5 in the standard manner.

Since $h(\omega)$ has no zeros for $|\omega| < 1$, it follows that $\log h(\omega)$ is analytic for $|\omega| < 1$; and further, by construction,

$$|h(\omega)| = \Delta d(\omega), \text{ for } \omega \in \Gamma_1, \tag{6A}$$

$$|h(\omega)| = M(\omega), \text{ for } \omega \in \Gamma_2. \tag{6B}$$

Hence, if we define a reduced real-analytic function

$$\tilde{f}(\omega) = f(\omega)/h(\omega) \tag{7}$$

for all ω in the unit disk, and a reduced data function,

$$\tilde{d}(\omega) = d(\omega)/h(\omega), \tag{8A}$$

for $\omega \in \Gamma_1$, and

$$\tilde{d}(\omega) = 0, \tag{8B}$$

for $\omega \in \Gamma_2$, then the two inequalities (2) and (3) can be subsumed under the single condition

$$|\tilde{f}(\omega) - \tilde{d}(\omega)| \leqslant 1, \tag{9}$$

for $\omega \in \Gamma_1 + \Gamma_2$, i.e. for $|\omega| = 1$. Our task now is to characterize all functions, $\tilde{f}(\omega)$, that are analytic in the unit disk, and which satisfy (9).

We make a Fourier expansion of the reduced data function:

$$\tilde{d}(e^{i\theta}) = \sum_{n=-\infty}^{\infty} d_n e^{in\theta}. \tag{10}$$

It is important to note that this series is valid only for real θ, i.e. for ω on the unit circle. Since $\tilde{d}(\omega)$ vanishes for $\omega \in \Gamma_2$ and not for $\omega \in \Gamma_1$, $\tilde{d}(\omega)$ cannot possibly have an analytic extension into the unit disk. Of course $\tilde{f}(\omega)$, per hypothesi, has such an extension, so that it can be written, for $|\omega| < 1$, as

$$\tilde{f}(\omega) = \sum_{n=0}^{\infty} f_n \omega^n. \tag{11}$$

We must now practise the delicate art of matching the positive frequencies of $\tilde{f}$ with the positive and negative frequencies of $\tilde{d}$, in such a way that (9) is observed, i.e. such that

$$\left| \sum_{n=-\infty}^{\infty} d_n e^{in\theta} - \sum_{n=0}^{\infty} f_n e^{i\theta} \right| \leqslant 1, \tag{12}$$

for $\theta \in (-\pi, \pi)$. This problem can be solved in completely general terms; but an important simplification can be made if we assume that $d(\omega)$ is sufficiently continuous to ensure the absolute convergence of the Fourier series (10). In that case we can truncate the negative frequencies at n = -N, writing

$$\hat{d}(e^{i\theta}) = \sum_{-N}^{\infty} d_n e^{i\theta}; \tag{13}$$

and the difference between the true data function, $\tilde{d}$, and the truncated data function, $\hat{d}$,

$$\left|\hat{d}(e^{i\theta}) - \tilde{d}(e^{i\theta})\right| = \left|\sum_{-\infty}^{-N-1} d_n e^{i\theta}\right| \leqslant \sum_{-\infty}^{-N-1} \left|d_n\right| \equiv \eta_N, \tag{14}$$

can be made as small as one likes by making N large enough.

We may replace (12) by

$$|\hat{d}(e^{i\theta}) - \tilde{f}(e^{i\theta})| \leqslant 1 + \eta_N, \tag{15}$$

where we choose N so large that $\eta_N << 1$. Finally, we make a minor renormalization,

$$\bar{f}(e^{i\theta}) = (1 + \eta_N)^{-1}\tilde{f}(e^{i\theta}); \tag{16A}$$

$$\bar{d}(e^{i\theta}) = (1 + \eta_N)^{-1}\hat{d}(e^{i\theta}), \tag{16B}$$

so that

$$|\bar{d}(e^{i\theta}) - \bar{f}(e^{i\theta})| \leqslant 1. \tag{17}$$

Now although $\hat{d}(\omega)$ did not have an extension into the unit disk, the artificially truncated $\bar{d}(\omega)$ does have such a meromorphic continuation. Clearly it has an Nth order pole at the origin. Hence the function

$$\psi_o(\omega) = \omega^N\left[\bar{d}(\omega) - \bar{f}(\omega)\right]$$

$$= (1+\eta_N)^{-1} \sum_{n=0}^{N-1} d_{n-N}\omega^n + (1+\eta_N)^{-1} \sum_{n=N}^{\infty} (d_{n-N} - f_{n-N})\omega^n, \tag{18}$$

is analytic in the unit disk; and so, by virtue of (17) and the maximum modulus principle,

$$|\psi_o(\omega)| \leqslant 1, \tag{19}$$

for $|\omega| \leqslant 1$.

The function $\psi_o(\omega)$ is not the most general analytic function that is bounded by unity within the unit disk, since its first N Taylor coefficients are determined by the negative frequency coefficients of $\bar{d}$, as we see from (18). Accordingly, we define successively

$$\psi_{k+1}(\omega) = \frac{1}{\omega} B\left[\psi_k(\omega), \psi_k(0)\right], \tag{20}$$

for k = 0, 1, 2, ..., N-1, where the Blaschke factor is given by

$$B[a,b] = \frac{a-b}{1-ab} \; . \tag{21}$$

Now one can show easily that $|\psi_k(\omega)| \leqslant 1$ implies $|\psi_{k+1}(\omega)| \leqslant 1$, and hence, for all $|\omega| \leqslant 1$, we may assert

$$|\psi_k(\omega)| \leqslant 1 \tag{22}$$

for k = 0, 1, 2, ..., N. Moreover, whereas the first N Taylor coefficients of $\psi_o(\omega)$ were determined by the negative frequency coefficients of $\overline{d}$, only the first N-1 coefficients of $\psi_1(\omega)$ are so determined. Similarly, only the first N-2 coefficients of $\psi_2(\omega)$ are prescribed, and so on, so that finally $\psi_N(\omega)$ has no prescribed coefficients at all.

Since $\psi_N(\omega)$ is analytic in the unit disk, and is bounded there in modulus by unity, the maximal member of the set of all permissible functions is simply the constant +1, and the minimal member is the constant -1. One can invert (20) to find $\psi_{N-1}(\omega)$ in terms of $\psi_N(\omega)$ and $\psi_{N-1}(0)$, which latter is a function only of the known negative frequency coefficients of $\overline{d}$. Moreover, $\psi_{N-1}(\omega)$ is a monotonically increasing function of $\psi_N(\omega)$, at fixed $\psi_{N-1}(0)$, so that its extremal values are assumed when $\psi_N(\omega)$ takes on its extremal values, namely $\pm$ 1. In a similar manner, one can proceed recursively, determining successively the extremal members of the set of all possible $\psi_k(\omega)$, for k = N-1, N-2, ...,0. From the relation

$$\overline{f}(\omega) = \overline{d}(\omega) - \omega^{-N}\psi_o(\omega), \tag{23}$$

one obtains the extremal members of the set of $\overline{f}(\omega)$, subject to (17) and the condition of analyticity within the unit circle. Hence the maximum and minimum possible values of $\overline{f}(0)$ can be calculated: the mean is then the predicted value of the extrapolated function, the difference twice the extrapolated uncertainty.

This method has been used in the design of a general computer program, Ciulli 77; and related techniques have been applied, Gensini 77, to study both the $\pi\pi$ and the πN systems.

Interior - Interior Extrapolations.

Here we understand that an analytic function is measured at certain points not on the cuts, and that its value is required at some other point, not on the cuts. An example of this kind of problem is the extrapolation of the differential cross-section for pion production ($\pi N \to \pi\pi N$), from the physically accessible region, $t < 0$, to the (second order) pion pole, at $t = m^2$, where t is the square of the difference of the initial and final nucleon four momenta. This cross-section, $\sigma(t)$, depends on the total energy, and on some angles, as well as t, but we shall suppress these extra variables. In addition to the double pole at $t = m^2$, $\sigma(t)$ also has a cut running from $t = 9m^2$ to $t = \infty$. In this case we need a stabilizing condition, in the form of an upper bound for $|\sigma(t)|$, for all points on the cut, since none of them are experimentally accessible. A suitably emasculated off-shell dual pion four-point function might be used; but such a model is so questionable that considerable latitude would have to be left in the allowed ranges of the various parameters. Nevertheless, the point must be made again, ad nauseam, that without such a stabilizing condition, the analytic continuation cannot meaningfully be made, since, Diabolus Volens, any value could be obtained for the extrapolate.

The first step (de rigueur is to map the cut t-plane into the unit disk in the ω-variable, in the standard manner, so that the cut is mapped onto $|\omega| = 1$, and such that the physical region, $t < 0$, is transformed into a part of the interval $-1 < \omega < 1$, the pion pole at $t = m^2$ becoming $\omega = \omega_o$, where ω_o lies to the right of the physical region, and to the left of $\omega = 1$ (the image of the branch-point, $t=9m^2$).

We define $f(\omega)$ by

$$\sigma(t) = \frac{f(\omega)}{(\omega-\omega_o)^2} , \tag{24}$$

so that $f(\omega)$ has no pole at $\omega = \omega_o$. We suppose that $f(\omega)$ has been measured at a certain set of experimental points, $\omega_1, \omega_2, \ldots, \omega_N$. Let us designate these values by $d(\omega_k)$, $k = 1, 2, \ldots, N$, where we impose the ordering

$$-1 < \omega_N < \omega_{N-1} < \ldots\ldots\ldots < \omega_1 < \omega_o < 1 \tag{25}$$

for convenience. For the moment we shall neglect experimental errors completely, so we require $f(\omega)$ to be real-analytic for $|\omega| < 1$, and to be such that

$$f(\omega_k) = d(\omega_k), \tag{26}$$

for $k = 1, 2, \ldots, N$. On the unit circle (i.e. the cut in the t-variable), we impose a stabilizing condition,

$$|f(\omega)| \leqslant M(\omega), \tag{27}$$

for $|\omega| = 1$.

Now we construct an exterior function, $h(\omega)$, as in the exterior-interior problem, except that, since (27) holds for all $|\omega| = 1$, we must impose (5B) on the whole circle, and thus deduce (6B) also for all points on the unit circle. We define a reduced function, $\tilde{f}(\omega)$, as in (7), and reduced data, $\tilde{d}(\omega_k)$, as in (8A), except that this is now done at the physical points, $\omega_1, \omega_2, \ldots, \omega_N$, which are on the real axis. The reduced problem consists in finding a real-analytic function, $\tilde{f}(\omega)$, such that

$$\tilde{f}(\omega_k) = \tilde{d}(\omega_k), \tag{28}$$

for $k = 1, 2, \ldots, N$, and for which

$$|\tilde{f}(\omega)| \leqslant 1, \tag{29}$$

for $|\omega| = 1$. By the maximum-modulus principle, (29) can be immediately extended to the complete unit disk, $|\omega| \leqslant 1$.

Let us define successively, in terms of the Blaschke factor (21),

$$\tilde{f}_{p+1}(\omega) = \frac{B[\tilde{f}_p(\omega), \tilde{f}_p(\omega_p)]}{B[\omega, \omega_p]}, \tag{30}$$

for p = 1, 2, ..., N, where $\tilde{f}_1(\omega) = \tilde{f}(\omega)$, $\tilde{f}_1(\omega_p) = \tilde{f}(\omega_p) = \tilde{d}(\omega_p)$, and where

$$\tilde{f}_{p+1}(\omega_{p+1}) = \frac{B[\tilde{f}_p(\omega_{p+1}), \tilde{f}_p(\omega_p)]}{B[\omega_{p+1}, \omega_p]}, \tag{31}$$

for p = 1, 2, ..., N-1. Now $\tilde{f}_1(\omega)$ is specified at the N points $\omega_1, \omega_2, \ldots, \omega_N$, according to (28), but $\tilde{f}_2(\omega)$ is specified only at the N-1 points $\omega_2, \omega_3, \ldots, \omega_N$, $\tilde{f}_3(\omega)$ at N-2 points, and so on, so that $\tilde{f}_N(\omega)$ is specified at only one point, ω_N, and finally $\tilde{f}_{N+1}(\omega)$ is not specified at any point. Moreover, all the functions, $\tilde{f}_p(\omega)$, p = 1, 2, ..., N+1, are analytic in $|\omega| < 1$, and it is easy to show, from the properties of the Blaschke factor, that $|\tilde{f}_p(\omega)| \leqslant 1$ implies $|\tilde{f}_{p+1}(\omega)| \leqslant 1$. Hence all the functions, $\tilde{f}_p(\omega)$, are bounded in modulus by unity; and in fact $\tilde{f}_{N+1}(\omega)$ may be any such real-analytic function, since it is not specified at any point.

As before, the constant +1 is the maximal, and -1 the minimal element in the set of all permissible functions, $\tilde{f}_{N+1}(\omega)$. One can invert (30) to obtain $\tilde{f}_p(\omega)$ in terms of $\tilde{f}_{p+1}(\omega)$ and $\tilde{f}_p(\omega_p)$, which latter is known in terms of the data values $\tilde{d}(\omega_k)$, via (31). Again one can show that the extremal functions, $\tilde{f}_p(\omega)$, are obtained, at constant $\tilde{f}_p(\omega_p)$, by inserting the extremal functions, $\tilde{f}_{p+1}(\omega)$. Hence one can obtain the extremal $\tilde{f}_N(\omega)$ by substituting $\pm$ 1 for $\tilde{f}_{N+1}(\omega)$, and thence obtain the extremal $\tilde{f}_k(\omega)$ successively, for k = N-1, N-2, ..., 1. The mean of the extremal values of $\tilde{f}_1(\omega)$, at the point $\omega = \omega_o$, constitutes the prediction for the one-shell pion-pion cross-section, and the difference is twice the uncertainty.

It should be noted carefully that the extrapolated uncertainty, to which we have just alluded, has nothing to do with the experimental error, which we have so far neglected. This uncertainty is solely a consequence of the freedom that is allowed by the stabilizing function, $M(\omega)$. Let us now suppose that the data function, $d(\omega_k)$, is in fact subject to an experimental error, $\Delta d(\omega_k)$. We define a reduced error by dividing by $h(\omega_k)$; and so now, instead of (28), we require only

$$|\tilde{f}(\omega_k) - \tilde{d}(\omega_k)| \leqslant \Delta\tilde{d}(\omega_k), \tag{32}$$

for k = 1, 2, ..., N. We wish to estimate the effect of these errors, in order to see how they are compounded with the uncertainties allowed by the stabilizing condition. Unfortunately, we have no elegant algorithm to do this. One can of course successively take different sets of trial values, $\tilde{f}(\omega_k)$, k = 1, 2, ..., N, which are consistent with (32), and repeat the method we have described for each set.

The ultimate extremal values of $\tilde{f}(\omega_o)$ would then be extracted from the complete set of results.

In fact, there is a serious practical difficulty, since, if one has a dozen experimental points or more, and a reasonable stabilizing condition, then for most choices of $\tilde{f}_1(\omega_k)$, k = 1, 2, ..., N, consistent with (32), there is simply no solution at all! A set of values selected by hand, at a few dozen points, can usually only be interpolated by the sort of hellish function for which the stabilizing condition was specifically invoked to cast out. This phenomenon manifests itself when one calculates $\tilde{f}_p(\omega_k)$, p = 2, 3, ..., N; k = p, p+1, ..., N. These numbers are functions only of the selected values, $\tilde{f}_1(\omega_k)$, k= 1, 2, ..., N, and they should all lie in the interval (-1,1). One has first to check all these numbers for each set of selected values; and if any of them falls outside (-1,1) one must throw the set away and try again.

A recent streamlining of this approach, due to Stefǎnescu, starts from the frank recognition that the unaided eye of man is incapable of distinguishing between an allowed set of values and a very slightly different set, that is nevertheless rejected by the stabilizing condition. Let us take the measured data set, $\tilde{d}(\omega_k)$, which will generally admit only infernal interpolations, and calculate

$$\chi_N^2(\tilde{f}) = \frac{1}{N} \sum_{k=1}^{N} \left[\frac{\tilde{f}(\omega_k) - \tilde{d}(\omega_k)}{\Delta\tilde{d}(\omega_k)} \right]^2 . \tag{33}$$

Now let $\tilde{f}$ range over all the allowed functions, i.e. all the real-analytic functions that are bounded in modulus by one within the unit circle. If we can calculate, or at least approximate well, the function, $\tilde{f}_o(\omega)$, in this allowed set, for which $\chi_N^2(\tilde{f})$ is minimum, then this minimal function is the best attempt we can make to match the data with the stabilizing condition, and $\chi_N^2(\tilde{f}_o)$ is a measure of the goodness of fit, in the usual statistical sense. In fact, the minimization of $\chi_N^2(\tilde{f})$ does much better justice to the world of experimental physics than does a rigid insistence on inequality (32), since errors are normally estimated as statistical standard deviations rather than as absolute uncertainties.

One can think of the set of all values of $\tilde{f}(\omega_k)$, k = 1, 2,..,N, that are compatible with the stabilizing condition as a certain body in an N-dimensional space, where the k-th coördinate is simply the value of $\tilde{f}(\omega_k)$. Each point of the body represents an allowed set of values, $\tilde{f}(\omega_k)$, k = 1, 2, ..., N. The data set $\tilde{d}(\omega_k)$, k = 1, 2, ..., N, will also be a point in the N-dimensional space; but it will usually lie outside the body. The task is to find the "distance" from the data point to the body of non-diabolical functions, and the point of this body that is closest to the data point.

In fact it is convenient, or at least amusing, to introduce two slight changes in the rules. It is the case that a given point in the body does not correspond to just one function, $\tilde{f}(\omega)$, but rather to an infinite family of them. We can remove this degeneracy by replacing the data points by a continuous function, $\tilde{d}(\omega)$, a fit to the measurements, as we did in the discussion of exterior-interior extrapolations. We suppose that $\Delta\tilde{d}(\omega)$, the standard deviation, is also given as a continuous function. Then we can replace the sum (33) by an integral. Before we do that, however, we shall indulge in a final contortion, namely the Cayley transformation:

$$\check{d}(\omega) = C\,[\tilde{d}(\omega)] \quad , \tag{34A}$$

$$\Delta\check{d}(\omega) = C\,[\Delta\tilde{d}(\omega)] \quad , \tag{34B}$$

$$\check{f}(\omega) = C\,[\tilde{f}(\omega)] \quad , \tag{34C}$$

where

$$C\,[z] = i\,\frac{1-z}{1+z} \quad . \tag{35}$$

Now $|\tilde{f}(\omega)| \leqslant 1$ implies $\mathrm{Im}\,\check{f}(\omega) \geqslant 0$, and that is the point of the Cayley transform, for a stabilizing condition of boundedness of the modulus has been traded for a condition of positivity of the imaginary part.

We generalize our picture of an N-dimensional space of function values to a space with a continuous infinity of dimensions, and imagine (or try to imagine!) the body of functions, $\check{f}(\omega)$, with non-negative imaginary parts. We define a distance function that generalizes (33):

$$\chi^2(\check{f}) = \frac{1}{b-a} \int_a^b d\omega \left[\frac{\check{f}(\omega) - \check{d}(\omega)}{\Delta\check{d}(\omega)} \right]^2 , \tag{36}$$

where (a,b) is the interval of the real ω-axis that corresponds to the experimentally accessible domain. In the topology induced by the distance function (36) on the set of analytic functions $\check{f}(\omega)$, the set of allowed functions (i.e. the set for which $\mathrm{Im}\,\check{f}(\omega) \geqslant 0$ for $|\omega| \leqslant 1$), is a closed, convex cone. As in the finite-dimensional case, the data function, $\check{d}(\omega)$, will generally not lie inside this hallowed body. Ştefănescu shows that there is a unique function, $\check{f}_o(\omega)$, in the cone that minimizes χ^2. Moreover, although the cone has infinitely many dimensions, it is an extremely flattened body in practice; and a good approximation to $\check{f}_o(\omega)$ and $\chi^2(\check{f}_o)$ can be found by working with the projection of the cone onto a finite, small number of dimensions.

A further field of application of the stabilized interior-interior extrapolation is in energy-dependent phase-shift analysis, Pietarinen 76. There are theoretical advantages in working in the unphysical region $0 \leqslant t < 4m^2$, Burkhardt 75; and a preliminary extrapolation in t from the physical region to this region of positivity, and a subsequent Pietarinen analysis for these unphysical t-values, should usefully complement the work in the physical region.

References

Atkinson 76. D. Atkinson, A.C. Heemskerk, and S.D. Swierstra, Nucl. Phys. B109, 322(1976).

Bowcock 75. J.E. Bowcock and H. Burkhardt, Rep. Prog. Phys. 38, 1099(1975).

Burkhardt 75. H. Burkhardt and A. Martin, Nuovo Cim. 29A, 141(1975).

Chew 59. G.F. Chew and F.E. Low, Phys. Rev. 113, 1640(1959).

Ciulli 75. S. Ciulli, C. Pomponiu and I.S. Stefănescu, Phys. Rep. 17C, 134 (1975).

Ciulli 77. M. Ciulli and S. Ciulli, Cern TH-2267, 2268(1977).

Gensini 77. P. Gensini (to be published in Phys. Rev.).

Hamilton 63. J. Hamilton and W.S. Woolcock, Rev. Mod. Phys. 35, 737(1963).

Pietarinen 76. E. Pietarinen, Nucl. Phys. B107, 21(1976).

STIELTJES FUNCTIONS AND APPROXIMATE SOLUTIONS OF AN INVERSE PROBLEM

G. Turchetti - C. Sagretti

Istituto di Fisica - Università di Bologna, Italy

Introduction.

A large class of inverse problems consists in finding a second order differential operator D when its spectral function or an equivalent set of data is given. There is rich mathematical literature for the Schrödinger and Sturm-Liouville operators: existence and uniqueness conditions and inversion equations have been given by Gel'fand-Levitan (1951), Marchenko (1955, 1963, 1974), Regge (1950), Newton (1962), Sabatier (1966), Borg (1946), Krein (1955a, 1955b) and Barcilon (1974a, 1974b). In practical applications only finite sets of data are available and consequently the inverse problem is no longer uniquely defined. Computational methods based on trial and error procedures are commonly used; however, the final results and the convergence rate depend on the initial guess of the differential operator.

In the case of second order Sturm-Liouville operators D one can also find, for any finite set of data, a unique finite difference operator Δ. The required formalism, related to positive continued fractions, was first developed by Stieltjes in his investigation of the oscillations of an elastic chain. The operators D and Δ are uniquely determined by the spectra associated to a different pair of boundary conditions, corresponding, in the mechanical analogue, to a string or a chain with both ends fixed, with one end fixed and the other free to move. The spectra of D are infinite sequences of eigenvalues $\{\lambda_i\}_{i=1}^{\infty}$, $\{\lambda_i'\}_{i=1}^{\infty}$ and any truncation at a finite order N does no longer allow the determination of D but rather of the finite difference operator of rank N Δ_N. A one-to-one correspondence also exists between D and a meromorphic Stieltjes function $R(\lambda)$ having λ_i, λ_i' as zeroes and poles, Δ_N and the corresponding Stieltjes rational fraction $R_N(\lambda)$.

Therefore, starting from a sequence of values of $R(\lambda)$ or a sequence of Taylor coefficients, we can construct converging sequences of Stieltjes fractions $R_N(\lambda)$ and the corresponding finite difference operators Δ_N. The procedure is unambiguously defined and, if a subsequence of Δ_N converges, the limit is D.

After a short review of the basic properties of Stieltjes functions in section I, we discuss the Sturm-Liouville equation of a transversally vibrating string in section II, and the S wave Schrödinger's equation for a finite range potential in section III.

Section I - Stieltjes functions.

A Stieltjes function f(x) is the Hilbert transform of a positive measure and can be written

$$(1.1) \qquad f(x) = \int_0^\infty \frac{d\mu(u)}{1+ux}$$

where $\mu(u)$ is a bounded non-descreasing function. Different analytic structures of f(x) are allowed: if $\mu(u)$ is monotonically increasing for $0 < u < 1/R$ and constant elsewhere, f(x) is analytic in the x plane cut along$]-\infty, -R]$, if $\mu(u)$ is discontinuous at $u = u_0$, f(x) has a pole with a positive residue at $x = -u_0^{-1}$. When the radius of convergence R is finite, f(x) is completely determined by its Taylor series or by its values in a numerable set of points and convergent sequences of rational fractions can be constructed.

The moment problem. The Taylor series of f(x) reads

$$(1.2) \qquad f(x) \sim \sum_{n=0}^{\infty} \mu_n(-x)^n \qquad \mu_n = \int_0^\infty \mu(u)u^n du$$

and is either convergent ($R > 0$) or asymptotic ($R = 0$) to f(x); the μ_n are known as moments of $\mu(u)$.

A necessary and sufficient condition for f(x) to be a Stieltjes function is the positivity of the Hadamard determinants

$$(1.3) \qquad \begin{vmatrix} \mu_0 & \mu_1 \cdots & \mu_n \\ \vdots & \vdots & \vdots \\ \mu_n & \mu_{n+1} \cdots & \mu_{2n} \end{vmatrix} > 0 \qquad \begin{vmatrix} \mu_1 & \mu_2 \cdots & \mu_n \\ \vdots & \vdots & \vdots \\ \mu_n & \mu_{n+1} \cdots & \mu_{2n-1} \end{vmatrix} > 0 \qquad n = 1, 2 \ldots$$

Consequently, f(x) can be expanded into a positive continued fraction

$$(1.4) \qquad f(x) = \cfrac{a_0}{1 + \cfrac{x\,a_1}{1 + \cfrac{x\,a_2}{1 + \ddots \cfrac{x\,a_n}{1 + x f_n(x)}}}} \qquad a_i > 0$$

where the remainder $f_n(x)$ is itself a Stieltjes function for any n, see Wall (1948) or Akhiezer (1965). The rational fraction obtained by truncating (1.4) at odd orders, that is by setting $f_{2N-1}(x) = 0$, is a Stieltjes function and can be written as the ratio of two polynomials of degree $N-1$ and N, respectively, known as the $[N-1/N]$

Padè approximant. Even order truncation, that is $f_{2N}(x) = 0$, produces the [N/N] Padè approximant.

More generally the [N/M] Padè approximant to an arbitrary function f(x) is defined by

$$(1.5) \qquad [N/M]_{f(x)} = \frac{P_N(x)}{Q_M(x)} \qquad P_N(x) - f(x)\, Q_M(x) = O(x^{N+M+1})$$

where $P_N(x)$, $Q_M(x)$ are polynomials of degree N, M respectively, see Baker (1970, 1975).

The positivity of a_i implies that $[N-1/N]_{f(x)}$ and $[N/N]_{f(x)}$ are for $x > 0$ monotonic sequences bounding f(x)

$$(1.6) \qquad [N-1/N]_{f(x)} < f(x) < [N/N]_{f(x)}$$

If $R > 0$ both sequences have the same limit; if $R - 0$ the limit is the same only if the μ_n do not grow faster than (2n)! The convergence is uniform in any compact of the x plane excluding the real negative axis. Indeed the poles and zeroes of these sequences lie on the negative real axis where they interlace to reproduce the singularities of f(x).

The interpolation problem. Let f(x) be a Stieltjes function with radius of convergence $R > 0$ and $\{x_i\}_{i=1}^{\infty}$ a bounded monotone increasing sequence of points

$$-R < x_1 < x_2 \ldots \qquad \lim_{i\to\infty} x_i = x^* < \infty$$

It is proven that f(x) has a continued fraction expansion

$$(1.7) \qquad f(x) = \cfrac{b_0}{1 + \cfrac{(x-x_1)\, b_1}{1 + \ddots \cfrac{(x-x_n)\, b_n}{1 + (x-x_{n+1})\, f_n(x)}}} \qquad b_i > 0$$

where $f_n(x)$ is still a Stieltjes function. We notice that the b_n can be computed with simple recursive formulae

$$(1.8) \qquad f_0(x) = f(x) \qquad f_n(x) = \frac{f_{n-1}(x_n) - f_{n-1}(x)}{x - x_n} \qquad b_n = f_n(x_{n+1})$$

The rational fraction obtained by truncating (1.7) at odd orders, $f_{2N-1}(x) = 0$, can be written as the ratio of two polynomials of degrees N - 1 and N respectively, known as the [N - 1/N] multipoint Padè approximant. Even order truncation, $f_{2N}(x) = 0$, produces a [N/N] Padè approximant. Barnsley (1973) has shown that

$$(1.9) \qquad [N-1/N]_{f(x)} < f(x) \qquad [N/N]_{f(x)} < f(x) \qquad -R < x < x_1$$

and consequently both the $N-1/N_{f(x)}$ and $[N/N]_{f(x)}$ give alternating bounds in the intervals $]x_1, x_2[$, $]x_2, x_3[$, ... since the difference with f(x) is a function with simple zeroes at $x = x_i$. All the poles and zeroes of $[N-1/N]_{f(x)}$ and $[N/N]_{f(x)}$ lie in the negative real axis $x < -R$, and if f(x) has p poles below the cut, $[N-1/N]_{f(x)}$ and $[N/N]_{f(x)}$ have for $N > p$ exactly p poles approaching the exact ones from below.

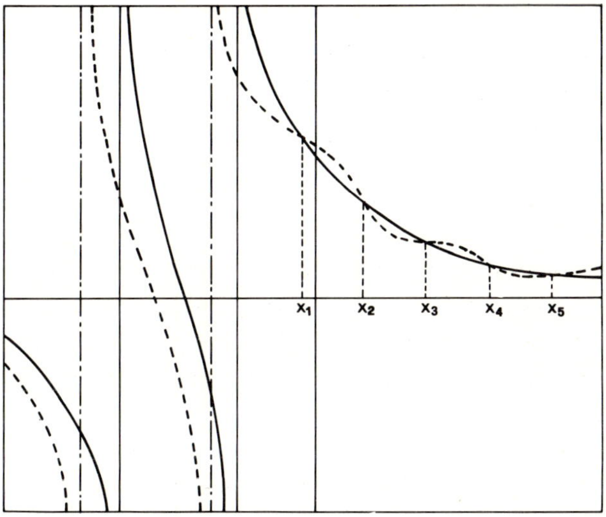

Fig. 1

In fig. 1 we sketch the behaviour of a f(x) with p = 2 (solid line) and the $[2/2]_{f(x)}$ (dashed line).

The following convergence theorem is given by Barnsley (1973). If $x_i \to x^* < \infty$ then

$$(1.10) \qquad \lim_{N\to\infty} [N-1/N]_{f(x)} = \lim_{N\to\infty} [N/N]_{f(x)} = f(x)$$

uniformly in any compact region of the x plane cut from $-\infty$ to $-R$.

When f(x) is meromorphic the poles and residues of $[N-1/N]_{f(x)}$ and $[N/N]_{f(x)}$ converge to the exact ones for $N \to \infty$. In this case it is helpful to consider the functions $\delta(x) = \arctan f(x)$, $\delta_{[N-1/N]}(x) = \arctan [N-1/N]_{f(x)}$ for which the bounding properties can be extended to the whole negative axis, as shown by Bessis (1975).

Since we shall deal with meromorphic functions we wish to point out that the standard expansion of f(x) in series of partial fractions

$$(1.11) \qquad f(x) = \sum_{i=1}^{\infty} \frac{\gamma_i}{x+\alpha_i}$$

or the representation of f(x) in the form of infinite product

$$(1.12) \qquad f(x) = \prod_{i=1}^{\infty} \frac{\left(1 + \dfrac{x}{\beta_i}\right)}{\left(1 + \dfrac{x}{\alpha_i}\right)}$$

is basically different from the expansion in terms of Padè approximants since the zeroes, poles and residues of the latter depend on the order of approximation

$$(1.13) \qquad f(x) = \lim_{N\to\infty} [N-1/N]_{f(x)} \qquad [N-1/N]_{f(x)} = \sum_{i=1}^{N} \frac{\gamma_{Ni}}{x+\alpha_{Ni}} = \prod_{i=1}^{N} \frac{\left(1 + \dfrac{x}{\beta_{Ni}}\right)}{\left(1 + \dfrac{x}{\alpha_{Ni}}\right)}$$

Section II - The vibrating string.

We examine the problem of the transverse vibrations of a string of length L with the left edge fixed at the origin. The string has unit elastic modulus and density $\rho(x)$ so that the equation for the free harmonic oscillations of frequency $\sqrt{\lambda}$ is given by

$$(2.1) \qquad \frac{d^2y}{dx^2} + \lambda\, \rho(x)\, y = 0$$

We consider two different boundary conditions, namely the fixed right end

$$(2.2) \qquad y(0) = 0 \qquad y(L) = 0$$

the right end free to move

$$(2.3) \qquad y(0) = 0 \qquad \frac{dy}{dx}(x = L) = 0$$

and label by $\sqrt{\lambda_i}$, $\sqrt{\lambda_i'}$ the corresponding eigenfrequencies.

If the right end is free and acted upon by a periodic force $F(t) = F_0 \sin\sqrt{\lambda}\, t$, then it undergoes a periodic motion $y(L, t) = R(\lambda)\, F_0 \sin\sqrt{\lambda}\, t$, where $R(\lambda)$, called the coefficient of dynamic yield, is given by

$$(2.4) \qquad R(t) = L \prod_{i=1}^{\infty} \frac{\left(1 - \dfrac{\lambda}{\lambda_i}\right)}{\left(1 - \dfrac{\lambda}{\lambda_i'}\right)}$$

It can be shown that the eigenfrequencies must interlace

$$(2.5) \qquad 0 < \lambda_1' < \lambda_1 < \ldots < \lambda_i' < \lambda_i < \ldots$$

and

$$(2.6) \qquad \lim_{n\to\infty} \frac{\lambda_n L^2}{n^2\pi^2} = 1 \qquad \lim_{n\to\infty} \frac{\lambda'_n L^2}{n^2\pi^2} = 1$$

so that $R(-\lambda)$ is a Stieltjes function, Krein (1955a).

The inverse problem consists in determining the density $\rho(x)$ from the eigenfrequencies or the function $R(\lambda)$. Borg (1946) showed that $\{\lambda_i\}_{i=1}^{\infty}$ and $\{\lambda'_i\}_{i=1}^{\infty}$ determine $\rho(x)$ uniquely while Krein proved the same result for $R(\lambda)$. These results were extended by Barcilon (1974a,b) to higher order Sturm-Liouville problems.

Consider a finite set of data given by the first N couples of frequencies λ_i, λ'_i or by the first 2N moments of $R(\lambda)$ or by the values of $R(\lambda)$ for 2N distinct frequencies smaller than λ'_1 (remember also that $R(0) = L$). In each case we obtain a unique Stieltjes rational fraction, either by cutting the infinite product (2.4) at order N or by computing the [N/N] Padè approximant.

Letting λ_{Ni} and λ'_{Ni} be the zeroes and poles $R_N(\lambda)$ we write

$$(2.7) \qquad R_N(\lambda) = \prod_{i=1}^{N} \frac{\left(1 - \dfrac{\lambda}{\lambda_{Ni}}\right)}{\left(1 - \dfrac{\lambda}{\lambda'_{Ni}}\right)}$$

Ordering the λ_{Ni} and λ'_{Ni} as non-decreasing sequences we have

$$(2.8) \qquad \lambda'_{N1} < \lambda_{N1} < \dots < \lambda'_{NN} < \lambda_{NN}$$

When $N\to\infty$ according to the theorems of section I $R_N(\lambda) \to R(\lambda)$ and $\lambda_{Ni} \to \lambda_i$, $\lambda'_{Ni} \to \lambda'_i$ for any fixed i.

For any finite N λ_{Ni}, λ'_{Ni} and $R_N(\lambda)$ are the eigenfrequencies and the coefficient of dynamic yield of a finite difference operator Δ_N, that is of a system of N oscillators with masses m_i and rest distances l_i as shown in fig. 2.

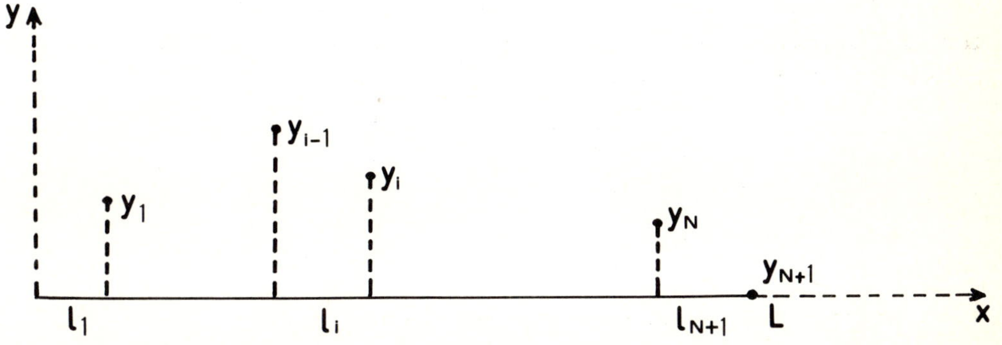

Fig. 2

Therefore $R_N(\lambda)$ which depends on 2N parameters λ_{Ni}, λ'_{Ni} determines unambiguously the 2N constants m_i, ℓ_1 for $i = 1, \dots, N$.

The N oscillators chain. This system first investigated by Stieltjes (1895) and revisited by Gantmakher (1960) and Barcilon (1976) in a more complex context, has a lagrangian

$$(2.9)\qquad \mathscr{L} = \sum_{i=1}^{N} \frac{m_i}{2}\, \dot{y}_i^2 - \sum_{i=1}^{N+1} \frac{(y_i - y_{i-1})^2}{2\,\ell_i}$$

so that the equations of motions for harmonic oscillations $y_i(t) = y_i \sin \sqrt{\lambda}\, t$ of frequency $\sqrt{\lambda}$ are given by

$$(2.10)\qquad \lambda m_i y_i + \frac{y_{i-1} - y_i}{\ell_i} + \frac{y_{i+1} - y_i}{\ell_{i+1}} = 0 \qquad i = 1, \dots, N$$

and the ℓ_i fulfil the constraint

$$(2.11)\qquad \sum_{i=1}^{N+1} \ell_i = L$$

By solving equations (2.10) with two distinct conditions corresponding to the last particle that is fixed

$$(2.12)\qquad y_0 = 0 \qquad y_{N+1} = 0$$

or free to move

$$(2.13)\qquad y_0 = 0 \qquad y_N = y_{N+1}$$

we find two distinct sets of eigenvalues λ_{Ni}, λ'_{Ni} $i = 1, \dots N$.

Since y_i is a polynomial of degree $i-1$ in λ as can be seen from (2.10), it is convenient to introduce two sets of polynomials $P_i(\lambda)$, $Q_i(\lambda)$ of degree i in λ according to

$$(2.14)\qquad y_i = P_{i-1}(\lambda) y_1$$

and

$$(2.15)\qquad Q_i(\lambda) = \frac{P_i(\lambda) - P_{i-1}(\lambda)}{\ell_{i+1}}$$

and it is easy to check, using (2.10), that they fulfil the following recursion relation

$$(2.16)\qquad \lambda m_i P_{i-1}(\lambda) - Q_{i-1}(\lambda) + Q_i(\lambda) = 0$$

From (2.10) and (2.15) multiplied by y_1 we obtain

$$(2.17)\qquad y_{N+1} = P_N(\lambda)y_1 \qquad y_{N+1} - y_N = \ell_{N+1}Q_N(\lambda)y_1$$

so that $P_N(\lambda)$, $Q_N(\lambda)$ are the characteristic polynomials for conditions (2.12) and (2.13). The normalization at $\lambda = 0$ is also fixed by (2.15) and (2.16) so that we can write

$$(2.18)\qquad P_N(\lambda) = \frac{L}{\ell_1}\prod_{i=1}^{N}\left(1 - \frac{\lambda}{\lambda_{Ni}}\right) \qquad Q_N(\lambda) = \frac{1}{\ell_1}\prod_{i=1}^{N}\left(1 - \frac{\lambda}{\lambda'_{Ni}}\right)$$

The ratio $R_N(\lambda) = P_N(\lambda)/Q_N(\lambda)$ is the coefficient of dynamic yield for the system and $R_N(-\lambda)$ is a Stieltjes function whose continued fraction expansion, (that geos into (1.4) with the change $\lambda \to \lambda^{-1}$) obtained from (2.15), (2.16) reads

$$(2.19)\qquad R_N(-\lambda) = \ell_{N+1} + \cfrac{1}{\lambda m_N + \cfrac{1}{\ell_N + \cfrac{1}{\lambda m_{N-1} + \ddots + \cfrac{1}{m_1\lambda + \cfrac{1}{\ell_1}}}}}$$

The inversion procedure. The inversion procedure for $R_N(\lambda)$ requires first the computation of $P_N(\lambda)$ and $Q_N(\lambda)$ in the form

$$(2.20)\qquad P_N(\lambda) = \sum_{i=1}^{N} p_{Ni}\lambda^i \qquad Q_N(\lambda) = \sum_{i=1}^{N} q_{Ni}\lambda^i$$

and then obtaining the m_i and ℓ_1 from the recurrence equations (2.15), (2.16).

To be more explicit we start from (2.15) for $i = N$ and compare the coefficients of λ^N to obtain

$$(2.21)\qquad \ell_{N+1} = \frac{p_{NN}}{q_{NN}} \qquad P_{N-1}(\lambda) = P_N(\lambda) - \ell_{N+1}\, Q_N(\lambda)$$

Since the polynomial $P_{N-1}(\lambda) = p_{N-1,N-1}\,\lambda^{N-1} + \ldots + p_{N-1,0}$ is known from (2.16) we obtain

$$(2.22)\qquad m_N = -\frac{q_{NN}}{p_{N-1,N-1}} \qquad Q_{N-1}(\lambda) = Q_N(\lambda) + \lambda m_N\, P_{N-1}(\lambda)$$

Iteration of this procedure determines the remaining parameters $\ell_N, \ldots \ell_2$ $m_{N-1}, \ldots, m_1$ and ℓ_1 is obtained from (2.11).

To the discrete set of parameters so far determined we associate the stepwise function $\mu_N(x)$

$$(2.23) \qquad \mu_N(x) = \sum_{i=1}^{N} m_i \, \vartheta(x - x_i) \qquad x_i = \ell_1 + \ell_2 + \dots + \ell_i$$

and claim that if $\mu_N(x)$ convergens to a continuos function $\tilde{\mu}(x)$ for $N \to \infty$ and if we define the exact mass distribution $\mu(x)$ by

$$(2.24) \qquad \mu(x) = \int_0^x \rho(u)\,du$$

then $\tilde{\mu}(x) = \mu(x)$. In fact, if this were not true, since $R_N(\lambda) \to R(\lambda)$, we would have the same coefficient of dynamic yield for two different mass distributions, contradicting Krein's uniqueness theorem.

A *numerical example.* In order to check the procedure described above we have applied it in the simplest case defined for L = 1 by the frequencies

$$(2.25) \qquad \lambda_n = n^2\pi^2 \qquad \lambda'_n = \left(n - \frac{1}{2}\right)^2 \pi^2$$

and a coefficient of dynamic yield

$$(2.26) \qquad R(\lambda) = \frac{\tan\sqrt{\lambda}}{\sqrt{\lambda}}$$

that correspond to $\rho(x) = 1$, i.e. a linear mass distribution $\mu(x)$.

In table 1 we show the distribution of masses m_i at points x_i corresponding to the coefficient of dynamic yield $R_N(\lambda)$ given by the first N couples of exact frequencies and by the [N/N] Padè approximant (computed from the Taylor series of $R(\lambda)$).

Table 1

	Exact λ_i, λ'_i $i \leq N$		Padè [N/N]	
	x_i	m_i	x_i	m_i
N = 3	.547	.371	.623	.421
	.859	.288	.933	.195
N = 4	.393	.258	.442	.299
	.602	.184	.689	.213
	.771	.160	.870	.146
	.925	.151	.978	.066

	Exact λ_i, λ'_i $i \leq N$		Padè [N/N]	
	x_i	m_i	x_i	m_i
N = 7	.299	.194	.334	.219
	.454	.134	.513	.157
	.575	.112	.657	.133
	.680	.100	.779	.111
	.776	.093	.879	.086
	.868	.090	.951	.057
	.956	.088	.992	.025

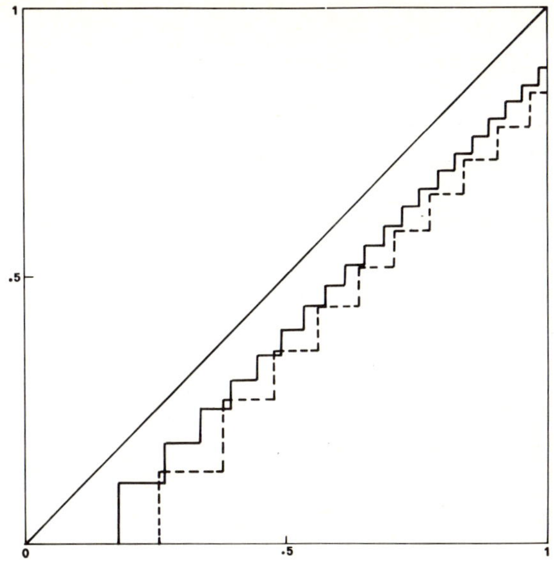

Fig. 3

The onset of a linear behaviour in $\mu_N(x)$ is evident but quite slow and one has to reach high values of N as shown by fig. 3 (where $\mu_{10}(x)$, the dotted line, and $\mu_{20}(x)$, the solid line, are computed from the exact λ_i, λ_i' for $i \leq 10$, and $i \leq 20$ respectively) in order to reproduce a straight line more accurately. Of course such high orders can hardly be reached with a Padè approximation to $R(\lambda)$ since the required input accuracy of $R(\lambda)$ increases with N. However, a mixed procedure based on the first set of eigenvalues λ_i $i = 1, \ldots, N$ and the values of $R(\lambda)$ at N points can probably be used up to reasonably high orders.

Section III - The Schrödinger equation.

The previous analysis applies also to the Schrödinger equation if we consider an inverse problem for fixed energy and variable strenght of the potential. The S wave Schrödinger equation for a short range potential $gV(r)$ vanishing for $r > a$ and normalized so that $V(a-0) = 1$ reads

$$(3.1) \qquad -\frac{1}{2m}\frac{d^2\varphi(r)}{dr^2} + gV(r)\,\varphi(r) = E\varphi(r)$$

with the standard boundary conditions

$$(3.2) \qquad \varphi(0) = 0 \qquad \underset{r>a}{\varphi(r)} = \sin\,[Kr + \delta(E,g)] \qquad K = (2mE)^{1/2}$$

The exact inversion equations when $\delta(E, g)$ is known for $E > 0$ and fixed $g > 0$ is somewhat simpler than in the general case and was developed by Regge (1958).

We consider the reaction matrix $R(E, g)$ first introduced by Wigner (1949), defined by

$$(3.3) \qquad R = \varphi(a) \left[a \frac{d\varphi}{dr} (r = a) \right]^{-1}$$

and related to the phase shift through

$$(3.4) \qquad R = \frac{\tan(Ka + \delta)}{Ka}$$

We notice that R(E, g) is a Stieltjes function of $-E$ for fixed $g > 0$, and of g for fixed $E < 0$. More generally R(E, g) is a meromorphic Hamburger function (that is a function with poles along the real axis with positive residues).

In fact, by letting u_n and $E_n(g)$ be the eigenfunctions and eigenvalues of the Sturm-Liouville equation obtained from (3.1) (for fixed g) with boundary conditions

$$(3.5) \qquad u_n(0) = u_n'(a) = 0$$

from the standard Wronskian relations and the completeness condition for u_n we get

$$(3.6) \qquad R(E, g) = \frac{1}{2ma} \sum_{n=1}^{\infty} \frac{u_n^2(a)}{E_n(g) - E}$$

In a similar way, by letting $v_n(x)$ and $g_n(E)$ be the eigenfunctions and eigenvalues obtained from (3.1) (for fixed E) with boundary conditions

$$(3.7) \qquad v_n(0) = v_n'(a) = 0$$

we obtain

$$(3.8) \qquad R(E, g) = \frac{1}{2ma} \sum_{n=1}^{\infty} \frac{v_n^2(a)}{g_n(E) + g}$$

where for $E < 0$ it is easy to show that $g_n(E) > 0$.

Any approximation to (3.6) or (3.8) gives a Stieltjes (or Hamburger) rational fraction R_N in E or g.

For the variable energy case one would like to identify R_N with the reaction matrix of a finite difference Schrödinger equation with equal steps a/N. However, such an inverse problem considered by Zacharev (1974) and Melnikov (1976) following Case (1973a,b) has in general no solution since the completeness relations involve additional constraints that are not fulfilled by an arbitrary Stieltjes fraction.

Conversely, let us consider R_N as the coefficient of dynamic yield of a system of oscillators with rest distances ℓ_i and masses $m_i = 2mv_i$ and submitted to a uniform elastic force of intensity $-2mE$ along the φ axis, and identify it with the reaction matrix of a finite difference Schrödinger equation with unequal space steps

$$(3.9)\qquad -\frac{1}{2m}\left(\frac{\varphi_{i+1}-\varphi_i}{\ell_{i+1}}+\frac{\varphi_{i-1}-\varphi_i}{\ell_1}\right)+gv_i\varphi_i-E\varphi_i=0$$

As in section II we introduce two polynomials $P_i(g, E)$, $Q_i(g, E)$ of degress i in both g and E according to

$$(3.10)\qquad \varphi_i=P_{i-1}\varphi_1 \qquad Q_i=\frac{P_i-P_{i-1}}{\ell_{i+1}}$$

and the matrix R_N is given by

$$(3.11)\qquad R_N(E, g)=a^{-1}\frac{P_N(E, g)}{Q_N(E, g)}=\varphi_{N+1}\left(a\,\frac{\varphi_{N+1}-\varphi_N}{\ell_{N+1}}\right)^{-1}$$

From the equations generalizing (2.19), (2.20) to this case, we easily see that the inverse problem for R_N, namely the constants v_i and ℓ_i, is uniquely determined when R_N is given as a Stieltjes fraction of g for fixed E, undetermined and in general with no solution when R_N is a Stieltjes fraction of E for fixed g.

To conclude we wish to stress that in both cases the inverse problem related to the exact reaction matrix R, that is to the Schrödinger equation (3.1), is uniquely defined.

REFERENCES.

Akhiezer N.I. (1965) "The classical moment problem", Oliver and Boyd, London.

Baker G., Gammel J. (1970) "The Padè Approximants in Theoretical Physics", Academic Press, New York.

Baker G. (1975) "Essentials of Padè Approximants", Academic Press, New York.

Barcilon V. (1973) J. Math. Phys. *15*, 429.

Barcilon V. (1974a) Geophys. J. R. Astr. Soc. *38*, 287.

Barcilon V. (1974b) Geophys. J. R. Astr. Soc. *39*, 143.

Barcilon V. (1976) Geophys. J. R. Astr. Soc. *44*, 61.

Barnsley M. (1973) J. of Math. Phys. *14*, 299.

Bessis D., Villani M. (1975) J. of Math. Phys. *16*, 462.

Borg G. (1946) Acta Math. *78*, 1.

Case K.M., Kac M. (1973a) J. of Math. Phys. *14*, 594.

Case K.M., Chin S.C. (1973b) J. of Math. Phys. *14*, 1943.

Gantmakher F.R., Krein M.G. (1960) "Oscillating Matrices and Kernels: Small Oscillations of Mechanical Systems", Academie Verlag, Berlin.

Gel'fand I.M., Levitan B.M (1951) Isv. Akad. Nauk. SSSR *15*, 309 Am. Math. Soc. Transl. *1*, 253 (1956).

Krein M.G. (1955a) Dokl. Akad. Nauk. SSSR *82*, 669.

Krein M.G. (1955b) Dokl. Akad. Nauk. SSSR *87*, 881.

Marchenko V.A. (1955) Dokl. Akad. Nauk. SSSR *104*, 695.

Marchenko V.A. (1963) "The Inverse Problem in Scattering Theory", Gordon and Breach, New York.

Marchenko V.A. (1974) "Spectral Theory of Sturm Liouville operators", Naukova Dumka, Kiev.

Melnikov V.N., Rudjak B.V., Ushakov I.B., Zakhariev B.N. (1976) " A Model of the Inverse Problem in Nuclear Physics", JINR, E4, Dubna.

Newton R.G. (1962) J. Math. Phys. *3*, 75.

Regge T. (1950) Nuovo Cimento *14*, 1251.

Regge T. (1958) Nuovo Cimento *9*, 491.

Sabatier P.C. (1966) J. Math. Phys. *7*, 1515.

Stieltjes T.J. (1895) "Recherches sur les fractions continues", Ann. Fac. Sci. Toulouse *8*, 1 Oevres complètes vol. *2*, 402, Groningen (1918).

Wall H.S. (1948) "Analytic thoery of continued fractions", Van Nostrand, Princeton.

Wigner E.P., Eisenbud L. (1949) Phys. Rev. *72*, 29.

Zakhariev B.N., Niyargulov S.A., Suzko A.A. (1974) Sov. J. Nucl. Phys. *20*, 667.

QUELQUES METHODES SUR LA RECHERCHE D'UN DOMAINE OPTIMAL

Jean CEA
Département de Mathématiques
Faculté des Sciences de NICE

INTRODUCTION :

Dans de nombreux problèmes, il est question de chercher la "meilleure forme" d'un ouvert Ω de $\mathbb{R}^n$; cependant ce problème peut prendre différents aspects :

1. recherche d'une meilleure forme (parmi un ensemble de formes admissibles). Ce peut être la forme d'un diélectrique, d'un disque de turbomachine,...

2. recherche d'une meilleure place : comme par exemple placer au mieux des électrodes, un tuyau, ... ; dans ce cas la forme est donnée.

3. Identifier la forme d'un objet non entièrement accessible comme par exemple la forme du fond d'un glacier, d'une nappe d'eau,...

4. Problème à frontière libre : un ouvert Ω a pour frontière $\Gamma \cup \Sigma$; Γ est donnée ; on donne 3 opérateurs A,B,C et on cherche u et Σ tels que :

$$\begin{array}{ll} Au = 0 & \text{dans } \Omega \\ Bu = 0 & \text{sur } \Gamma \cup \Sigma \\ Cu = 0 & \text{sur } \Sigma \end{array}$$

Si Σ était donnée, la 3ième condition serait surabondante ; cette condition supplémentaire permet de trouver la frontière libre Σ.

5. Problème de contrôle optimal - domaine optimal : on désigne par Ω le support du contrôle dans un problème de contrôle optimal. Le contrôle optimal dépend bien entendu de Ω ; la fonction économique du problème de contrôle pour le choix du contrôle optimal peut être encore "meilleure" si le support de Ω est bien choisi ; il s'agit de trouver le "meilleur" support du "meilleur contrôle".

1. Formalisme général ; les 2 problèmes modèles.

D'une façon générale, on dispose d'une équation d'état qui permet d'obtenir un état y à partir d'un domaine Ω :

$$\Omega \longrightarrow y_\Omega$$

on dispose aussi d'une fonction coût :

$$\Omega, y_\Omega \longrightarrow J(\Omega)$$

on disposera enfin d'une classe $\mathcal{G}$ d'ouverts Ω admissibles. Le problème est le suivant : on cherche Ω_{opt} tel que

$$\begin{cases} \Omega_{opt} \in \mathcal{G} \\ J(\Omega_{opt}) \leq J(\Omega) \quad \forall \Omega \in \mathcal{G} \end{cases}$$

Les questions d'existence, d'unicité et d'approximation se posent. Nous allons nous limiter ici à l'approximation.

Problème modèle 1 : on donne un ouvert fixe $D \subset \mathbb{R}^n$ de fonction Σ ; on partage D en $\Omega \cup \Gamma \cup B$ où Ω est un ouvert tel que $\overline{\Omega} \subset D$, $\Gamma = \partial\Omega$.

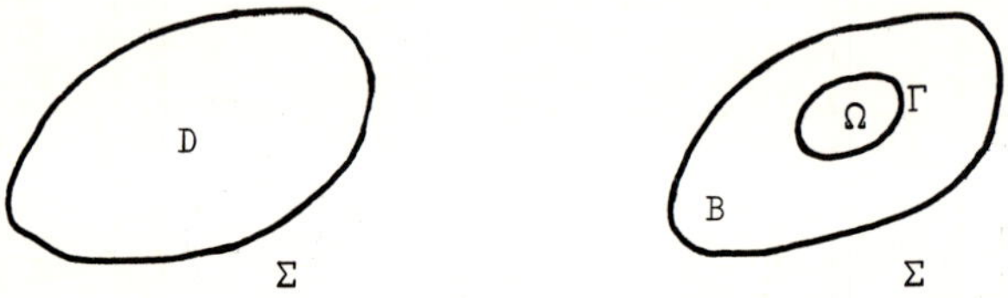

on désigne par u la fonction caractéristique de Ω . On pose :

$$V = H^1(D) = \{v | v \in L^2(D), \frac{\partial v}{\partial x_i} \in L^2(\Omega) \quad i = 1,\dots,n\}$$

$$a(u,y,\varphi) = \int_D u[\nabla y \, . \, \nabla\varphi + y \, \varphi] \, dx + \int_D (1-u) \, [k \, \nabla y \, \nabla\varphi + y \, \varphi] \, dx$$

$$L(\varphi) = \int_D f.\varphi \, dx$$

k est un nombre donné positif.

Equation d'état :

$$\begin{cases} a(u,y_u,\varphi) = L(\varphi) & \forall\varphi \in V \\ y_u \in V \end{cases}$$

Fonction coût :

$$J(u) = \frac{1}{2} \|y_u - y_d\|^2_{L^2(D)}$$

où y_d est donnée dans $L^2(D)$.

On cherche le meilleur Ω (ou le meilleur u) afin que J soit minimum et que y_u soit le plus voisin possible de y_d.

Si on pose $y_1 = y_u/_\Omega$ $\quad y_2 = y_u/_B$, l'équation d'état peut s'écrire

$$\begin{cases} - \Delta y_1 + y_1 = f/_\Omega & (\Omega) \\ - k \, \Delta y_2 + y_2 = f/_B & (B) \\ y_1 = y_2 & (\Gamma) \\ \dfrac{\partial y_1}{\partial x_1} = k \, \dfrac{\partial y_2}{\partial x_2} & (\Gamma) \\ \dfrac{\partial y_2}{\partial x_2} = 0 & (\Sigma) \end{cases}$$

où x_1 et x_2 désignent les normales extérieures à Ω et B.

On cherche donc Ω pour que la solution y_u d'un problème de transmission dans $\bar{\Omega} \cup \bar{B}$ soit la plus voisine possible d'une fonction donnée y_d.

Problème modèle 2 : on donne dans R^n un compact fixe B contenant un ouvert fixe ω. On introduit un ouvert Ω de frontière Γ tel que

$$\omega \subset \Omega \subset B$$

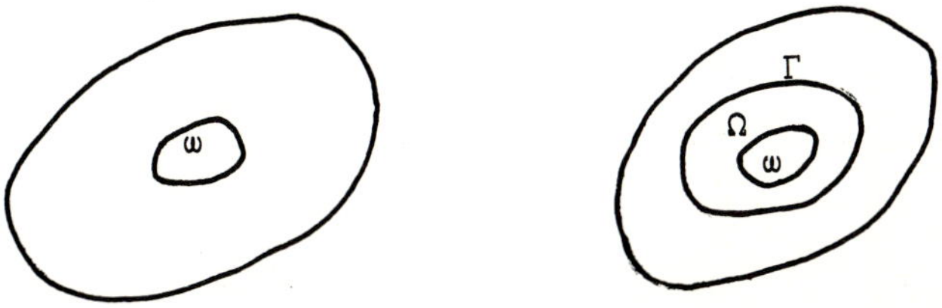

espace $V = H^1(\Omega)$; f donnée dans $L^2(B)$

Equation d'état :

$$\begin{cases} y \in V \\ \int_\Omega (\nabla y \,.\, \nabla\varphi + y\,\varphi)\,dx = \int_\Omega f.\varphi\,dx \qquad \forall\varphi \in V \end{cases}$$

ce qui peut s'écrire :

$$\begin{cases} -\Delta y + y = f & (\Omega) \\ \frac{\partial y}{\partial u} = 0 & (\Gamma) \end{cases}$$

Fonction coût :

$$J(\Omega) = \tfrac{1}{2}\,\|y_\Omega - y_d\|^2_{L^2(\omega)}$$

on cherche à minimiser J c'est-à-dire qu'on cherche Ω pour que la solution d'un problème de Neumann soit "le plus voisin possible" d'une fonction y_d donnée.
Nous allons donner 2 familles de méthodes d'approximation du Ω optimal (ou de construction d'une suite Ω_n "minimisante").

1. Méthode du type POINT FIXE : on décrira cette méthode sur le problème modèle n°1.

2. Méthode du type GRADIENT : on décrira cette méthode sur le problème modèle n°1.

2. APPROXIMATION par une méthode du type POINT FIXE.

On revient au problème modèle 1. On associe à Ω ou à sa fonction caractéristique u une fonction G_u définie sur D et telle que :

$$u = u_{opt} \Leftrightarrow \begin{cases} G_u(x) \geq 0 & \forall x \in B_{opt} \\ G_u(x) \leq 0 & \forall x \in \Omega_{opt} \end{cases}$$

On définit T par :

$$\Omega \to T\Omega \qquad \text{ou} \qquad u \to Tu$$

$$Tu(x) = 1 \Leftrightarrow G_u(x) < 0$$

$$Tu(x) = 0 \Leftrightarrow G_u(x) \geq 0$$

Notons que Ω_{opt} vérifie :

$$u_{opt} = Tu_{opt}$$

donc u_{opt} est un point fixe de T.

On utilise alors la méthode des approximations successives :

$$\begin{cases} u_o \ \text{donné} \\ u_{n+1} = T\, u_n \end{cases}$$

En résumé : Ω_{n+1} est l'ensemble des points x où $G_{\Omega_n}(x)$ est négatif. La mise en oeuvre de cette méthode est extrêmement facile.

Il ne nous reste plus qu'à exhiber la fonction G.

Recherche de la fonction G : Rappelons que $V = H^1(D)$.

$$(2.1) \quad a(u,y_u,\varphi) = L(\varphi) \qquad \forall \varphi \in V$$

avec

$$a(u,y,\varphi) = \int_D u(\nabla y\ \nabla\varphi + y\ \varphi)\ dx + \int_D (1-u)\ (k\ \nabla y\ \nabla\varphi + y\ \varphi)\ dx$$

$$= \int_D (k\ \nabla y\ \nabla\varphi + y\ \varphi)\ dx + \int_D ((1-k)u\ \nabla y\ \nabla\varphi\ dx$$

$$= b(y,\varphi) + c(u,y,\varphi)$$

avec des notations évidentes ; et (2.1) s'écrit :

$$(2.1)' \quad b(y_u,y) + c(u,y_u,\varphi) = L(\varphi) \qquad \forall \varphi \in V$$

changeant le domaine Ω en changeant u en $u+\delta u$:

$$(2.2) \quad a(u + \delta u, y_u + \delta y, \varphi) = L(\varphi) \qquad \forall \varphi \in V$$

avec (2.1)' et (2.2) on a :

$$(2.3) \quad a(u,\delta y,\varphi) = - c(\delta u, y_u,\varphi) - c(\delta u,\delta y,\varphi)$$

d'autre part :

$$(2.4) \quad J(u+\delta u) = J(u) + (y_u - y_d,\delta y) + \tfrac{1}{2}\|\delta y\|^2$$

Etat adjoint : on introduit p_u par :

$$(2.5) \begin{cases} p_u \in V \\ a(u,\Psi,p_u) = (y_u - y_d,\Psi) \qquad \forall \Psi \in V \end{cases}$$

$$J(u+\delta u) = J(u) + a(u,\delta y,p_u) + \tfrac{1}{2}\,\|\delta y\|^2$$
$$= J(u) - c(\delta u,y_u,p_u) - c(\delta u,\delta y,p_u) + \tfrac{1}{2}\,\|\delta y\|^2$$

(2.6) $J(u+\delta u) = J(u) + T_1(u,\delta u) + T_2(u,\delta u,\delta y)$

avec

$$T_1(u,\delta u) = -c(\delta u,y_u,p_u)$$
$$T_2(u,\delta u,\delta y) = -\,c(\delta u,\delta y,p_u) + \tfrac{1}{2}\,\|\delta y\|^2$$

T_1 contient les termes d'ordre 1 et T_2 ceux d'ordre 2.

Rappelons

(2.7) $$T_1(u,\delta u) = -\,(1-k)\int_D \delta u\,.\,\nabla y_u\,.\,\nabla p_u\,dx$$

On pose (2.8) $$G_u = -\,(1-k)\,\nabla y_u\,.\,\nabla p_u$$

et alors

$$T_1(u,\delta u) = \int_D \delta u\,.\,G_u\,dx$$

On pose (en élargissant la signification de + et -)

$$\Omega + \delta\Omega = \Omega + \delta\Omega^+ - \delta\Omega^-$$

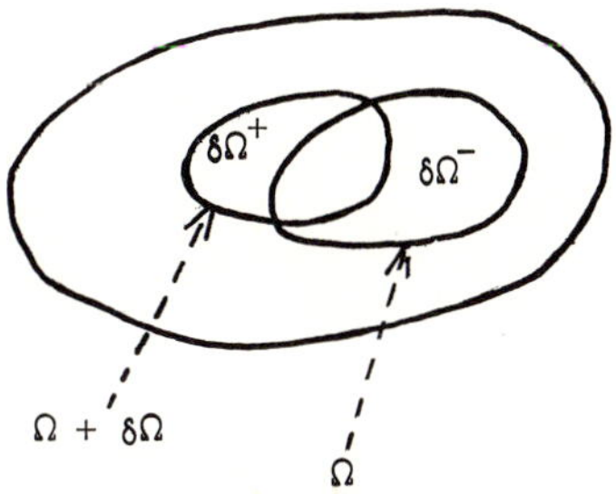

Alors on a :

(2.9) $$T_1(u,\delta u) = \int_{\delta\Omega^+} G_u(x)\,dx - \int_{\delta\Omega^-} G_u(x)\,dx$$

Si Ω est optimal alors $T_1(u,\delta u) \geq 0$ pour tout $\delta\Omega$, donc pour tous $\delta\Omega^+$, $\delta\Omega^-$ et donc $G_u(x) \geq 0$ $\forall x \in \complement\Omega$, $G_u(x) \leq 0$ $\forall x \in \Omega$, c'est bien ce qui était annoncé.

Remarque 2.1: pour obtenir G il faut résoudre 2 problèmes de transmissions, dans lesquels seuls les second membres changent. Le changement $\Omega_n \to \Omega_{n+1}$ ($x \in \Omega_{n+1} \Leftrightarrow G_{u_n}(x) < 0$) est trés facile à réaliser. Cette méthode s'est révélée trés efficace, on a obtenu le domaine optimal en quelques itérations.

On pourra consulter la publication CEA, GIOAN, MICHEL [5] au sujet de cette méthode.

3. APPROXIMATION par une méthode de type GRADIENT

3.1. Rappel : Soit à résoudre le problème $\min_{u \in \mathbb{R}^n} J(u)$ on va construire une famille $u(t)$, $t \geq 0$, telle que $u(\infty) = u_{opt}$

$u(t)$ vérifie une équation différentielle :

$$\begin{cases} u'(t) = v(u(t),t) \\ u(0) = u_o \end{cases}$$

Le problème réside dans le choix de la "vitesse" v ; posons $j(t) = J(u(t))$ et alors

$$j'(t) = J'(u(t),u'(t)) = (G(u(t)),u'(t)) = (G(u(t)),v(u(t),t))$$

si on choisit

$$(3.1) \qquad v(u,t) = -\,G(u)$$

alors

$$j'(t) = -\,\|G(u(t))\|^2$$

et

$$j(t) = j(0) - \int_o^t \|G(u(s))\|^2\,ds$$

J étant bornée inférieurement (en général), l'intégrale $\int_o^{+\infty} \|G(u(s))\|^2\,ds$ est convergente, et en général $\lim_{s\to+\infty} \|G(u(s))\|^2 = 0$ et cela montre que pour t assez grand, j' est assez petit, ce qui conduit à $\lim_{t\to+\infty} u(t) = u_{opt}$.

Dans le cas où la variable u devient un domaine Ω, on va utiliser les mêmes principes pour construire une famille $\Omega(t)$ qui va converger vers Ω_{opt} lorsque $t \to +\infty$.

3.2. Le principe de l'approximation du domaine optimal :

On construit une famille Ω_t, $t \geq 0$; Ω_t est supposé être la position à l'instant t d'un milieu continu qui se déforme ; on suit la trajectoire des points du milieu en introduisant la vitesse particulaire v :

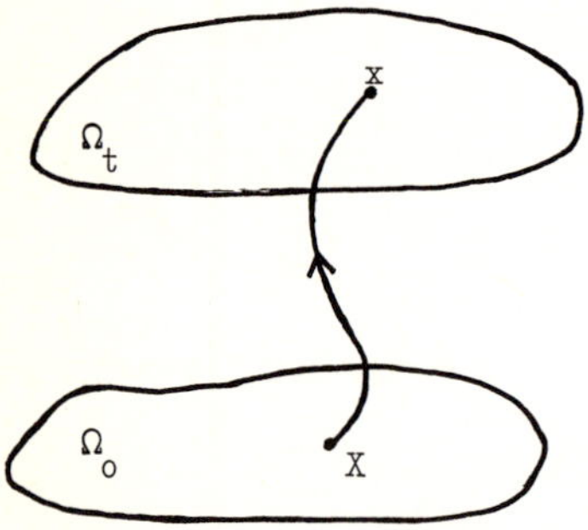

$$\begin{cases} x'(t) = v(x(t),t) \\ x(0) = X \end{cases}$$

(v dépend de X et il faudrait écrire $x_X(t)$ pour rappeler l'origine de la trajectoire.)

v est la vitesse de déformation du milieu Ω_t et constitue la vraie inconnue du problème. Nous allons, comme dans le rappel 3.1, montrer comment il faut choisir v pour que $\lim_{t\to\infty} \Omega_t = \Omega_{opt}$.

3.3. Dérivation par rapport à t :

En général l'équation d'état est du type

$$\int_\Omega A(y,\varphi)\,dx + \int_\Gamma B(y,\varphi)\,d\sigma = 0 \qquad \forall\varphi \in V$$

et si $\Omega = \Omega_t$ alors on a

$$\int_{\Omega_t} A(y_t,\varphi)\, dx + \int_{\Gamma_t} B(y_t,\varphi)\, d\sigma = 0$$

La fonction coût introduit des fonctions du type :

$$j(t) = J(\Omega_t) = \int_{\Omega_t} C(y_t)\, dx + \int_{\Gamma_t} D(y_t)\, d\sigma$$

pour chercher la dérivée $j'(t)$, on a besoin de dériver les intégrales qui sont toutes de 2 types suivants

$$\int_{\Omega_t} E(x,t)\, dx \quad \text{et} \quad \int_{\Gamma_t} F(x,t)\, d\sigma$$

Cela se fera conformément à des techniques de la Mécanique des milieux continus : par exemple si :

$$K(t) = \int_{\Omega_t} C(x,t)\, dx$$

alors

$$\frac{d\,K(t)}{dt} = \int_{\Omega_t} \frac{\partial C}{\partial t}(x,t)\, dx + \int_{\Omega_t} \operatorname{div} C \vec{v}\, dx$$

ou

$$\frac{d\,K(t)}{dt} = \int_{\Lambda t} \frac{\partial C}{\partial t}(x,t)\, dx + \int_{\Gamma_t} C < \vec{v},\vec{n} >_{\mathbb{R}^n} d\sigma$$

$< a,b >$ indique le produit scalaire dans $\mathbb{R}^n$ $\vec{v}$ est la vitesse de déformation du milieu.

3.4. Application au problème modèle n°2 :

Rappelons la géométrie

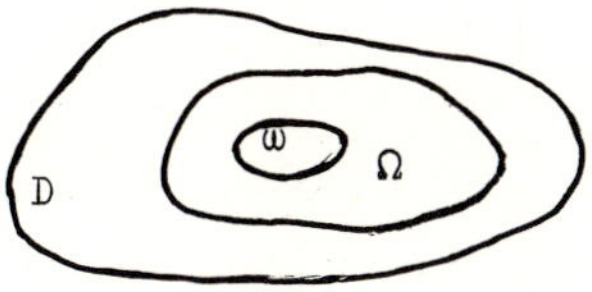

l'équation d'état :

$$\int_\Omega \{\nabla y\, \nabla\varphi + y\varphi - f\varphi\}\, dx = 0 \qquad \forall\varphi \in V = H^1(\Omega)$$

et la fonction coût

$$J(\Omega) = \tfrac{1}{2} \int_\omega |y - y_d|^2\, dx$$

En indiçant tout par t il vient :

$$(3.1) \quad \int_{\Omega_t} \{\nabla y_t \cdot \nabla\varphi + y_t\varphi - f\varphi\}\, dx = 0$$

$$(3.2) \quad j(t) = J(\Omega_t) = \tfrac{1}{2} \int_\omega |y_t - y_d|^2\, dx$$

et en dérivant par rapport à t, dérivation notée . ,

(3.3) $\int_{\Omega_t} \{\nabla\dot{y}_t \cdot \nabla\varphi + \dot{y}_t\, \varphi\}\, dx + \int_{\Omega_t} \operatorname{div}\, \{\nabla y_t \cdot \nabla y + y_t\, \varphi - f\, \varphi\}\, \vec{v}\, dx = 0$

(3.4) $\dot{j}\,(t) = \int_\omega (y_t - y_d)\, \dot{y}_t\, dx$

on introduit alors <u>l'état adjoint</u> $p_t \in H^1(\Omega) = V$

(3.5) $\int_{\Omega_t} \{\nabla\Psi\, \nabla p_t + \Psi\, p_t\} = - \int_\omega (y_t - y_d)\, \Psi\, dx \qquad \forall\Psi \in V$

et alors

$$\dot{j}(t) = \int_\omega (y_t - y_d)\dot{y}_t\, dx = - \int_{\Omega_t} \{\nabla \dot{y}_t\ \nabla p_t + \dot{y}_t\, p_t\}\, dx$$

(3.6) $\dot{j}(t) = \int_{\Omega_t} \operatorname{div}\, \{\nabla y_y \cdot \nabla p_t + y_t\, p_t - f\, p_t\}\, \vec{v}\, dx$.

ce qui peut s'écrire dans un espace convenable

$$\dot{j}(t) = \langle G, v \rangle$$

ce qui conduit à choisir $v = - G$; de façon plus précise soit W un espace de Hilbert où se trouve $\vec{v}$ notons que $[w,v]$ le produit scalaire dans W et $[\![v]\!]$ la norme et choisissons v tel que

(3.7) $[v,\Psi] = - \int_{\Omega_t} \operatorname{div}\, \{\nabla y_t \cdot \nabla p_t + y_t\, p_t - f\, p_t\}\, \Psi\, dx$

alors en faisant $\Psi = v$ il vient :

$$\dot{j}(t) = - [\![v]\!]^2$$

Si $v = 0$ c'est que

$$- \int_{\Omega_t} \operatorname{div}\{\nabla y_t\, \nabla p_t + y_t p_t - f\, p_t\}\, \Psi\, dx = 0 \qquad \forall\Psi$$

et donc $\dot{j} = 0$ pour toute vitesse de déformation ; on est donc en un domaine critique.

En résumé, on obtient des relations du type suivant :

$$(3.8) \quad \begin{cases} E(\Omega_t, y_t) = 0 \\ F(\Omega_t, y_t, p_t) = 0 \\ H(\Omega_t, y_t, p_t, v_t) = 0 \\ \Omega_t = \{x | x = x(t)\ ,\ x'(s) = v(x(s),s),\ x(0) = X,\ \ \forall X \in \Omega.\} \end{cases}$$

Il y a dans (3.8) un couplage

$$\Omega_t \Rightarrow y_t \Rightarrow p_t \Rightarrow v_t \Rightarrow \Omega_t$$

On introduit un pas de discrétisation du temps : τ , on désigne par $\Omega_n = \Omega_{n\tau}$, $\Omega_{n+1} = \Omega_{(n+1)\tau}$, $y_n = y_{n\tau}$ ... d'où le <u>changement de domaine</u>

$\underline{\Omega_n \to \Omega_{n+1}}$ par

$$(3.9)\quad \begin{cases} E(\Omega_n, y_n) = 0 \\ F(\Omega_n, y_n, p_u) = 0 \\ H(\Omega_n, y_n, p_u, v_n) = 0 \\ \Omega_{n+1} \text{ est l'ensemble des points } x_{n+1} \text{ de la forme} \\ \qquad x_{n+1} = x_n + \tau\, v\,(x_n, n\,\tau) \qquad x_n \in \Omega_n \end{cases}$$

(on a donc discrétisé les équations différentielles des trajectoires par la méthode d'Euler

3.5. Point de vue numérique :

Dans tous les exemples cités dans l'introduction, les équations d'états sont des équations aux dérivées partielles ; on résoudra les équations d'états, états adjoints,... par la méthode des éléments finis (en général). A chaque itéré Ω_n il faudra associer une triangulation $\mathcal{T}_n$; cela peut coûter cher ; en fait on va transporter les triangulations de la manière suivante : si a,b,c constitue un triangle de $\mathcal{T}_n$, on va lui associer un triangle A,B,C dans $\mathcal{T}_{n+1}$.

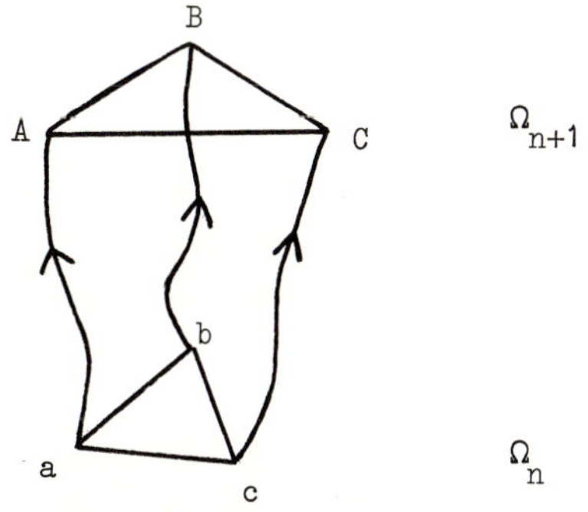

par :
$$\begin{cases} x_A = x_a + \tau\, v(x_a, n\tau) \\ x_B = x_b + \tau\, v(x_b, n\tau) \\ x_C = x_c + \tau\, v(x_c, n\tau) \end{cases}$$

Naturellement il faudra veiller à ce que la triangulation transportée ne dégénère pas (écrasement des triangles, superposition ...) quitte à reconstruire de temps à autre une triangulation par un autre procédé que celui indiqué.

Remarque

La méthode exposée plus haut est trés générale, son défaut est celui des méthodes de gradient simple (choix du pas τ, convergence lente ...)

Une mise en oeuvre plus sophistiquée (type gradient conjugué par exemple) serait souhaitable.

3.6. Cas particuliers :

On a envisagé plus haut le cas général d'un ouvert Ω ; cependant il y a des cas particuliers importants.

Cas particulier 1 :

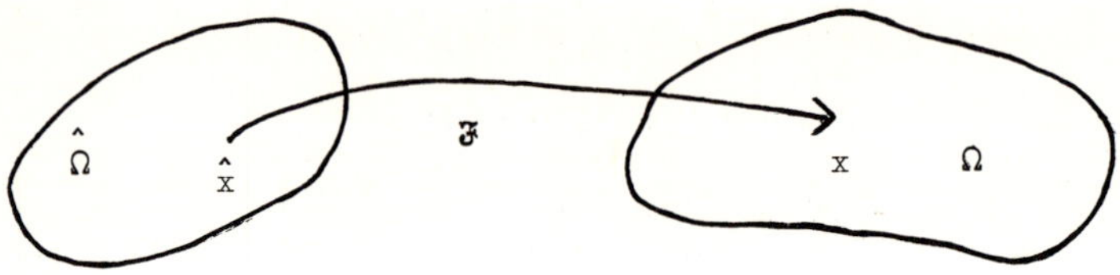

$$\Omega = \hat{\mathfrak{F}}\,\hat{\Omega} \qquad \begin{cases} x = \hat{\mathfrak{F}}(\hat{x},a) \\ a \in K \end{cases}$$

Ω est l'image d'un domaine fixe par une famille de transformations qui dépendent d'un paramètre a (de dimension finie ou infinie).

On emploie la même méthode : on cherche $t \to a(t)$ pour que Ω_{opt} = transformé de $\hat{\Omega}$ pour a_∞ : autrement dit $a_{opt} = a_\infty$.

On écrira :

$$\begin{cases} a'(t) = W(a(t),t) \\ a(0) = a_0 \end{cases}$$

et on cherchera W pour que $a_{opt} = a_\infty$.

Il est intéressant de se ramener au cas général en liant les vitesses de déformations w et v de a et de Ω . En général, on obtiendra une relation du type

$$v = Aw$$

si bien que

$$j(t) = \langle G,v \rangle = \langle G,AW \rangle$$
$$= \langle\langle A^*G,W \rangle\rangle$$

et il faudra choisir $W = - A^*G$.

Cas particulier 2 : On peut appliquer la même technique au cas d'un problème où le domaine Ω est donné mais où les opérateurs dépendent d'un paramètre a (d'un espace de dimension finie ou infinie):

y_a est solution de léquation d'état : $E(\Omega,a,y_a) = 0$ et $j(a) = J(a,y_a)$.

On procède comme dans le cas particulier précédent ; les formules de différentiation seront plus simples, le domaine ne dépendant pas de t .

BIBLIOGRAPHIE SOMMAIRE :

[1] BEGIS D., GLOWINSKI R. : Application de la méthode des éléments finis à la résolution d'un problème de domaine optimal.
Springer Verlag - Lecture Notes in Computer Science, 11, 1974.

[2] BENSON D.C., : An elementary Solution of a variational Problem of Aerodynamics.
J. of Opt. Theory and Appli. Vol.1,n°2, 1967.

[3] BENDALI A., DJAADANE A. : Thèse - Universié Alger, 1975

[4] CEA J. : Identification de domaines.
Springer Verlag - Lecture Notes in Computer Science, 3, 1973

[5] CEA J., GIOAN A., MICHEL J. : Quelques résultats sur l'identification de domaines.
CALCOLO, III-IV, 1973

[6] CEA J., GIOAN A., MICHEL J. : Adaptation de la méthode du gradient à un problème d'identification de domaine.
Springer Verlag - Lecture Notes in Computer Science, 11, 1974

[7] CHENAIS D. : On the existence of a solution in a domain identification problem.
J. of Math. Anal. and Appli., 52,2,1975

[8] DANILJUK I.I. : Sur une classe de fonctionnelles intégrales à domaine variable d'intégration.
Actes, Congrès Internat. Math., 1970, Tome 2, p.703-à-715.

[9] DERVIEUX A., PALMERIO B. : Une formule de Hadamard dans les problèmes d'optimal design.
Springer Verlag-Lecture Notes in Computer Science, 40, 1976

[10] DERVIEUX A., PALMERIO B. : Thèse Université de Nice, 1974

[11] GARABEDIAN P.R., SCHIFFER M. : Convexity of domain functionals.
J. d'Analyse Math. 3, 246-344, 1953

[12] GERMAIN P. : Cours de Mécanique des milieux continus.
Masson 1973

[13] HADAMARD J. : Leçons sur le calcul des variations.
Gauthiers-Villars 1910 et Oeuvres Complètes.

[14] JOSEPH D.D. : Parameter and Domain Dependance of Eigenvalues of elliptic Partial Differential Equations.
Arch.Rat.Mech. Anal., vol 24, 1967

[15] KAC : Can one hear the shape of a drum ?
Ann. Math. Monthly 73, 1-23, 1966

[16] KAGIWADA H.H., KALABA R.E. : A practical method for determining Green's functions using Hadamard's variational formula.
J. of Opt. Theory and Appli., Vol. 1, n°1, 1967

[17] LIONS J.L. : On the optimal control of distributed parameter systems, pp.137-158 : Techniques of Optimization, A. Balakrishnan, Academic Press, 1972.

[18] MIELE A.n HULL D.G. : Sufficiency Proofs for the problem of the Optimum transversal contour.
SIAM J. Appl. Math. 15, 2, 1967

[19] MOREL P. : Utilisation en analyse numérique de la formule de variation d'Hadamard.
RAIRO, R-2, 115-119, 1973

[20] PIRONNEAU O. : Optimisation de structure, Application à la mécanique des fluides.
Springer Verlag - Lecture Notes in Economics and Mathematical Systems, 107, 1974 .

[21] PIRONNEAU O. : On optimum profiles in Stokes flow.
J. Fluid. Mech., 59, 117-128, 1973

[22] PIRONNEAU O. : On optimum design in fluids mechanics.
J. Fluid. Mech., 64, 97-III, 1974

[23] MURAT R., SIMON J. : Quelques résultats sur le contrôle par un domaine géométrique.
Rapport n° 74003. Labo d'Analyse Numérique, Université Paris VI, 1974

[24] MURAT R., SIMON J. : Etude des Problèmes d'optimal design.
Springer Verlag - Lecture Notes in Computer Sciences, 40, 1976

[25] RAYLEIGH J.W. The theory of Sound.
2nd Ed., Cambridge 1894-1896

[26] ROUSSELET B. : Problèmes Inverses de Valeurs Propres.
Springer Verlag - Lecture Notes in Computer Sciences, 40, 1976

[27] SCHIFFER M. : Variation of domain functional.
Bull. A.M.S. 60, 303-328, 1954.

[28] TROESCH B.A. : Elliptical Membranes with Smallest Second eigenvalue.
Math. of Comp., 27, 124, 1973

[29] TROESCH B.A. , TROESCH H.R. : Eigenfrequencies of an Elliptic Membrane.
Math. of Computation, 27, 124, 1973

[30] ZOLESIO J.P. : Sur un problème d'identification de domaine.Localisation du support d'un contrôle obtimal (à paraître).

UNIVERSITE DE BORDEAUX I

U.E.R. DE MATHEMATIQUES ET INFORMATIQUE

351, Cours de la Libération

33405 TALENCE

DIVERSES DONNEES SPECTRALES POUR LE PROBLEME INVERSE DISCRET DE STURM-LIOUVILLE

P. MOREL

DIVERSES DONNEES SPECTRALES POUR LE PROBLEME INVERSE DISCRET DE STURM-LIOUVILLE

Résumé : On envisage le problème (P) suivant : Reconstruire à partir de données de type spectral l'opérateur aux différences L, qui est l'analogue discret de l'opérateur de Sturm-Liouville.

Dans un premier temps on emploie la théorie des polynomes orthogonaux pour obtenir et décrire d'une manière élémentaire la mesure spectrale de L. On rappelle alors que la connaissance de celle-ci est suffisante pour résoudre le problème (P).

On montre ensuite que dans certains problèmes inverses discrets envisagés par Anderson, Hochstadt, Hald on sait déterminer la mesure spectrale et par suite reconstruire l'opérateur. Ce point de vue permet d'unifier et de généraliser les résultats obtenus par les auteurs précédemment cités.

On obtient en particulier une interprétation de deux suites entrelacées comme les spectres d'une même expression aux différences associées à deux systèmes de conditions aux limites ; on obtient également l'interprétation d'une suite comme le spectre d'un opérateur vérifiant certaines propriétés de symétrie. On examine aussi quelques questions concernant la stabilité, et on donne pour terminer quelques résultats numériques d'un algorithme simple et efficace.

1 - INTRODUCTION ET NOTATIONS.

Considérons sur l'intervalle fini $[0,\pi]$ le problème de Sturm-Liouville :

$$-\frac{d}{dx}\left(p(x)\frac{dy}{dx}\right) + q(x)\, y = \lambda y$$

$$\alpha_1\, y(0) + \beta_1\, y'(0) = \alpha_2\, y(\pi) + \beta_2\, y'(\pi) = 0$$

où p et q sont des fonctions continues, p étant à valeurs positives. On appelle problème inverse à donnée spectrale la recherche des fonctions p et q connaissant par exemple la mesure spectrale de l'opérateur (cf Guelfand-Levitan [1]) ou bien les deux suites de valeurs propres obtenues en donnant deux valeurs distinctes à β_2 dans la condition limite en $x=\pi$. (cf Marchenko [2] , Lévitan [3], Borg [4]).

Beaucoup de résultats ont été obtenus pour ce problème inverse continu. Mais pour des raisons dues au calcul effectif des fonctions p et q c'est l'analogue discret de ce problème qui nous intéresse ici. Pour le problème inverse discret, on consultera d'abord Hald [5, 6] , puis Hochstadt [7] et Anderson [9]. Dans notre travail nous faisons jouer un rôle primordial à la mesure spectrale de l'opérateur ; ce point de vue permet d'unifier et de généraliser les résultats obtenus par les auteurs précédemment cités.

On introduit le pas $h=\pi/(n+1)$ et les noeuds $x_i = ih$ $i=0, 1, \ldots, n, n+1$. On notera f_i la valeur en $x=x_i$ de la fonction f. Les opérateurs de différence avant et arrière du premier ordre seront notés :

$$D_+ y_i = (y_{i+1}-y_i)/h \qquad D_- y_i = (y_i-y_{i-1})/h$$

La relation de récurrence linéaire du deuxième ordre qui est l'analogue discret de l'expression différentielle de Sturm-Liouville sur $[0,\pi]$; s'écrit :

$$(\mathcal{L}y)_i = D_-(p_i\, D_+ y_i) + q_i\, y_i \qquad i=1, 2, \ldots, n$$

$p_0, p_1, \ldots, p_n \,;\, q_1, q_2, \ldots, q_n$ sont les $2n+1$ coefficients de l'expression aux différences et $(y_0, y_1, \ldots, y_{n+1})$ est le vecteur sur lequel $\mathcal{L}$ agit.

Pour alléger l'écriture, on pose :

$$b_i = p_i/h^2 \quad i=0,1,\ldots,n \qquad a_i = p_i/h^2 + p_{i-1}/h^2 + q_i \quad i=1,2,\ldots,n$$

Il est clair que l'on retrouve les p_i, q_i connaissant les a_i, b_i et réciproquement. La relation de récurrence s'écrit alors :

$$(\mathcal{L}y)_i = -b_{i-1}\, y_{i-1} + a_i\, y_i - b_i\, y_{i+1} \qquad i=1,2,\ldots,n$$

et on adoptera pour conditions aux limites :

$$y_0 + ky_1 = y_{n+1} + Ky_n = 0$$

Dans le calcul de $(\mathcal{L}y)_i$ pour $i=1$ on pose $y_0 = -ky_1$ et pour $i=n$ on pose $y_{n+1} = -Ky_n$. Dans $\mathbb{R}^n$ on définit alors l'opérateur L par $(Ly)_i = (\mathcal{L}y)_i$ $i = 1, 2, \ldots, n$ et on notera pour rappeler les conditions limites $\{L;k,K\}$.

On appellera problème 1 (Pb1) la détermination de L ; c'est à dire des $(2n+1)$ coefficients b_i $i=0,1,\ldots,n$; a_i $i=1,2,\ldots,n$ connaissant la mesure spectrale $d\tau_{\{L;k,K\}}$ de $\{L;k,K\}$ dont on précisera la définition au paragraphe suivant.

A l'opérateur $\{L;k,K\}$ on peut associer une matrice $n \times n$ symétrique tridiagonale J. En effet le vecteur $Ly \in \mathbb{R}^n$ peut être considéré comme l'image de $y \in \mathbb{R}^n$ par la matrice J.

$$J = \begin{vmatrix} a_1 + kb_0 & -b_1 & & & \\ -b_1 & a_2 & -b_2 & & O \\ & -b_2 & \ddots & \ddots & \\ & & \ddots & \ddots & -b_{n-1} \\ & O & & -b_{n-1} & a_n + Kb_n \end{vmatrix}$$

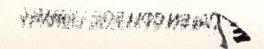

Le problème 1' (Pbl') correspond à la détermination de la matrice J, connaissant $d\tau_{\{L;k,K\}}$. Remarquons que le problème 1' est à priori moins fin que le problème 1 : il n'y a que (2n - 1) inconnues.

C'est un fait bien connu que la théorie de matrices de ce type, dite de Jacobi, est liée à la fois à la théorie des polynomes orthogonaux et au problème des moments. (cf Stone [10], Szegö [11], Akhiezer [12], Bérézanskii [13]). Nous proposons en quelque sorte de revenir à ce fait fondamental, et de l'exploiter pour la résolution du problème inverse discret.

Dans le paragraphe 2 nous développons sous une forme directement accessible des résultats très classiques sur l'opérateur $\{L;k,K\}$ et sur sa mesure spectrale. Le paragraphe 3 sera consacré au problème de la reconstruction de l'opérateur connaissant explicitement sa fonction spectrale ; nous examinerons l'existence, l'unicité et la stabilité d'une telle reconstruction. Dans les paragraphes 4, 5, 6 nous envisagerons trois manières particulières de se donner de manière implicite la mesure spectrale. Chaque fois que nous poserons le problème de l'existence et de l'unicité, et le problème de la reconstruction par un algorithme efficace.

2 - L'OPERATEUR $\{L;k,K\}$ ET SA MESURE SPECTRALE $d\tau_{\{L;k,K\}}$.

Considérons la relation de récurrence $(\mathcal{L}y) = -b_{i-1}\, y_{i-1} + a_i\, y_i - b_i\, y_{i+1}$ $\quad i = 1,2,\ldots,n$ dans laquelle $b_i > 0$ pour tout i.

Il est immédiat d'établir la relation de Green :

$$\sum_{j=1}^{u} (\mathcal{L}u)_j\, v_j - u_j\, (\mathcal{L}v)_j = b_n(u_{n+1}\, v_n - u_n v_{n+1}) - b_0(u_1 v_0 - u_0 v_1)$$

En tenant compte des conditions aux limites, on en déduit que $\{L;k,K\}$ est un opérateur auto-adjoint de $\mathbb{R}^n$ dans $\mathbb{R}^n$, et donc en particulier que ses valeurs propres sont réelles.

Considérons maintenant l'équation récurrente linéaire du deuxième ordre $(\mathcal{L}y)_i = \lambda y_i$ $\quad i = 1,2,\ldots,n$. Elle permet de déterminer y_{j+1} connaissant

y_j et y_{j-1}, car $b_j > 0$. On pose $y_0(\lambda) = -k$, $y_1(\lambda) = 1$ et on détermine $y_{i+1}(\lambda)$ par $(\mathcal{L}y)_i = \lambda y_i(\lambda)$ $i = 2 ; n$.

On obtient pour $y_i(\lambda)$ des polynômes en λ, à des coefficients réels dont le terme de degré le plus élévé est $(-1)^{i-1}\lambda^{i-1}/(b_1, b_2, \ldots, b_{i-1})$; suivant Berezanskii [13], Ahiezer [12] on les appelle polynômes de première espèce associés à la récurrence.

Cette solution de l'équation $(\mathcal{L}y)_i = \lambda y_i$ $i=1,2,\ldots,n$ vérifie $y_0 + ky_1 = 0$. Elle vérifiera $y_{n+1} + ky_n = 0$ si et seulement si λ est une racine du polynôme de degré n $\varpi_k(\lambda) = y_{n+1}(\lambda) + Ky_n(\lambda)$. Ainsi les valeurs propres de $\{L;k,K\}$ sont les zéros de $\varpi_k(\lambda) = 0$. Le vecteur propre associé à λ_r est $(y_1(\lambda_r) = 1, y_2(\lambda_r), \ldots, y_n(\lambda_r))$; notons qu'il est normalisé par $y_1(\lambda_r) = 1$. Les valeurs propres et vecteurs propres de $\{L;k,K\}$ sont également les valeurs et vecteurs propres de la matrice de Jacobi associée à $\{L ; k,K\}$.

Les polynômes $y_i(\lambda)$ $i=1,2,\ldots,n$ forment une suite de polynômes orthonormée pour la mesure spectrale de $\{L;k,K\}$. Nous allons caractériser celle-ci d'une manière élémentaire en mettant en relief certains détails utiles pour la suite.

LEMME 1 - Identité de Christoffel-Darboux.

Pour les polynômes de première espèce $\{y_i(\lambda)\}_1^n$ associés à la récurrence $(\mathcal{L}y)_i = \lambda y_i$ $i=1,2,\ldots,n$ on a les deux identités.

$$\forall p=1,2,\ldots,n \quad (\mu-\lambda)\sum_{i=1}^{p} y_i(\lambda)y_i(\mu) = b_p(y_{p+1}(\lambda)\, y_p(\mu) - y_{p+1}(\mu)\, y_p(\lambda))$$

$$\sum_{i=1}^{p} y_i(\lambda)^2 = b_p(y'_{p+1}(\lambda)\, y_p(\lambda) - y_{p+1}(\lambda)\, y'(\lambda)).$$

Démonstration : Szegö [11] page 42.

Ce lemme permet d'affirmer que les zéros de $\varpi_k(\lambda)$ sont non seulement tous réels, mais sont également simples.

LEMME 2 - Si λ_r et λ_s sont deux racines distinctes de $\varpi_k(\lambda) = 0$ et si les $\{y_i(\lambda)\}_i^n$ sont les polynômes de l'espèce associés à la récurrence $(\mathcal{L}y)_i = \lambda y_i$ $i=1,2\ldots,n$. On a :

$$\sum_{p=1}^{n} y_p(\lambda_r)\, y_p(\lambda_s) = \rho_r\, \delta_{rs} \quad \text{avec } \rho_r = \sum_{p=1}^{n} y_p(\lambda_r)^2 = b_n y_n(\lambda_r)\, \varpi_k'(\lambda_r)$$

Démonstration : On applique le lemme 1 avec $\lambda=\lambda_r$, $\mu=\lambda_s$ et $p=n$; il vient :

$$(\lambda_s-\lambda_r) \sum y_i(\lambda_r)\, y_i(\lambda_s) = b_n \begin{vmatrix} y_{n+1}(\lambda_r) & y_{n+1}(\lambda_s) \\ y_n(\lambda_r) & y_n(\lambda_s) \end{vmatrix}$$

On multiplie la seconde ligne du déterminant par K et on l'ajoute à la première. On tient compte du fait que $\varpi_k(\lambda_r) = \varpi_k(\lambda_s) = 0$. Pour $r=s$, comme dans la démonstration du lemme 1, c'est une application de la règle de l'Hospital.

Il est facile de démontrer le lemme d'algèbre suivant :

LEMME 3 - Soit $Y = (y_{rs}) \in \mathcal{M}_{nn}(\mathbb{R})$ telle que :

$$\sum_r a_r\, y_{rs}\, y_{rt} = \rho_s\, \delta_{st} \qquad s,t = 1,2,\ldots,n \qquad a_r, \rho_s > 0$$

Alors on a :

$$\sum_r (y_{sr}\, y_{tr})/\rho_r = \delta_{st}/a_s \qquad s,t = 1,2,\ldots,n$$

Des lemmes 2 et 3, on tire la proposition suivante qui permet de définir la fonction spectrale.

THEOREME 1 - Soit $\{y_i(\lambda)\}_1^n$ les polynômes de première espèce associés à l'opérateur $\{L\,;k,K\}$ et soit λ_r une racine de $\varpi_k(\lambda) = y_{n+1}(\lambda) + k y_n(\lambda) = 0$. Alors pour tout p et $q=1,2\ldots,n$ on a :

$$\sum_{r=1}^{n} y_p(\lambda_r)\, y_q(\lambda_r)\, \frac{1}{\rho_r} = \delta_{pq}$$

où :

$$\int_{-\infty}^{\infty} y_p(\lambda)\, y_q(\lambda)\, d\tau_{\{L;k,K\}}(\lambda) = \delta_{pq}$$

avec :

$$\tau_{\{L;k,K\}}(\lambda) = \begin{cases} \sum_{0<\lambda\leq\lambda_r} \rho_r^{-1} & \text{si } \lambda \geq 0 \\ -\sum_{\lambda<\lambda_r\leq 0} \rho_r^{-1} & \text{si } \lambda < 0 \end{cases}$$

DEFINITION - On appelle $\tau_{\{L;k,K\}}$ la fonction spectrale de $\{L;k,K\}$ et la mesure $d\tau_{\{L;k,K\}} = \sum_i \rho_i^{-1} \delta_{\lambda_i}$ la mesure spectrale de l'opérateur.

Dans ce contexte très simple, la mesure spectrale a son support concentré sur le spectre $\{\lambda_i\}_1^n$ de $\{L;k,K\}$. Les masses correspondantes $\{\rho_i^{-1}\}_1^n$ apparaissent comme les inverses des normes euclidiennes des vecteurs propres. Ainsi la mesure spectrale est caractérisée par 2n nombres, les n masses étant par ailleurs telles que la masse totale vaut 1.

3 - RECONSTRUCTION DE L'OPERATEUR A PARTIR DE LA MESURE SPECTRALE.

Dans le paragraphe précédent nous avons rappelé qu'à un opérateur $\{L;k,K\}$ on sait associer une mesure $d\tau_{\{L,k,K\}}$. Réciproquement une mesure $d\tau = \sum_{i=1}^{n} \alpha_i \delta_{\lambda_i}$ de support n points distincts et dont les masses α_i vérifient $\alpha_i > 0$, $\sum_{i=1}^{n} \alpha_i = 1$ peut elle être considérée comme la mesure spectrale d'un certain opérateur aux différences ? et si la réponse est affirmative peut-on en calculer explicitement les coefficients ?

LEMME 4 - Soit $d\tau = \sum_{i=1}^{n} \alpha_i \delta_{\lambda_i}$ une mesure de probabilité ayant pour support n points distincts. Alors il existe un système unique de n polynômes $g_1(\lambda)$ $g_2(\lambda), \ldots, g_n(\lambda)$ tel que :

i) $\int_{-\infty}^{\infty} g_i(\lambda)\, g_j(\lambda)\, d\tau(\lambda) = \delta_{ij}$ $\quad 1 \leq i, j \leq n$

ii) $g_i(\lambda) = (-1)^{i-1} C_i \{\lambda^{i-1} + \ldots\}$ $\quad c_i > 0.$

on orthogonalise par le procédé de Gram-Schmidt la suite $1, \lambda, \lambda^2, \ldots, \lambda^{n-1}$.

THEOREME 2 - Soit $d\tau$ une mesure de probabilité de la forme $d\tau = \sum_{i=1}^{n} \alpha_i \delta_{\lambda_i}$ $\alpha_i > 0$ dont le support est constitué par n points distincts. Alors on peut construire une famille à deux paramètres α, β d'opérateurs $\{L(\alpha, \beta) ; k, K\}$ telle que la mesure spectrale $d\tau_{\{L(\alpha,\beta);k,K\}}$ égale la mesure donnée $d\tau$.

Démonstration : Soit $\{g_i(\lambda)\}_1^n$ la suite des polynômes orthonormée par la mesure $d\tau$. Ces polynômes doivent vérifier une relation de récurrence à trois termes dont on va déterminer les coefficients.

Comme $g_1(\lambda) = 1$, pour que la condition limite à gauche soit vérifiée on est amené à poser $g_0(\lambda) = -k$. $\lambda g_1 = \lambda$ est un monôme du premier degré que l'on peut écrire :

$$\lambda g_1(\lambda) = \alpha_2^1 g_2(\lambda) + \alpha_1^1 g_1(\lambda) + \alpha_0^1 g_0(\lambda).$$

Les α_1^1 ne seront pas déterminés de manière unique les $g_2(\lambda), g_1(\lambda), g_0(\lambda)$ n'étant pas linéairement indépendants. Pour déterminer α_2^1 on identifie les termes de plus haut degré , il vient :

$$\lambda = \alpha_2^1 \times (-C_2\lambda + \ldots\ldots.) \qquad \text{d'où } \alpha_2^1 = -C_2^{-1} < 0$$

On posera $b_1 = -\alpha_2^1 = C_2^{-1} > 0$. Pour obtenir α_1^1 et α_0^1 multiplions les deux membres par $p_1(\lambda)$ et intégrons par rapport à $d\tau$; il vient :

$$\int \lambda g_1^2(\lambda)\, d\tau\, (\lambda) = \alpha_1^1 + \alpha_0^1 \int g_0(\lambda)\, g_1(\lambda)\, d\tau(\lambda) = \alpha_1^1 - \alpha_0^1 k.$$

On fixe arbitrairement la valeur de α_0^1 soit $\alpha_0^1 = \alpha$; k étant donné on obtient $\alpha_1^1 = \int \lambda g_1^2(\lambda)\, d\tau(\lambda) + \alpha k$. On pose :

$$a_1 = \alpha_1^1 \quad \text{et} \quad b_0 = -\alpha = -\alpha_0^1$$

Pour $i = 2, 3, \ldots, n-1$ on considère le polynôme $\lambda g_i(\lambda)$ dont le coefficient de tête est $(-1)^{i-1} C_i \lambda^i$. Son écriture sur la base des $g_1(\lambda), g_2(\lambda), \ldots, g_{i+1}(\lambda)$ est $\lambda g_i(\lambda) = \alpha_{i+1}^i g_{i+1}(\lambda) + \alpha_1^i g_i(\lambda) + \ldots + \alpha_1^i g_1(\lambda)$

L'examen du signe du coefficient de tête prouve que $\alpha_{i+1}^i < 0$. Les relations d'orthogonalité permettent de calculer les divers coefficients ; il vient :

$$\alpha_{i+1}^i = \int_{-\infty}^{\infty} \lambda g_i(\lambda)\, g_{i+1}(\lambda) d\tau(\lambda) \; ; \; \alpha_{i-1}^i = \int \lambda g_i(\lambda)\, g_{i-1}(\lambda)\, d\tau(\lambda)$$

$$\alpha_i^i = \int \lambda g_i(\lambda)^2\, d\tau(\lambda) \quad ; \text{ tous les autres } \alpha_k^i \text{ étant nuls.}$$

On pose $b_i = -\alpha_{i+1}^i > 0$ $b_i = -\alpha_{i-1}^i > 0$ et $a_i = \alpha_i^i$.

Pour $i = n$ on doit déterminer a_n, $b_n > 0$ et $g_{n+1}(\lambda)$ tels que d'une part $b_n g_{n+1}(\lambda) = (a_n - \lambda)\, g_n(\lambda) - b_{n-1}\, g_{n-1}(\lambda)$ d'autre part $g_{n+1}(\lambda_i) + K g_n(\lambda_i) = 0$ pour $i = 1, 2, \ldots, n$. On pose arbitrairement $g_{n+1}(\lambda) = \beta\, (-1)^n \prod_i (\lambda - \lambda_i) - K y_n(\lambda)$ β étant une constante positive. Le polynôme $\lambda g_n(\lambda)$ de degré n possède une écritureunique sur la base $g_{n+1}(\lambda)\; g_n(\lambda), \ldots, g_1(\lambda)$; soit $\lambda g_n(\lambda) = \alpha_{n+1}^n g_{n+1}(\lambda) + \ldots + \alpha_1^n g_1(\lambda)$. Le coefficient α_{n+1}^n est déterminé par l'identification des coefficients de λ^n ; il vient $(-1)^{n-1} C_n = \alpha_{n+1}^n . \beta . (-1)^n$. d'où $\alpha_{n+1}^n = -C_n . \beta^{-1}$. On pose $b_n = -\alpha_{n+1}^n = C_n \beta^{-1} > 0$ ainsi b_n dépend du paramètre β ; on pourra lui assigner de même qu'à b_0 une valeur arbitraire. Les autres α_i^n sont déterminés par la méthode d'orthogonalité, en remarquant que $\int g_{n+1}(\lambda)\, g_s(\lambda)\, d\tau(\lambda) = -K \int g_n(\lambda)\, g_s(\lambda)\, d\tau(\lambda)$.

On constate alors que $\alpha_1^n = \alpha_2^n = \cdots = \alpha_{n-2}^n = 0$ et que $\alpha_n^n = \int \lambda g_n(\lambda)^2\, d\tau(\lambda) + K\alpha_{n+1}$. On pose $a_n = \alpha_n^n = \int \lambda g_n(\lambda)^2 d\tau - K C_n^{-1} \beta$.

On vérifie que $\alpha_{n-1}^{n} = -b_{n-1}$; ce dernier terme ayant été déterminé à l'étape $i=n-1$.

Il est facile de vérifier que pour tout α et β les opérateurs $\{L(\alpha,\beta)\ ;\ k,K\}$ ainsi obtenus ont pour mesure spectrale $d\tau_{\{L(\alpha,\beta);\ k,K\}}$ la mesure fixée $d\tau$.

Le théorème 2 rend cohérente la définition suivante :

DEFINITION - Soit $d\tau$ une mesure de probabilité de la forme $d\tau = \sum_{1}^{n} \alpha_i \delta_{\lambda_i}$ $\alpha_i > 0$ dont le support est constitué de n points distincts.

On notera $\{L;k,K\}_1$ l'unique opérateur aux différences de mesure spectrale $d\tau$, et qui vérifie de plus $b_0 = b_n = 1$.

Il est intéressant d'examiner la dépendance des divers coefficients vis à vis des paramètres α, β, k, K.

Remarque 1 : Revenons à la notation en p_i, q_i initiale. On a :

$$\begin{cases} b_0 = -\alpha = p_0/h^2 \\ a_1 = (p_0+p_1)/h^2 + q_1 = \int_{-\infty}^{\infty} \lambda g_1^2(\lambda)\, d\tau(\lambda) - kb_0 \end{cases}$$

$$\begin{cases} b_i = -p_i/h^2 = \int_{-\infty}^{\infty} \lambda g_{i-1}(\lambda)\, g_i(\lambda)\, d\tau(\lambda) \\ a_i = (p_i+p_{i-1})/h^2 + q_i = \int_{-\infty}^{\infty} \lambda g_i^2(\lambda)\, d\tau(\lambda) \end{cases}$$

$$\begin{cases} b_n = p_n/h^2 = -C_n \beta^{-1} \\ a_n = (p_n + p_{n-1})/h^2 + q_n = \int_{-\infty}^{\infty} \lambda g_n^2(\lambda)\, d\tau(\lambda) - Kb_n . \end{cases}$$

De sorte que : $q_1 = \int \lambda g_1^2(\lambda)\, d\tau(\lambda) - \int \lambda g_1(\lambda)\, g_2(\lambda)\, d\tau - b_0(1+k)$

et que : $$q_n = \int \lambda g_n^2(\lambda)\, d\tau(\lambda) - \int \lambda g_n(\lambda)\, g_{n-1}(\lambda) d\tau - b_n(1+K)$$

Ainsi la mesure $d\tau$ étant fixée, le vecteur $p = (p_0, p_1, \ldots, p_n)^T$ est fixé indépendement de k, K ; on peut par le choix des paramètres α, β donner une valeur arbitraire à p_0 et p_n. De même pour le vecteur $q=(q_1, q_2, \ldots, q_n)^T$, $d\tau$ etant fixée seules les composantes q_1 et q_n dépendent des paramètres α, β et des constantes k, K.

Remarque 2 - Traduisons la remarque précédente sur la matrice de Jacobi : J associé à l'opérateur $\{L(\alpha, \beta) ; k, K\}$. On a :

$$J = \begin{vmatrix} a_1 + kb_0 & -b_1 & & \\ -b_1 & a_2 & & \\ & & a_{n-1} & -b_{n-1} \\ & & -b_{n-1} & a_n + Kb_n \end{vmatrix}$$

J semble dépendre de k, K et de α, β par l'intermédiaire de b_0, et b_n ; il n'en est rien. En effet d'après les formules explicitées dans la démonstration du théorème 2, on a :

$$a_1 + kb_0 = \int \lambda g_1^2(\lambda)\, d\tau(\lambda)$$

$$a_n + Kb_n = \int \lambda g_1^2(\lambda)\, d\tau(\lambda)$$

Ainsi à une mesure de probabilité $d\tau = \sum_1^n \alpha_i \delta_{\lambda_i}$ $\alpha_i > 0$ ayant pour support n points distincts correspond une seule matrice de Jacobi J. Quelque soit k, K et quelque soit α, β l'opérateur $\{L(\alpha, \beta); k, K\}$ déterminé par $d\tau$ est associé à J.

Remarque 3 - Si $d\tau_{\{L;k,K\}} = d\tau_{\{\tilde{L}\,;\,\tilde{k}, \tilde{K}\}_1}$ alors nécessairement

$a_1 - \tilde{a}_1 = \tilde{k} - k$, $a_i = \tilde{a}_i$ $i=2, 3, \ldots, n-1$, $a_n - \tilde{a}_n = \tilde{K} - K$ et $b_i = \tilde{b}_i$ $i=0, 1, \ldots n$.

Notons ici une légère différence avec le cas continu. En effet, si les fonctions spectrales de deux opérateurs de Sturm-Liouville sont multiples l'une de l'autre, et si de plus on a coïncidence des conditions limites à gauche alors on en déduit l'égalité des coefficients et l'égalité des conditions aux limites à droite (cf Levitan [3]).

THEOREME 3 - Sur l'ouvert des mesures de probabilité de la forme $d\tau = \sum_1^n \rho_i \delta_{\lambda_i}$ $\rho_i > 0$ portée par n points distincts on considère l'application φ qui à une mesure $d\tau$ associe la matrice de Jacobi : $J(d\tau)$. φ est continue.

Démonstration : On considère les moments successifs des diverses mesures associées aux fonctions τ et $\tau^{(k)}$. On notera :

$$m_i = \int_{-\infty}^{\infty} \lambda^i \, d\tau(\lambda) \quad \text{et} \quad m_i^{(k)} = \int_{-\infty}^{\infty} \lambda^i \, d\tau^{(k)}(\lambda).$$

les ièmes moments de $d\tau$ et $d\tau^{(k)}$ respectivement. Montrons que pour tout $i = 0, 1, 2, \ldots$ $m_i^{(k)} \to m_i$.

On introduit les fonctions $f_i^{(k)}(\lambda)$ définies par :

$$f_i^{(k)}(\lambda) = \begin{cases} 0 & \text{si } -\infty < \lambda < \lambda_1^{(k)} \\ [\lambda_p^{(k)}]^i \, \rho_p^{(k)}/\rho_p & \text{si } \lambda_p^{(k)} \leq \lambda < \lambda_{p+1}^{(k)} \\ & p = 1, 2, \ldots, n-1 \\ [\lambda_n^{(k)}]^i \, \rho_n^{(k)}/\rho_n & \text{si } \lambda_n^{(k)} \leq \lambda < +\infty \end{cases}$$

Il est alors clair que :

$$m_i^{(k)} = \int_{-\infty}^{\infty} \lambda^i \, d\tau^{(k)}(\lambda) = \sum_{p=1}^{n} [\lambda_p^{(k)}]^i \, \rho_p^{(k)} = \sum_{p=1}^{n} f_i^{(k)}(\lambda_p) \, \rho_p$$

$$m_i^{(k)} = \int f_i^{(k)}(\lambda) \, d\tau(\lambda).$$

Il est clair que pour chaque i, et pour k tendant vers l'infini les $f_i^{(k)}(\lambda)$ tendent vers la fonction $f_i(\lambda)$ constante par morceaux définie par :

$$f_i(\lambda) = \begin{cases} 0 & -0 < \lambda < \lambda_1 \\ \lambda_p^i & \lambda_p^i \leq \lambda < \lambda_{p+1}^i \\ \lambda_n^i & \lambda_n \leq \lambda < +\infty \end{cases}$$

Ainsi $m_i^{(k)} = \int f_i^{(k)}(\lambda)\, d\tau(\lambda) \to \sum_p \lambda_p^i \rho_p = \int \lambda^i\, d\tau(\lambda) = m_i$

Les polynômes orthonormaux par rapport à la mesure $d\tau(\lambda)$ [resp $d\tau^{(k)}(\lambda)$] sont en fonction des moments donnés par :

$$g_i(\lambda) = (-1)^{i-1} \det_{p,q=0,\dots,i-2} \{m_{p+q+1} - \lambda m_{p+q}\} \sqrt{D_{i-2} D_{i-1}}$$

$$\text{resp } [g_i^{(k)}(\lambda) = (-1)^{i-1} \det_{p,q=0,\dots,1-2} \{m_{p+q+1}^{(k)} - \lambda m_{p+q}^{(k)}\} / \sqrt{D_{i-2}^{(k)} D_{i-1}^{(k)}}]$$

où $D_n^{(k)} = \det_{p,q=0,\dots,n} \{m_{p+q}^{(k)}\}$.

On tient compte maintenant des formules de calcul des éléments des matrices J et $J^{(k)}$ associées respectivement à $d\tau(\lambda)$ et $d\tau^{(k)}(\lambda)$.

$$a_{m,n}^{(k)} = \int_{-\infty}^{\infty} \lambda g_n^{(k)}(\lambda)\, g_m^{(k)}(\lambda)\, d\tau^{(k)}(\lambda)$$

$$a_{m,n} = \int_{-\infty}^{\infty} \lambda g_n(\lambda)\, g_m(\lambda)\, d\tau(\lambda).$$

De la première partie assurant la convergence du $m_i^{(k)}$ vers m_i on en tire que $a_{m,n}^{(k)} \to a_{m,n}$ pour k tendant vers l'infini.

Notons que ce résultat de continuité, n'est pas un résultat de stabilité numérique ; des essais numériques hélas le confirment. On peut l'expliciter par le fait que le calcul des divers moments est numériquement très instable.

Nous allons maintenant aborder des problèmes inverses pour lesquels nous ne connaissons pas explicitement la mesure spectrale. Nous séparerons en général les problèmes d'unicité et d'existence. Les résultats de deux types seront obtenus en montrant que les hypothèses permettent de retrouver la mesure spectrale. L'algorithme pour le calcul effectif des coefficients ne suivra pas cette voie.

4 - RECONSTRUCTION DE L'OPERATEUR CONNAISSANT SON SPECTRE ET LES ZEROS DE SON DERNIER POLYNOME DE PREMIERE ESPECE.

Nous voulons montrer que les n valeurs propres $\{\lambda_i\}_1^n$ et les (n-1) zéros $\{\nu_i\}_1^{n-1}$ de $y_n(\lambda)$ dernier polynôme de première espèce associé à $\{L;k,K\}$ déterminent de manière unique $d\tau_{\{L;k,K\}}$; de la sorte, nous serons amenés au problème original du paragraphe 2. La solution est contenue dans le lemme 2.

THEOREME 4 - Supposons que l'opérateur aux différences $\{L;k,K\}$ ait pour spectre les $\{\lambda_i\}_1^n$ et pour zéros de son dernier polynôme de première espèce les $\{\nu_i\}_1^{n-1}$. Alors sa mesure spectrale $d\tau_{\{L;k,K\}}$ est proportionnelle à la mesure $d\tau = \sum_1^n r_i\delta_{\lambda_i}$ où $r_i^{-1} = \prod_{j=1}(\lambda_i - \nu_j) \prod_{j\neq 1}(\lambda_i - \lambda_j)$

De la mesure spectrale $d\tau_{\{L;k,K\}}$ nous connaissons à priori le support : $\{\lambda_i\}_1^n$. Les masses correspondantes $\{\rho_i\}_1^n$ vérifient $\rho_i^{-1} = b_n y_n(\lambda_i)\varpi_k(\lambda_i)$ $i=1,2,\ldots,n$. D'après la définition des $\{\lambda_i\}_1^n$ le polynôme $\varphi(\lambda) = \prod_{j=1}^{n}(\lambda - \lambda_j)$ est un multiple de $\varpi_k(\lambda)$ de même le polynôme $\psi(\lambda)\ \prod_{j=1}^{n-1}(\lambda - \nu_j)$ est un multiple de $y_n(\lambda)$. Ainsi les $r_i^{-1} = \psi(\lambda_i)\wedge\varphi'(\lambda_i)$ sont ils proportionnels aux ρ_i^{-1} $i=1,2,\ldots,n$. On connait donc la mesure $d\tau_{\{L;k,K\}}$.

Nous décrirons un algorithme efficace après avoir envisagé le problème de l'existence.

THEOREME 5 - Soient les réels $\{\lambda_i\}_1^n$ et $\{\nu_i\}_1^{n-1}$

Pour tout k, K il existe un et un seul opérateur $\{L;k,K\}_1$ vérifiant :

i) $Sp\{L;k,K\}_1 = \{\lambda_1\}_1^n$

ii) $y_n(\nu_i) = 0 \quad i=1,2,\dots,n-1$ où $y_n(\lambda)$ est le dernier polynôme de première espèce associé.

si et seulement si

$$\lambda_1 < \nu_1 < \lambda_2 \dots\dots\dots\dots < \nu_{n-1} < \lambda_n$$

Démonstration : Soit k, K et soit $\{L;k,K\}_1$. On a nécessairement l'entrelacement $\lambda_1 < \nu_1 < \lambda_2 < \dots\dots < \nu_{n-1} < \lambda_n$. En effet, les $\{\lambda_i\}_1^n$ sont les valeurs propres de la matrice de Jacobi associée à $\{L;k,K\}_1$ et les $\{\nu_i\}_1^n$ sont les (n-1) valeurs propres du premier mineur principal de la même matrice. On en déduit l'entrelacement d'après le principe du min-max (cf. Wilkinson [14]).

Réciproquement, considérons à priori la mesure $d\tau = \sum_{i=1}^n \rho_i \delta_{\lambda_i}$ avec :

$$\rho_i^{-1} = \prod_{j=1}^{n=1} (\lambda - \nu_j)|_{\lambda=\lambda_i} \times \frac{d}{d\lambda} \prod_{j=1}^{n} (\lambda - \lambda_j) = \prod_{j=1}^{n} (\lambda_i - \lambda_j) \prod_{j\neq 1} (\lambda_i - \lambda_j).$$

On vérifie que ces masses sont positives grâce à l'hypothèse d'entrelacement. Choisissons k, K et considérons l'opérateur $\{L;k,K\}_1$ déterminé par ces constantes et la mesure $d\tau(\lambda)$. Il est clair d'après le théorème 2 qu'il possède pour spectre les $\{\lambda_i\}_1^n$. Soit $y_n(\lambda)$ le dernier polynôme de première espèce associé à cet opérateur. Par l'absurde supposons que $y_n(\lambda)$ ne soit pas proportionnel à $\prod_1^{n-1} (\lambda - \nu_j)$. Comme $y_n(\lambda)$ est un polynôme de degré n-1 c'est à dire que quelque soit la constante c on a $y_n(\lambda_i) \neq c \prod_{j=1}^{n-1} (\lambda_i - \nu_j) \quad i=1,2,\dots n$

D'après le lemme 2 la mesure spectrale de $\{L;k,K\}$ est $\sum_{1}^{n} \rho_i \delta_{\lambda_i}$ avec $\rho_i^{-1} = b_n y_n(\lambda_i)\varpi_n'(\lambda_i)$ et d'après le théorème 4 elle est proportionnelle à $\sum_{1}^{n} r_i \delta_{\lambda_i}$ avec $r_i^{-1} = \prod_{j=1}^{n-1} (\lambda_i - \nu_j)\varpi_n'(\lambda_i)$; ce qui est contradictoire.

L'expression de ce résultat en termes de matrices de Jacobi fournit le corollaire suivant :

<u>COROLLAIRE 1</u> - (Hochstadt [8], Hald [6]).

Soient $\lambda_1 < \nu_1 < \lambda_2 < \ldots\ldots\ldots < \nu_{n-1} < \lambda_n$. Alors il existe une et une seule matrice $n \times n$ de Jacobi admettant les $\{\lambda_i\}_1^n$ pour valeurs propres et les $\{\nu_i\}_1^n$ pour les valeurs propres du premier mineur principal.

Du résultat de stabilité énoncé dans le théorème 3, on tire aisément un résultat semblable pour ce problème.

<u>COROLLAIRE 2</u> - Sur l'ouvert $\mathcal{O} = \{ v \in \mathbb{R}^{2n-1} \mid v_1 < v_2 < - < v_{2n-1} \}$ on considère ψ qui à $v \in \mathcal{O}$ fait correspondre la matrice $J(v)$; le spectre de $J(v)$ et du premier mineur principal étant respectivement constitué par les $\{ v_{2i-1} \}_{i=1}^n$ et les $\{ v_{2i} \}_{i=1}^{n-1}$. Alors ψ est continue.

5 - PROBLEME INVERSE DONT LA DONNEE EST DEUX SPECTRES.

Considérons l'équation aux différences

$$(\mathcal{L}y)_i - \lambda y_i = -b_{i-1} + (a_i - \lambda)y_i \quad -b_i y_{i+1} = 0 \qquad i=1,2,\ldots\ldots,n$$

et les deux systèmes de conditions aux limites suivants :

$$y_o + ky_1 = y_{n+1} + k_1 y_n = 0 \qquad y_o + ky_1 = y_{n+1} + k_2 y_n = 0$$

avec $k_1 \neq k_2$. Les opérateurs $\{L;k,K_1\}$ et $\{L;k,K_2\}$ ont pour valeurs propres $\{\lambda_i\}_1^n$ et $\{\mu_i\}_1^n$ respectivement. Ce sont les zéros de $\varpi_{k_i}(\lambda) = y_{n+1}(\lambda) + K_i y_n(\lambda) = 0 \quad i=1,2.$

Le problème est de retrouver les coefficients de l'expression aux différences connaissant les deux spectres. Enonçons d'abord une caractérisation de la mesure spectrale, d'où nous déduirons un résultat d'unicité, qui est un analogue discret d'un théorème de Borg [4].

THEOREME 6 - Supposons que $Sp\{L;k,K_1\} = \{\lambda_i\}_1^n$ et que $Sp\{L;k,K_2\} = \{\mu_i\}_1^n$ avec $K_2 \neq K_2$. Alors la mesure spectrale de $\{L;k,K_1\}$ est proportionnelle à la mesure:

$$\sum_{i=1}^{n} r_i \delta_{\lambda_i} \quad \text{où} \quad r_i^{-1} = \prod_{j=1}^{n} (\lambda_i - \mu_j) \prod_{j \neq i} (\lambda_i - \lambda_j)$$

Démonstration : Considérons les polynômes $\varphi(\lambda) = \prod_i (\lambda - \lambda_i)$ et $\psi(\lambda) = \prod_i (\lambda - \mu_i)$.

On a :

$$\varphi(\lambda) = c\,\varpi_{K_1}(\lambda) = c(y_{n+1}(\lambda) + K_1\, y_n(\lambda))$$

$$\psi(\lambda) = c\,\varpi_{K_2}(\lambda) = c(y_{n+1}(\lambda) + K_2 y_n(\lambda))$$

où on remarque que le coefficient de proportionnalité c est le même pour les deux relations ; en l'occurence égale le coefficient de tête de $g_{n+1}(\lambda)$.

On peut donc calculer le polynôme $y_n(\lambda)$ à une constante près.

$$y_n(\lambda) = \frac{c}{K_1 - K_2} \{\varphi(\lambda) - \psi(\lambda)\}$$

La mesure $d\tau_{\{L;k,K_1\}}$ qui a pour support les $\{\lambda_i\}_1^n$ et pour masses correspondantes les $\{\rho_i\}_1^n$ où $\rho_i^{-1} = b_n y_n(\lambda_i)\varpi'_{K_1}(\lambda_i)$ est donc proportionnelle à la mesure $\sum_1^n r_i \delta_{\lambda_i}$ où les $r_i^{-1} = \psi(\lambda_i)\varphi'(\lambda_i) = \prod_j (\lambda_i - \mu_j) \prod_{j \neq i} (\lambda_i - \lambda_j)$

COROLLAIRE - Si $Sp\{L;k,K_1\}_1 = Sp\{\widetilde{L};k,K_1\}_1 = \{\lambda_i\}_1^n$

et si $Sp\{L;k,K_2\}_1 = Sp\{\widetilde{L};k,K_2\}_1 = \{\mu_i\}_1^n$

avec $K_1 \neq K_2$ alors $L = \widetilde{L}$ ie $a_i = \widetilde{a}_i$ $i=1,2,\ldots,n$ et $b_i = \widetilde{b}_i$ $i=0,1,\ldots,n$.

Notons que l'on connaissait déjà plusieurs démonstrations de ce résultat cf. Hald [ ] page 91. Nous donnons ici une rédaction qui emploie la notion de mesure spectrale ce qui réduit à notre avis la longueur et les difficultés de la démonstration.

Considérons maintenant le problème de l'existence, plus précisément, deux suites $\{\lambda_i\}_1^n$ et $\{\mu_i\}_1^n$ peuvent elles être interprétées comme les spectres d'une même équation aux différences $\mathcal{L}$ pour deux systèmes de conditions aux limites ?

THEOREME 7 - Soient $\{\lambda_i\}_1^n$ et $\{\mu_i\}_1^n$ fixés, ainsi que les constantes k, K_1. Une condition nécessaire et suffisante pour qu'il existe une expression aux différences : $\mathcal{L}$ et une constante $K_2 > K_1$ telles que

$$Sp\{L;k,K_1\}_1 = \{\lambda_i\}_1^n \quad \text{et} \quad Sp\{L;k,K_2\}_1 = \{\mu_i\}_1^n$$

est que :

$$\lambda_1 < \mu_1 < \lambda_2 < \ldots\ldots\ldots < \lambda_n < \mu_n.$$

Démonstration : La condition d'entrelacement est nécessaire. On sait que les racines de $\varpi_{k_i}(\lambda) = 0 \quad i=1,2$ sont réelles et simples. Donc si λ_i et λ_{i+1} désignent deux zéros consécutifs de $\varpi_{K_1}(\lambda) = 0$.

On a :

$$\varpi'_{K_1}(\lambda_i)\, \varpi'_{K_1}(\lambda_{i+1}) < 0$$

Pour tout λ on a :

$$\varpi'_{K_1}(\lambda)\, \varpi_{K_2}(\lambda) - \varpi_{K_1}(\lambda)\, \varpi'_{K_2}(\lambda) =$$

$$= (K_2 - K_1)\,(y'_{n+1}(\lambda)\, y_n(\lambda) - y_{n+1}(\lambda)\, y'_n(\lambda)).$$

Cette expression est positive d'après le lemme 1. On en déduit que $\varpi'_{K_1}(\lambda_i)\, \varpi_{K_2}(\lambda_i) > 0$. Cela implique que ϖ_{K_2} change de signe entre deux racines consécutives de ϖ_{K_1} ; d'où l'entrelacement car on pourrait mener un raisonnement analogue en permutant les roles de ϖ_{K_1} et ϖ_{K_2}. Pour placer la dernière racine on peut étudier $y_{n+1}(\lambda)/y_n(\lambda)$.

On a :

$$\lim_{\lambda\to\infty} y_{n+1}(\lambda)/y_n(\lambda) = -\infty \quad \text{avec } -K_1 > -K_2 \,;$$

cela impose que la plus grande racine est celle pour laquelle $y_{n+1}(\lambda)/y_n(\lambda) = -K_2$ soit donc μ_n.

La condition d'entrelacement est suffisante. Considérons à priori la mesure $d\tau(\lambda)$ qui pour support aura les $\{\lambda_i\}_1^n$ et pour masses correspondantes les $\{\rho_i\}_1^n$ avec :

$$\rho_i^{-1} = \prod_{\ell=1}^{n} (\lambda_i - \mu_\ell) \prod_{\ell \neq i} (\lambda_i - \lambda_\ell) \qquad i=1,2,\ldots\ldots,n$$

Les ρ_i sont positifs grâce à l'entrelacement des λ_i et des μ_i. D'après le corollaire 1 de la proposition 2 on peut connaissant la mesure $d\tau(\lambda)$ et les constantes k, K_1 construire un opérateur $\{L;k,K_1\}_1$ dont la mesure spectrale soit $d\tau(\lambda)$. Soit $y_1(\lambda), y_2(\lambda), \ldots\ldots\ldots, y_n(\lambda)$ la suite des polynômes de première espèce associée à $\{L;k,K_1\}_1$ et soit $y_{n+1}(\lambda)$ le polynôme de degré n obtenu par la récurrence. Par construction on a

$$y_{n+1}(\lambda_i) + K_1 y_n(\lambda_i) = 0 \qquad i=1,2,\ldots\ldots\ldots,n$$

et donc :

$$\prod_i (\lambda - \lambda_i) = c\{y_{n+1}(\lambda) + K_1 y_n(\lambda)\}$$

c étant une constante que l'on peut déterminer : c'est l'inverse du coefficient de tête de $y_{n+1}(\lambda)$. Nous voulons maintenant prouver qu'il existe une constante K_2 telle que :

$$\prod_i (\lambda - \mu_i) = c\{y_{n+1}(\lambda) + K_2\, y_n(\lambda)\}$$

car cette identité exprime que $Sp\{L;k,K_2\} = \{\mu_i\}_1^n$

$\prod_i (\lambda - \mu_i)$ est un polynôme de degré n qui possède une écriture unique sur la base $y_1 \ldots\ldots\ldots y_n\, y_{n+1}$

On a :

$$\prod_i (\lambda - \mu_i) = \alpha_{n+1}\, y_{n+1}(\lambda) + \alpha_n y_n(\lambda) + \alpha_{n-1} y_{n-1}(\lambda) + \ldots + \alpha_1 y_1(\lambda).$$

Par identification du terme de plus haut degré on a $d_{n+1} = c$. Montrons en utilisant l'orthogonalité des $y_i(\lambda)$ par rapport à $d\tau(\lambda)$ que les α_i $i=1,2,..,n-1$ sont nuls. Pour cela montrons que pour $0 \leq k \leq n-2$

$$\int_{-\infty}^{\infty} \prod_i (\lambda-\mu_i)\,\lambda^k d\tau(\lambda) = 0.$$

Il vient :

$$\int_{-\infty}^{\infty} \prod_i (\lambda-\mu_i)\,\lambda^k\, d\tau(\lambda) = \sum_{r=1}^{n} \frac{\prod_{\ell}(\lambda_r-\mu_\ell)\,\lambda_r^k}{\prod_{\ell\neq r}(\lambda_r-\lambda_\ell)\prod_{\ell}(\lambda_r-\mu_\ell)} = \sum_r \frac{\lambda_r^k}{\prod_{\ell\neq r}(\lambda_r-\lambda_\ell)}$$

Considérons maintenant l'identité polynomiale obtenue en décomposant λ^{k+1} sue les n polynômes de Lagrange de degré $n-1$ construit sur les noeuds $\{x_i\}_1^n$ tous distincts ; il vient pour $k \leq n-2$:

$$\lambda^{k+1} = \sum_{r=1}^{n} \frac{x_r^{k+1}\prod_{\ell}(\lambda-x_\ell)}{(\lambda-x_r)\prod_{\ell\neq r}(x_r-x_\ell)}$$

Pour $\lambda = 0$ on obtient entre les $\{x_i\}_1^n$, l'identité scalaire :

$$0 = \sum_{r=1}^{n} \frac{x_r^k}{\prod_{\ell\neq r}(x_r-x_\ell)} \times (-1)^{n-1}\prod_{\ell} x_\ell$$

Cette identitée est vraie dès que les $\{x_i\}_1^n$ sont distincts deux à deux.
Si tous les $\{x_i\}_1^n$ sont distincts de zéro on en tire :

$$(R) \quad \sum_{r=1}^{n} \frac{x_r^k}{\prod_{\ell\neq r}(x_r-x_\ell)} = 0 \qquad k \leq n-2$$

Si l'un des $\{x_i\}_1^n$ disons x_1 vaut zéro, tous les autres sont différents de zéro. Alors en appliquant le premier résultat sur les $(n-1)$ derniers, il vient :

$$\sum_{r=2}^{n} \frac{x_r^k}{\prod_{\substack{\ell=2\\ \ell\neq r}}^{n}(x_r-x_\ell)} = 0$$

d'où encore la relation (R) puisque $x_1=0$. (R) est donc vraie dès que les $\{x_i\}_1^n$ sont deux à deux distinctes.

Appliquons ce résuttat aux $\{\lambda_i\}_1^n$; on en tire $\int \prod_i (\lambda-\mu_i)\,\lambda^k \, d\tau(\lambda) = 0$ pour $k \leq n-2$.

On peut naturellement encore utiliser ici le résultat de stabilité du théorème 3. En effet, il est clair que les masses $\{\rho_i\}_1^n$ avec $\rho_i^{-1} = \prod_j (\lambda_i-\mu_j) \prod_{j\neq i} (\lambda_i-\lambda_j)$ sont des fonctions continues des $\{\lambda_i\}_1^n$ et $\{\mu_i\}_1^n$ d'où le corollaire.

COROLLAIRE . Sur l'ouvert $\mathcal{O} = \{u \in \mathbb{R}^{2n} \mid u_n < u_2 < \ldots\ldots < u_{2n}\}$ on considère la fonction ψ qui à $u \in \mathcal{O}$ associe la matrice de Jacobi $J(u)$ déterminée par $\{u_{2i-1}\}_{i=1}^n$ et $\{u_{2i}\}_{i=1}^n$; ψ est continue.

Nous allons maintenant décrire un algorithme qui permettra le calcul des coefficients de l'opérateur connaissant les deux suites $\{\lambda_i\}_1^n$ et $\{\mu_i\}_1^n$ ainsi que les constantes k, K_1 , cela directement c'est à dire sans recalculer la mesure spectrale.

On considère la matrice $n \times n$ de Jacobi J_i associée à l'opérateur $\{L; k, K_i\}_1$. Comme $b_o = b_n = 1$ elle s'écrit :

$$J_i = \begin{vmatrix} a_1+k & -b_1 & & & 0 \\ -b_1 & a_2 & \ddots & & \\ & \ddots & \ddots & \ddots & \\ & & \ddots & a_{n-1} & -b_{n-1} \\ 0 & & & -b_{n-1} & a_n+K_i \end{vmatrix}$$

Il est facile de vérifier que les polynômes

$$p_o(\lambda) = 1$$

$$p_1(\lambda) = a_1 + k - \lambda$$

$$\begin{cases} p_i(\lambda) = (a_i - \lambda)\; p_{i-1}(\lambda) - b_{i-1}^2 \; p_{i-2}(\lambda) \\ i = 2, 3, \ldots\ldots, n-1 \end{cases}$$

sont respectivement proportionnels aux polynômes de première espèce

$$y_i(\lambda) \; y_2(\lambda) \ldots\ldots y_{i+1}(\lambda) \qquad i=2,3,\ldots\ldots,n-1$$

et que

$p_n(\lambda) = [(a_n + K_1) - \lambda]\; p_{n-1}(\lambda) - b_n^2\; p_{n-2}(\lambda)$ est proportionnel

à $y_{n+1}(\lambda) + K_1 y_n(\lambda)$.

Les polynômes $p_i(\lambda)$ ont l'avantage sur les $y_{i+1}(\lambda)$ d'avoir un coefficient de tête connu, en l'occurence $(-1)^{i-1}$. Ils sont donc parfaitement déterminés dès que l'on possède leur racines.

Ainsi : $p_n(\lambda) = (-1)^{n-1} \prod_i (\lambda - \lambda_i)$ est connu. Comme dans le cours de la démonstration de la proposition 6, on a :

$$\varphi(\lambda) = \prod (\lambda - \lambda_i) = c(y_{n+1}(\lambda) + K_1\, y_n(\lambda))$$

$$\psi(\lambda) = \prod (\lambda - \mu_i) = c\{g_{n+1}(\lambda) + K_2\, y_n(\lambda)\}$$

d'où $y_n(\lambda) = \frac{c}{K_1 - K_2} \{\varphi(\lambda) - \psi(\lambda)\}$. Connaissant φ et ψ on sait former un multiple de $y_n(\lambda)$; on peut donc déterminer complètement $p_{n-1}(\lambda)$: il suffit de ramener à $(-1)^{n-2}$ le coefficient de tête de $\prod_i (\lambda - \lambda_i) - \prod_i (\lambda - \mu_i)$.

On a franchi un stade essentiel dans l'algorithme quand on connait deux polynômes successifs. En effet, l'algorithme de la division euclidienne de $p_n(\lambda)$ par $p_{n-1}(\lambda)$ à une écriture unique sont :

$$p_n(\lambda) = q(\lambda)\, p_{n-1}(\lambda) + r(\lambda)$$

Or :

$$p_n(\lambda) = [(a_n + K_1) - \lambda]\, p_{n-1}(\lambda) - b_{n-1}^2\, p_{n-2}(\lambda)$$

De l'examen du quotient $q(\lambda) = q_1 \lambda + q_2$ on tire :

$$a_n = q_2 - K_1$$

De l'examen du reste $r(\lambda) = r_1 \lambda^{n-2} + r_2 \lambda^{n-3} + \ldots\ldots$ on tire :

$$b_{n-1} = \sqrt{|r_1|}$$

et :

$$p_{n-2}(\lambda) = \frac{1}{b_{n-1}^2}\, r(\lambda)$$

On itère alors ce processus de la division. La division de $p_{n-1}(\lambda)$ par $p_{n-2}(\lambda)$, donc a_{n-1}, b_{n-2} et $p_{n-3}(\lambda)$........ On continue jusqu'à la division de $p_2(\lambda)$ par $p_1(\lambda)$ qui fournit a_2 et b_1.

Enfin, l'examen des coefficients de $p_1(\lambda)$ permet d'obtenir a_1.

Pour obtenir K_2 on calcule la trace des matrices J_1 et J_2 ; il vient

$$\Sigma\lambda_i = \sum_1^n a_i + k + K_1 \qquad\qquad \Sigma\,\mu_i = \Sigma\, a_i + k + K_2$$

d'où $K_2 = K_1 - \Sigma\lambda_i + \Sigma\mu_i$. Résumons les règles de calcul dans l'organigramme suivant :

Algorithme.

1 - Lecture de n,

Lecture de $\{\lambda_i\}_1^n$, $\{\mu_i\}_1^n$

Lecture de k, K_1

2 - Calcul de $\varphi(\lambda) = \prod(\lambda - \lambda_i)$, $\psi(\lambda) = \prod(\lambda - \mu_i)$

3 - Calcul de $p_n(\lambda) = (-1)^{n-1}\,\varphi(\lambda)$

$$p_{n-1}(\lambda) = \varphi(\lambda) - \psi(\lambda) = c_{n-1}\,\lambda^{n-1} + \ldots + c_o$$

$$p_{n-1}(\lambda) = (-1)^{n-2}\,\frac{\Pi_{n-1}(\lambda)}{c_{n-1}}$$

4 - Division de $p_n(\lambda)$ par $p_{n-1}(\lambda)$

$$q(\lambda) = q_1\lambda + q_2 \qquad \text{d'où} \quad q_n = q_2 - K_1$$

$$r(\lambda) = r_1\,\lambda^{n-2} + r_2\lambda^{n-3} + \ldots\ldots + r_{n-1} \qquad \text{d'où} \quad b_{n-1} = \sqrt{|r_1|}$$

$$\text{et} \quad p_{n-2}(\lambda) = \frac{1}{b_{n-1}^2}\; r(\lambda)$$

5 - Pour $i = 1, 2, \ldots\ldots, N-2$ on effectue 5. 1 à 5. 4

5.1 Division de $p_{n-i}(\lambda)$ par p_{n-i-1}

5.2 $q(\lambda) = q_1 \lambda + q_i$ d'où $a_{n-i} = q_2$

5.3 $r(\lambda) = r_1 \lambda^{n-i-2} + \ldots + r_{n-i-2}$ d'où $b_{n-i-1} = \sqrt{|r_1|}$

5.4 et $p_{n-i-1}(\lambda) = \dfrac{-1}{b^2_{n-i-1}} r(\lambda)$

6 - On examine $p_1(\lambda) = p_1\lambda + p_2$

d'où $a_1 = p_2 - k$

7 - On calcule $K_2 = K_1 - \Sigma \lambda_i + \Sigma \mu_i$

Il est facile de voir comment l'on peut modifier cet algorithme pour résoudre les problèmes inverses précédemment cités. Si l'on connait le spectre et les zéros de $y_n(\lambda)$ cela revient pour l'essentiel à rentrer au point 4.

Si on a les deux spectres et les constantes k, K_1, K_2 on peut modifier un peu l'algorithme pour travailler avec la suite des polynômes de première espèce ; cela évite une partie des calculs préparatoires mais n'apporte rien à la qualité numérique.

Présentons maintenant quelques expériences numériques relatives à ce problème et à cet algorithme. Les calculs ont été effectués sur un IRIS 80 de la CII à l'Université de Bordeaux I. Le programme est écrit en FORTRAN IV avec l'option de la double précision. Les jeux d'essais ont été construit de la manière suivante. Pour n fixé on considère la matrice de Jacobi $n \times n$ qui possède des -2 sur la diagonale et des 1 sur les deux codiagonales. On choisit les constantes k, K_1 et on calcule alors le premier spectre par l'algorithme tql 2 de Wilkinson Reinsh [15] page 227. On procède de même pour k, K_2 et le deuxième spectre. Comme tests on prend l'erreur relative sur la norme euclidienne de la diagonale, de la codiagonale, des spectres des deux opérateurs, c'est à dire les nombres .

$$A = \frac{\| \text{diag. initiale - diag. recalculée} \|}{\| \text{diag. initiale} \|} \quad , \quad B = \frac{\| \text{codiag. initiale - codiag. recal.} \|}{\| \text{codiag. initiale} \|}$$

$$C = \frac{\| Sp_1 \text{ initial - } Sp_1 \text{ recalculé} \|}{\| Sp_1 \text{ initial} \|} \quad , \quad D = \frac{\| Sp_2 \text{ initial - } Sp_2 \text{ recalculé} \|}{\| Sp_2 \text{ initial} \|}$$

Dans le tableau 1, on examine le comportement de l'algorithme lorsque la dimension s'accroit.

Tableau 1 - $k = 0$; $K_1 = 0$; $K_2 = 1.0$.

n	A	B	C	D
5	0.11 E-27	0.11 E-27	0.55 E-29	0.65 E-29
10	0.53 E-21	0.58 E-21	0.23 E-23	0.13 E-22
15	0.76 E-15	0.81 E-15	0.41 E-16	0.55 E-16
20	0.75 E-07	0.79 E-07	0.11 E-8	0.30 E-9
25	0.75 E+04	0.15 E+05	0.98 E+04	0.100 E+5
30				

Nous avons mené le même expérience pour d'autres valeurs de k, K_1, K_2. Le comportement est essentiellement le même : les résultats sont très bons jusqu'à la taille 20, puis brutalement inacceptables. Nous avons vérifié que la reconstitution des polynômes à partir de leur racine est excellente. Les erreurs se produisent lors de la différence des deux polykômes et des divisions. En effet, même pour la taille 30 les six premiers coefficients diagonaux et codiagonaux calculés ont tous leurs chiffres exacts. Puis la détérioration intervient très rapidement.

Dans le tableau 2, on examine le comportement **numérique** de la solution, pour un problème de taille 20, mais pour les deux jeux d'essais obtenus pour $k=0$, et K_1 tendant vers $K_2=1.0$. Nous avons fait varier K_1

par pas de O. 1 entre 0 et 0. 9 puis par pas de 0. 01 entre 0,9 et 1. 0. Nous présentons seulement quelques lignes caractéristiques.

Tableau 2 - N=20 , K=0 ; K_2=1.

K1	A	B	C	D
	0. 31 E-07	0. 33 E-07	0. 11 E-08	0. 12 E-08
0.3	0. 32 E-06	0. 34 E-06	0. 40 E-08	0. 47 E-08
0.9	0. 10 E+01	0. 77 E-01	0 . 43 E-00	0. 433 E+00
0.91	0. 43 E-04	0. 45 E-04	0. 49 E-06	0. 56 E-06
0.98	0. 27 E+01	0. 13 E+01	0. 15 E+01	0. 15 E+01
0.99	0. 16 E-03	0. 17 E-03	0. 35 E-05	0. 35 E-05

Pour K_1 variant entre 0. 0 et 0. 8 les résultats sont bons. La variation de K_1 vers K_2 produit une instabilité dans les résultats à partir de 0. 9 Les valeurs propres du premier problème convergent vers les valeurs propres du second ; cependant les polynômes sont reconstitués avec une très bonne précision. Les erreurs se produisent de nouveau au niveau des différences et du processus de division.

Si l'on a les deux spectres et les constantes k, K_1, K_2 on peut modifier un peu l'algorithme pour travailler avec la suite des polynômes de première espéce ; cela n'apporte rien numériquement.

6 - PROBLEME INVERSE DONT LA DONNEE EST UN SPECTRE.

Le problème est de retrouver les coefficients de l'opérateur aux différences connaissant son spectre. Pour avoir existence et unicité il faudra une classe d'opérateurs telle que le support de la mesure spectrale permette de retrouver les masses. D'un point de vue très grossier on a maintenant n données, les $\{\lambda_i\}_1^n$ et à priori le double d'inconnues les $\{a_i\}_1^n$ et les $\{b_i\}_1^n$. On va imposer à ces inconues de vérifier certaines relations supplémentaires sous forme d'hypothèses de symétrie.

Considérons l'expression aux différences $(\ell y)_i = -b_{i-1}y_{i-1}+a_iy_i - b_iy_{i+1}$ dans laquelle $a_i = a_{n-i}$ $i = 1, 2, \dots, n$

et

$$b_i = b_{n-i+1} \qquad i = 0, 1, \dots\dots, n$$

Soit $\{L;k,k\}_1$ l'opérateur attaché à ℓ et aux conditions limites $y_o + ky_1 = y_{n+1} + ky_n = 0$. J la matrice $n \times n$ de Jacobi associée est symétrique par rapport à la seconde diagonale ; nous dirons qu'elle est 2-symétrique. Pour caractériser algébriquement la 2-symétrique introduisons la matrice $n \times n$ orthogonale et involutive $S = (\delta_{i,n-j+i})$.

Alors M est 2-symétrique si et seulement si $M = SMS$.

Un opérateur aux différences sera dit 2-symétrique si sa matrice de Jacobi associée est 2-symétrique.

THEOREME 8 - Si le spectre de l'opérateur aux différences 2-symétrique $\{L,k;k\}$ est constitué par les $\{\lambda_i\}_1^n$ alors la mesure spectrale de celui ci est proportionnelle à la mesure :

$$\sum_{i=1}^{n} \rho_i \,\delta\lambda_{\lambda_i} \qquad \text{avec} \quad \rho_i^{-1} = \Big|\prod_{\substack{\ell=1\\ \ell\neq i}}^{n} (\lambda_i - \lambda_\ell)\Big|$$

Considérons la matrice $n \times n$ J associée à $\{L;k,k\}$. Montrons que la connaissance du spectre de J implique la connaissance à un multiple scalaire près des modules de ses vecteurs propres normalisés en mettant leur première composante à l'unité. Le lemme 2 permettra de conclure.

Remarquons d'abord que si v est un vecteur propre de J alors Sv est aussi un vecteur propre ; en effet, on a $Jv = JSSv = \lambda v$

d'où $(SJS)(Sv) = \lambda(Sv) = J(Sv)$

Ainsi v et Sv sont 2 vecteurs propres associés à la valeur propre λ, qui est simple puisque J est de Jacobi. Cela implique que $Sv = kv$, mais S étant orthogonale on a $k = \pm 1$. Une conséquence est que le module de la première composante v égale le module de la dernière composante de v ; ie $|(v,e_i)| = |(v,e_n)|$.

Pour mettre en évidence les modules des vecteurs propres de J considérons la résolvante $G_\lambda = (J-\lambda I)^{-1} = (g_{rs}(\lambda))$.

Si $x=(J-\lambda I)^{-1}y$ pour $\lambda \notin Sp(J)$ on a l'identité :

$$x = (J-\lambda I)^{-1}y = \sum_i \frac{(y,v_i)}{\|v_i\|^2} \frac{v_i}{\lambda_i - \lambda}$$

où les $\{v_i\}_1^n$ sont les vecteurs propres de J. On les supposera normalisés par $(v_i, e_1) = 1$. De cette identité on tire en particulier :

$$g_{n1}(\lambda) = ((J-\lambda I)^{-1} e_1, e_n) = \sum_1^n \frac{(v_i, e_n)}{\|v_i\|^2(\lambda_i - \lambda)}$$

d'où $\displaystyle g_{n1}(\lambda) = \sum_{i=1}^n \frac{k_i}{\|v_i\|^2(\lambda_i-\lambda)}$ $\qquad k_i = \pm 1$ en utilisant la 2- symétrie de J. Une autre façon classique d'avoir G_λ est de considérer les solutions de problèmes de Cauchy pour l'expression aux différences définies par

$$y_o(\lambda) = -k \quad y_1(\lambda) = 1 \qquad \text{et} \quad w_{n+1}(\lambda) = -k \qquad w_n(\lambda) = 1$$

respectivement. On sait alors que

$$g_{rs}(\lambda) = \begin{cases} y_r(\lambda)\varpi_s(\lambda) \,/\, \varpi_k(\lambda) & r \leq s \\ y_s(\lambda)\varpi_r(\lambda) / \varpi_k(\lambda) & r \geq s \end{cases}$$

Le terme $g_{n1}(\lambda)$ est par construction directement accessible ; on a

$$g_{n_1}(\lambda) = \frac{1}{\varpi_k(\lambda)} = \frac{1}{c\prod(\lambda-\lambda_i)}$$

c étant une constante inconnue. On compare les deux écritures de $g_{n1}(\lambda)$ obtenues en les mettant toutes les deux sous forme de somme de fractions rationnelles , il vient :

$$g_{n_1}(\lambda) = \sum_1^n \frac{1}{\varpi'_k(\lambda_i)(\lambda-\lambda_i)} = \sum_1^n \frac{k_i}{\|v_i\|^2(\lambda_i-\lambda)}$$

Par identification il vient $\|v_i\|^2 = |\varpi'_k(\lambda_i-\lambda)|$ ou encore $\|v_i\|^2 = |c| . |\prod_{\ell\neq i}(\lambda_i-\lambda_e)|$ d'où le résultat.

Notons que le théorème 8 est l'analogue discret d'un théorème de Borg [3] page 81. O. Hald [5] page 70 en a donné des démonstrations dans des cas particuliers en utilisant un analogue discret de la formule intégrale de Cauchy.

<u>THEOREME 9</u> - Soit $\{\lambda_i\}_1^n$. Pour tout k il existe un opérateur aux différences 2-symétrique $\{L;k,k\}_1$ tel que $Sp\{L;k,k\}_1 = \{\lambda_i\}_1^n$. De plus il est unique et on sait le reconstruire.

<u>Démonstration</u> . On considère la mesure $d\tau(\lambda)$ de support les $\{\lambda_i\}_1^n$ et de masses correspondantes les $\{\rho_i\}_1^n$ où $\rho_i^{-1} = |\prod_{\ell\neq i}(\lambda_i-\lambda_e)|$. Cette mesure et la constante k permettent de déterminer l'opérateur $\{L;k,k\}_1$. On a bien $Sp\{L;k,k\}_1 = \{\lambda_i\}_1^n$. Vérifions que $\{L;k,k\}_1$ est 2-symétrique.

Soit J la matrice de Jacobi associée à $\{L;k,k\}_1$. Si $Jv^i = \lambda_i v^i$ alors on sait que $v^i = (y_1(\lambda_i) = 1, y_2(\lambda_i), \ldots, y_n(\lambda_i))^T$; où $y_1(\lambda_i) = 1$; montrons que pour $i=1,2,\ldots,n$ on a $|y_n(\lambda_i)| =$ constante. En effet la mesure spectrale de $\{L;k,k\}$ a pour masses les $\{\rho_i\}_1^n$ avec

$$\rho_i^{-1} = \bar{c}|\prod_{\ell\neq i}(\lambda_e-\lambda_i)| = b_n y_n(\lambda_i)\varpi'_k(\lambda_i) = \check{c}\, y_n(\lambda_i)\prod_{\ell\neq i}(\lambda_e-\lambda_i)$$

où $\bar{c}$, $\check{c}$ sont des constantes. D'où $y_n(\lambda_i) = \pm\frac{\bar{c}}{\check{c}}$ $i=1,2,\ldots,n$.

Soit $\tilde{J} = SJS$ et $\{\tilde{L};k,k\}_1$ l'opérateur associé : c'est par construction le 2-symétrique de $\{L;k,k\}_1$. La mesure spectrale de $\{\tilde{L};k,k\}$ a pour support le spectre de $\tilde{J}$, c'est à dire encore le spectre de J, donc les $\{\lambda_i\}_1^n$. Les masses de la mesure $d\tau_{\{\tilde{L};k,k\}_1}$ sont proportionnelles aux $\{r_i\}_1^n$ où

$r_i^{-1} = \|S v^i\| = \|v^i\|$ $i=1,2,\ldots.n$. En effet les vecteurs propres de $\tilde{J} = SJS$ sont les Sv^i, et d'après la première partie les Sv^i ont tous une première composante de même module.

Enfin $r_i^{-1} = \|v^i\|$ car S est orthogonale. Ainsi $\{L;k,k\}$ et $\{\tilde{L};k,k\}$, sont identiques car ils ont même mesure spectrale.

[1] I. M. GUELFANT - B. M. LEVITAN - The determination of a differential equation from its spectral function.
Amer. Math. Soc Transl. Series 2, 1 (1955) p 253, 3 A.

[2] MARCHENKO - Some problems of the theory of Second order differential operators. Dokl. 72 (1950) p 457-460.

[3] B. M. LEVITAN - Generalized Translation Operators.
Israel Program for Scientific Translations. Jerusalem, 1964.

[4] G. BORG - On the completeness of somme sets of functions Acta. Math 81 (1949), p 2?5-283.

[5] O. HALD - On discrete and numerical inverse Sturm-Liouville problem.
Uppsala University - Report n° 42, 1972.

[6] O. HALD - Inverse eigenvalue problems for Jacobi matrices (à paraître)

[7] H. HOCHSTADT - On some inverse problems in matrix theory. Arch. Math 18 (1967), p 201-207.

[8] H. HOCHSTADT - On the construction of a Jacobi matrix from spectral data. J. of Linear algebra and application.

[9] L. ANDERSON - On the effective determination of the wave operator from given spectral data in the case of a difference equation corresponding to a Sturm-Liouville differential equation.
J. Math Anal. Appl. 29 (1970) p 467-497.

[10] M.H. STONE - Linear transformations in Hilbert space and their applications to analysis.
An. Math. Soc. Colloq. public 15 1932.

[11] G. SZEGO - Orthogonal polynomials. An. Math Soc. Colloq. publ. 23 1939.

[12] N.I. AHIEZER - The classical moment problem and some related questions in analysis English Translat. Hafner, New-York 1965.

[13] J.M. BEREZANSKII - Expansions in eigen functions of self adjoint operators. Ame. Math Soc. Translations 17 1968.

[14] WILKINSON - The algebric eigenvalue problems. London, 1965 . Oxford University press.

[15] WILKINSON-REINSH - Linear algebra, Hand book for automatic computation II, Springer Verlag Berlin 1971.

ON THE REGULARIZATION OF LINEAR INVERSE PROBLEMS IN FOURIER OPTICS

M. Bertero*, C. De Mol+ and G.A. Viano*

* Istituto di Scienze Fisiche dell'Università and Istituto Nazionale di Fisica Nucleare, Genova.

+ Aspirant F.N.R.S., Département de Mathématique, Université Libre de Bruxelles.

Abstract. Least squares regularization methods for ill-posed problems are reviewed and applied to image extrapolation and object restoration in optics. The stabilizing constraints and the kind of continuity they ensure are discussed from a physical point of view.

1. Introduction

In the frame of Fourier optics, an isoplanatic imaging system is completely characterized by its impulse response $\tau(s)$, called the "spread function". A one-dimensional coherent object $x(s)$, identically zero outside the space interval $[-1,1]$, gives through this system the following image $y(s)$:

$$y(s) = \int_{-1}^{1} \tau(s-s')\, x(s')\, ds' \quad . \tag{1.1}$$

In Fourier space this convolution equation becomes a simple multiplication between Fourier transforms: $\hat{y}(\omega) = \hat{\tau}(\omega)\,\hat{x}(\omega)$. For a diffraction limited system the transfer function $\hat{\tau}(\omega)$ and hence $\hat{y}(\omega)$ vanish outside some frequency band $[-c/2\pi, c/2\pi]$ (we say that $\tau(s)$ and $y(s)$ are band-limited). So $y(s)$ is an entire analytic function. On the other side, $x(s)$ is zero outside the space interval $[-1,1]$, so that $\hat{x}(\omega)$ is entire and contains all spatial frequencies up to infinity.

Suppose we measure a finite portion of the image $y(s)$, for instance over $[-1,1]$. As the entire function $y(s)$ is in principle uniquely determined by its values over $[-1,1]$, a first problem investigated in optics is to extrapolate $y(s)$ over the whole real line, i.e. to perform an analytic continuation of the measured data (Frieden 1971). A second problem is to reconstruct the object $x(s)$, i.e. to solve the first kind Fredholm integral equation (1.1) (Frieden 1971, 1975). This is equivalent to extrapolate $\hat{x}(\omega)$ beyond the frequency band, i.e. to go beyond the Rayleigh resolution limit. However this program cannot be realized in a simple intuitive setting. Indeed a small modification of y(s) over $[-1,1]$, for instance due to noise, may be amplified in a catastrophic way by the inversion procedure (Viano 1976). The pathology encountered here, in the specific frame of Fourier optics, is in fact more general and typical of linear inverse problems which consist in solving an equation like $y = Ax$, where y is the data vector and A is some linear continuous operator. Whenever A^{-1} is not continuous, the inverse problem is unstable. Problems of this kind are said to be improperly posed in the sense of Hadamard.

In order to restore continuity one could restrict and equip with an ad hoc norm the space of data vectors y. However this procedure is not allowed since the data are always affected by errors. One could say (Talenti 1978) that the concept of ill-posed problem relies on the following fact: one regards as admissible only those topologies (such as the topology of C^0 or of L^2) that allow for an adequate representation of experimental errors. However it is still possible to restore continuous dependence on the data by means of supplementary constraints on the set of admissible solutions. For instance, stability is guaranteed when searching for a solution in a compact set (John 1955, Pucci 1955, 1958, Tikhonov 1963). This follows from an elementary theorem of functional analysis, i.e. that the inverse of a continuous operator with compact domain is continuous. The a priori knowledge about the solution is the price to be paid for controlling error propagation in the inversion procedure. Nevertheless, this extra-information has to be prescribed according to the physical character of the problem one considers. For instance, one assumes some realistic upper bound for the energy, for the depth of some earth-crust anomaly, for the density gradients in the atmosphere, etc.

Many works have been devoted to the regularization theory of linear ill-posed problems (Lavrentiev 1956, Miller 1970, Payne 1973, 1975, Tikhonov & Arsénine 1976). The basic features of all proposed methods are quite similar and we review them in Sect. 2.1, devoted to the so-

-called Tikhonov-Miller method. Less attention has been paid up to now to the probabilistic regularization method; Sect. 2.2 is devoted to this approach.

Now, let us emphasize the fact that restoring mere continuity on the data is not necessarily enough for practical purposes. The level of continuity should be sufficiently high to allow numerical computations. However this is not always true; for instance in the case of analytic continuation, if one pretends to go up to the boundary of the analyticity domain, then an error ε on the data induces an error on the solution proportional to $|\ln\varepsilon|^{-1}$ (John 1960). Such a poor continuity is called logarithmic continuity. The problem of obtaining precise stability estimates is often overlooked in the literature and we will focus on it in Sect. 3.

A nice particular case for illustrating these results is provided by an ideal diffraction limited imaging system. The transfer function is then given by: $\hat{\tau}(\omega) = 1$ for $-c/2\pi < \omega < c/2\pi$, $\hat{\tau}(\omega) = 0$ elsewhere, and eq. (1.1) becomes:

$$y(s) = \int_{-1}^{1} \frac{\sin[c(s-s')]}{\pi(s-s')} x(s')ds' \quad , \quad -1 \leqslant s \leqslant 1 . \qquad (1.2)$$

The eigenfunctions of this integral operator are the well-known "linear prolate spheroidal functions" (Frieden 1971). Their properties enable us to obtain more precise results for object restoration and image extrapolation in the special case of the ideal diffraction limited optical system. In particular we prove that, unless imposing very restrictive conditions to the object, the restored continuity is but a logarithmic one.

2. Regularization methods

2.1 . Tikhonov-Miller method - A) The general case. This method has been proposed firstly by Tikhonov (Tikhonov 1963) for the resolution of first kind Fredholm integral equations and extended by Miller (Miller 1970). Let us write our basic equation as follows:

$$Ax + z = y \qquad (2.1)$$

where x is the unknown vector, y the data vector and z the noise. A is a linear continuous operator from the solution space X to the data space Y, which are both Hilbert spaces. Besides we assume the existence of the inverse operator A^{-1}. Our problem is to estimate the unknown vector x. When A^{-1} is not continuous, the following bound for the noise:

$$\|z\|_Y = \|Ax - y\|_Y \leq \varepsilon \tag{2.2}$$

is not sufficient to construct an approximation of x. Hence we suppose to have some further a priori knowledge of x under the form of the constraint:

$$\|Bx\|_Z \leq E \tag{2.3}$$

where E is some positive given number and B is the "constraint operator" having dense domain in X, range in the Hilbert space Z, and bounded inverse B^{-1}. We consider any $\bar{x}$ satisfying both constraints (2.2) and (2.3) as an approximation to the unknown x.

As usual in least squares methods, let us combine quadratically the constraints (2.2) and (2.3) into the single one

$$\|Ax - y\|_Y^2 + \left(\frac{\varepsilon}{E}\right)^2 \|Bx\|_Z^2 \leq 2\varepsilon^2 \quad . \tag{2.4}$$

The vector $\tilde{x}$ which minimizes the l.h.s. of eq. (2.4) is given by:

$$\tilde{x} = \left[A^*A + \left(\frac{\varepsilon}{E}\right)^2 B^*B\right]^{-1} A^*y \tag{2.5}$$

If we choose some norm or seminorm $\langle \cdot \rangle$ in X for measuring the accuracy of the approximation,one can prove that (Miller 1970):

$$\langle \tilde{x} - \bar{x} \rangle \leq \sqrt{2}\, M(\varepsilon , E) \tag{2.6}$$

where $\bar{x}$ is any vector satisfying (2.2),(2.3) and $M(\varepsilon , E)$ is the following quantity:

$$M(\varepsilon , E) = \sup\left\{\langle x \rangle \mid x \in X, \|Ax\|_Y \leq \varepsilon , \|Bx\|_Z \leq E \right\} . \tag{2.7}$$

The problem is stable with respect to $\langle \cdot \rangle$ when $M(\varepsilon , E)$ tends to zero with ε , for fixed E . Then $M(\varepsilon , E)$ is called the best-possible stability estimate and $\tilde{x}$ is a nearly-best-possible approximation, since the error $< \tilde{x} - \bar{x} >$ is bounded by $M(\varepsilon , E)$ up to an irrelevant numerical factor (Miller 1970). Let us remark that we have assumed here both ε and E to be a priori known. This however is not necessary and one can assume, for instance, that only ε is known (Morozov 1967 , Miller 1970, Miller & Viano 1973).

B) Eigenfunction expansions. Suppose now that the operator A is compact. Let us denote by $\{\alpha_k\}_{k=0}^{+\infty}$ the set of singular values of the operator A, by $\{u_k\}_{k=0}^{+\infty}$ the set of eigenvectors of A^*A, and write

$$A^*Ax = \sum_{k=0}^{+\infty} \alpha_k^2 x_k u_k \quad , \quad x_k - (x , u_k)_X \quad . \tag{2.8}$$

Let us further assume that B^*B and A^*A commute and hence may be simultaneously diagonalized:

$$B^*B\,x = \sum_{k=0}^{+\infty} \beta_k^2 x_k u_k \quad . \tag{2.9}$$

In this case, solution (2.5) reduces to

$$\tilde{x} = \sum_{k=0}^{+\infty} \frac{\alpha_k}{\alpha_k^2 + (\frac{\varepsilon}{E})^2 \beta_k^2} y_k u_k \quad , \quad y_k = (y , v_k)_Y \tag{2.10}$$

where $\{ v_k = \alpha_k^{-1} A u_k \}_{k=0}^{+\infty}$ are eigenvectors of AA^* .

Another approximation $\approx\!\!x$ can be defined as follows. Let $I_{\varepsilon/E}$ be the set of those indices k such that $\alpha_k \geqslant \frac{\varepsilon}{E}\beta_k$. It can be shown that (Miller 1970):

$$\approx\!\!x = \sum_{k \in I_{\varepsilon/E}} \frac{y_k}{\alpha_k} u_k \tag{2.11}$$

is also a nearly-best-possible approximation. With a precise cut-off we recover here the well-known empirical truncation method for eliminating the noise amplification due to eigenvalues very near to zero.

C) Application to Fourier optics. For the object restoration problem, a natural choice for the spaces X and Y is $L^2(-1,1)$; the norm of a vector represents then its total energy. In the special case of the ideal diffraction limited system the operator A, given by eq. (1.2), is compact, selfadjoint and non-negative in $L^2(-1,1)$. Its eigenvectors are given by:

$$u_k(s) = \lambda_k^{-1/2} \psi_k(c,s) \ , \quad |s| \leqslant 1 \quad ; \quad k = 0,1,2, \ldots \qquad (2.12)$$

where $\psi_k(c,s)$ are the linear prolate spheroidal functions. The corresponding eigenvalues $\{\lambda_k\}_{k=0}^{+\infty}$ have a step behaviour: monotonically decreasing from 1 to 0 , they present a rapid fall for k around $2c/\pi$.

Formulae of the preceding section may be applied here. The a priori constraint (2.3) with $Z = L^2(-1,1)$, and B*B defined by eqs. (2.9) and (2.12) , has a simple physical interpretation. Indeed ψ_o is the L^2-function with support $[-1,1]$ which is the most concentrated in the frequency band $[-c/2\pi, c/2\pi]$. In the subspace orthogonal to ψ_o, ψ_1 has the same property, and so on for the other ψ_k's (Frieden 1971). The more rapidly the weights β_k grow to infinity, the more the object is concentrated on the band and the better the restoration. It is clear that we may not hope a good restoration for objects whose Fourier spectral components are mainly out of the band and thus completely lost in the imaging process.

For the image extrapolation problem a similar treatment can be done; thanks to the fact that the prolate spheroidal functions are also orthogonal in $L^2(-\infty,+\infty)$, they are suitable for the extrapolation of band-limited functions (Frieden 1971). As stabilizing constraint one can prescribe the total energy of the image to be bounded (Viano 1976). A stronger condition is to require that the image corresponds to an object whose energy is bounded (Bertero, De Mol & Viano 1978).

2.2 . <u>Probabilistic method</u> - A) <u>The general case.</u> The functions x, y , z in our basic equation (2.1) will be viewed here as samples of some Hilbert space valued random variables ξ , η, ζ related by

$$A\xi + \zeta = \eta \qquad (2.13)$$

To include processes like white noise, we only assume that they are weak random variables, i.e. they induce on the Hilbert space cylinder measures (not necessarily σ-additive probability measures). Here the a priori knowledge will be that of the joint measure of the object ξ and the noise ζ (Franklin 1970, Bertero & Viano 1978). In order to simplify the problem let us make some further assumptions which are reasonable in many practical situations. Suppose that ξ and ζ are gaussian, independent and of zero mean. Hence their joint measure is completely determined by the covariance operators R_ξ and R_ζ. The covariance operator of η is given by:

$$R_\eta = AR_\xi A^* + R_\zeta \tag{2.14}$$

and the cross-covariance operator $R_{\xi\eta}$ by

$$R_{\xi\eta} = R_\xi A^* \quad . \tag{2.15}$$

We also assume that R_ζ^{-1} exists, i.e. physically that all components of the data vector are affected by noise. Remark that the a priori knowledge of R_ξ corresponds to that of the bound E in the Tikhonov-Miller method. Similarly we see the correspondence between the knowledge of R_ζ and that of the noise bound (2.2) by putting $R_\zeta = \varepsilon^2 N$ where N, independent of ε, is a linear bounded operator.

Now, given a sample value y of the image η, how can we estimate a sample value x of the object ξ? Thanks to the gaussian assumption, this problem reduces to a linear estimation problem. If $L : Y \to X$ is any linear continuous operator, we call the weak random variable $\tilde{\xi}_L = L\eta$ a linear estimate of ξ. Then, for any $u \in X$, the reliability of the estimate for the random variable $(\xi, u)_X$ is measured by the mean square error:

$$\delta^2(u, \varepsilon; L) = E\left\{ \left| (\xi - L\eta, u)_X \right|^2 \right\} \tag{2.16}$$

(E denotes the mathematical expectation). If and only if the operator $R_{\xi\eta} R_\eta^{-1}$ is bounded over the range of R_η, then there exists a unique linear continuous operator L_o minimizing $\delta^2(u, \varepsilon ; L)$ for any $u \in X$. It is given by:

$$L_o = R_{\xi\eta} R_\eta^{-1} = R_\xi A^* \left[A R_\xi A^* + \varepsilon^2 N \right]^{-1} \tag{2.17}$$

and $\tilde{\xi} = L_o \eta$ is called the best linear estimate of ξ. Under further assumptions (Bertero & Viano 1978), L_o minimizes also the global mean square error

$$\delta^2(\varepsilon ; L) = E\left\{ \| \xi - L\eta \|_X^2 \right\} . \tag{2.18}$$

By putting $R_\xi = E^2(B^* B)^{-1}$ and $N = I$ (white noise), with the help of the identity

$$(A^* R_\xi^{-1} A + R_\xi^{-1}) R_\xi A^* = A^* R_\xi^{-1} (A R_\xi A^* + R_\xi) \tag{2.19}$$

the estimated solution $\tilde{x} = L_o y$ is seen to be formally equivalent to the Tikhonov-Miller one (see (2.5)).

B) <u>Eigenfunction expansion</u>. Using the notations of Sect. 2.1.B, assuming that A is compact and that A^*A commutes with R_ξ, whereas AA^* commutes with $R_\zeta = \varepsilon^2 N$, we write

$$R_\xi x = \sum_{k=0}^{+\infty} \rho_k x_k u_k \quad , \quad R_\zeta y = \varepsilon^2 \sum_{k=0}^{+\infty} \nu_k y_k v_k . \tag{2.20}$$

The operator L_o is bounded iff $\sup (\alpha_k \rho_k \nu_k^{-1}) < +\infty$. Equation (2.17) then gives

$$\tilde{x} = \sum_{k=0}^{+\infty} \frac{\alpha_k \rho_k}{\alpha_k^2 \rho_k + \varepsilon^2 \nu_k} y_k u_k \quad , \quad y_k = (y, v_k)_Y \tag{2.21}$$

which we may also put in the following form

$$\tilde{x} = \sum_{k=0}^{+\infty} (1 - e^{-2J_k}) \frac{y_k}{\alpha_k} u_k \tag{2.22}$$

where

TABLE 1

For reader's convenience we summarize both methods by the following scheme:

Tikhonov - Miller method	
data	the vector y ; $y = Ax + z$
a priori knowledge	$\|z\| = \|Ax - y\| \leq \varepsilon$, $\|Bx\| \leq E$ knowledge of ε , E and of the operator B
requirement	an estimate of the vector x
least squares solution	$\tilde{x} = \left[A^*A + (\frac{\varepsilon}{E})^2 B^*B\right]^{-1} A^*y$

Probabilistic method	
data	a value y of the r.v. $\eta = A\xi + \zeta$
a priori knowledge	the r.v. ξ, ζ are gaussian and independent; knowledge of the covariances R_ξ , R_ζ ; mean values $m_\xi = 0$, $m_\zeta = 0$
requirement	an estimate of a value of ξ
least squares solution	$\tilde{x} = R_\xi A^* \left[AR_\xi A^* + R_\zeta\right]^{-1} y$

$$J_k = \frac{1}{2} \ln \left[1 + \frac{\alpha_k^2 \rho_k}{\varepsilon^2 \nu_k} \right] \tag{2.23}$$

is the average mutual information of the random variables $\xi_k = (\xi, u_k)_X$ and $\eta_k = (\eta, v_k)_Y$. Solution $\tilde{x}$ is thus obtained from the unstable formal solution $A^{-1}y$ by penalizing those components of y that contain too little information about x. By defining $I_\varepsilon = \left\{ k \mid J_k \geqslant \frac{1}{2} \ln 2 \right\}$ we also get a truncated series similar to (2.11):

$$\tilde{\tilde{x}} = \sum_{k \in I_\varepsilon} \frac{y_k}{\alpha_k} u_k \quad . \tag{2.24}$$

C) Application to Fourier optics. With the help of Section 2.1.C , it is straightforward to apply these general results to the object restoration and image extrapolation problems (Bertero, De Mol & Viano 1978).

3. Stability estimates

3.1 . Tikhonov - Miller method. Under the mere assumption that the constraint operator B admits a bounded inverse, only weak stability is guaranteed. This means that the problem is stable with respect to the seminorm $\langle x \rangle = |(x, u)_X|$, where $u \in X$ is some fixed element. The corresponding best stability estimate (2.7) is bounded by (Miller 1970)

$$M_u(\varepsilon, E) \leqslant \sqrt{2}\, \varepsilon \left(\left[A^*A + \left(\frac{\varepsilon}{E}\right)^2 B^*B\right]^{-1} u, u \right)_X \tag{3.1}$$

and one can prove that $M_u(\varepsilon, E)$ tends to zero with ε . Moreover, when u belongs to the range of A^* (Bertero, De Mol & Viano 1978):

$$M_u(\varepsilon, E) = O(\varepsilon) \tag{3.2}$$

which is a very satisfactory kind of continuity since the error on the solution is proportional to the error on the data. However, when being satisfied with weak stability, one has to renounce to a punctual restoration of the function. One only pretends to know some weighted averages of it; for instance, in order to estimate x in a neighbourhood of a point $s_o \in [-1,1]$, we can take as smearing function u, a sharp gaussian curve peaked on s_o, and estimate the value of the scalar product:

$$(x, u(\cdot, s_o))_X = \int_{-1}^{1} \frac{1}{\sqrt{2\pi}\delta} e^{-\frac{(s_o-s)^2}{2\delta^2}} x(s)ds \qquad (3.3)$$

where δ is some small quantity.

To ensure <u>strong stability</u>, i.e. stability with respect to the norm of X, we require the operator B to have compact inverse. Thanks to this requirement, constraint (2.3) guarantees the compactness of the set of admissible solutions and hence, according to the theorem recalled in the Introduction, the problem is stable with respect to the strong topology of X. Under this general assumption, we only know that $M(\varepsilon, E)$ tends to zero with ε and not how fast it decreases. However, in view of practical computations, one is interested in knowing some precise estimation of $M(\varepsilon, E)$. To our knowledge, this latter problem has been solved only when A^*A and B^*B commute. In this case, stability estimates have been obtained with various techniques (Arsenin & Ivanov 1968 a), b), John 1960, Lavrentiev 1956, 1967, Miller 1964, Pucci 1958). However these methods present many common features and, in our opinion, this should allow for a unified presentation. Here, we try to illustrate this in the eigenfunction expansion case (see Sect. 2.1.B). Let us firstly recall that the instability of the inverse problem is due to the fact that the singular values of A, $\alpha_0 \geq \alpha_1 \geq \cdots \geq \alpha_k \geq \cdots$ tend to zero for $k \rightarrow +\infty$. The weights β_k are introduced in order to neutralize this effect (see eq. (2.10)). In particular, when B^{-1} is compact, $\beta_k \rightarrow +\infty$ for $k \rightarrow +\infty$ and we expect that the quantitative level of the restored continuity should depend essentially on the relationship between the decreasing α_k's and the increasing β_k's. To precise this intuitive statement, let us further assume the following inequality:

$$\alpha_k^2 \geq \beta_k^2 \, P(\frac{1}{\beta_k^2}) \qquad (3.4)$$

where the function $p(r)$, $r \geqslant 0$ satisfies the conditions:

i) $p(r)$ is convex and monotonically increasing;

ii) $p(0+) = 0$;

iii) $r\, p^{-1}(\frac{1}{r})$ is increasing.

Then (Talenti 1978) we obtain the following majoration for $M(\varepsilon, E)$:

$$M(\varepsilon, E) \leqslant E \sqrt{p^{-1}(\frac{\varepsilon^2}{E^2})} \quad . \tag{3.5}$$

This inequality follows from the Jensen inequality for convex functions p :

$$p(\sum_{k=0}^{+\infty} a_k b_k) \leqslant \sum_{k=0}^{+\infty} a_k\, p(b_k) \ ; \qquad \sum_{k=0}^{+\infty} a_k = 1 \quad . \tag{3.6}$$

Indeed, using (3.4) we get:

$$p(\frac{\|x\|^2}{\|Bx\|^2}) \leqslant \sum_{k=0}^{+\infty} \frac{\beta_k\, |x_k|^2}{\|Bx\|^2}\, p(\frac{1}{\beta_k^2}) \leqslant \frac{\|Ax\|^2}{\|Bx\|^2} \quad . \tag{3.7}$$

Hence:

$$\|x\| \leqslant \|Bx\| \sqrt{p^{-1}(\frac{\|Ax\|^2}{\|Bx\|^2})} \tag{3.8}$$

which gives (3.5) using (2.2) (with $y = 0$), (2.3) and assumption (iii) on p. When we specify the following choice for $p(r)$:

$$p(r) = r^{1/\alpha} \quad , \qquad 0 < \alpha < 1 \tag{3.9}$$

(3.5) becomes:

$$M(\varepsilon, E) \leqslant E(\frac{\varepsilon}{E})^{\alpha} \tag{3.10}$$

and one speaks of Hölder continuity, which is quite good in practice.

In the case of the ideal diffraction limited system (Sect. 2.1.C) the eigenvalues behave asymptotically, for $k \longrightarrow \infty$, like :

$$\alpha_k = \lambda_k = O\left[\frac{1}{k} \left(\frac{ce}{k} \right)^{2k} \right] . \qquad (3.11)$$

In order to get Hölder continuity, we have to take (according to eqs. (3.4), (3.9)) weights β_k's such that $\beta_k \geqslant \lambda_k^{-\mu}$ $(\mu = \alpha/(1-\alpha) > 0)$. Hence the β_k's need to increase faster than $\exp(2\mu k \ln k)$, and the components x_k of the object to decrease at inverse rate. This leads only to very smooth objects, which are restrictions to $[-1,1]$ of band-limited functions. So Hölder continuity is ensured by an assumption which appears too restrictive from the physical point of view. Let us show however that to deal with a reasonable class of objects, we have to take weights β_k growing only like some power of k.

More precisely, let us characterize the class of objects which can be restored when taking $\beta_k = O(k^{\mu})$, $\mu > 0$. For this purpose, let us observe that the linear prolate spheroidal functions $u_k(s)$ (of unit norm in $L^2(-1,1)$ - see eq. (2.12)) are solutions of the differential equation:

$$(Du_k)(s) = - \left[(1 - s^2) u_k'(s) \right]' + c^2 s^2 u_k(s) = \chi_k u_k(s) \qquad (3.12)$$

where $\chi_k = k(k+1) + \frac{1}{2}c^2 + O(k^{-2}) = O(k^2)$ for $k \longrightarrow \infty$. Now let us take as regularization operator B, the operator $D^{1/2}$. For any x in $L^2(-1,1)$ such that $\| D^{1/2} x \| < \infty$, we have:

$$\| D^{1/2} x \|^2 = \int_{-1}^{1} (1 - s^2) |x'(s)|^2 ds + c^2 \int_{-1}^{1} s^2 |x(s)|^2 ds \ ; \qquad (3.13)$$

besides, from eq. (3.12) we get:

$$\| D^{1/2} x \|^2 = \sum_{k=0}^{+\infty} \chi_k \left| (x, u_k) \right|^2 \qquad (3.14)$$

so we may conclude that $\| D^{1/2} x \| < +\infty$ iff :

$$\sum_{k=0}^{+\infty} k^2 \left| (x,u_k) \right|^2 < +\infty \quad . \tag{3.15}$$

Taking β_k proportional to k, we recover the class of objects for wich (3.13) is bounded. Similarly, taking β_k proportional to k^{μ}, we recover the class of objects for which $\| D^{\mu/2} x \|$ is bounded.

We obtain a stability estimate for the case $\beta_k = k^{\mu}$, $\mu > 0$, remarking that the eigenvalues $\alpha_k = \lambda_k$ satisfy the inequality: $\lambda_k^2 \geqslant \gamma^2 \exp(-2\beta_k^{2\nu})$, where $\nu = (1+\delta)/2\mu$, γ and δ being some positive constants. The function $p(r) = \gamma^2 r \exp(-2r^{-\nu})$ satisfies the conditions i) - iii) if $\nu < 1$ (i.e. $\mu > 1/2$; otherwise p is convex only in a neighbourhood of r = 0) ; besides, for $s \longrightarrow 0$, $p^{-1}(s)$ has the following behaviour:

$$p^{-1}(s) \sim 4^{\mu(1-\eta)} \left| \ln \left(\frac{s}{\gamma^2} \right) \right|^{-2\mu(1-\eta)} \tag{3.16}$$

where $1-\eta = (1+\delta)^{-1}$. Formula (3.5) becomes then:

$$M(\varepsilon, E) \leqslant 2^{\mu(1-\eta)} E \left| 2 \ln \left(\frac{\varepsilon}{\gamma E} \right) \right|^{-\mu(1-\eta)} \tag{3.17}$$

Hence, for reasonable classes of objects, we only get a logarithmic continuity which is in practice very poor. Note that the bound (3.17) may not be improved, since we also find for $M(\varepsilon, E)$ a logarithmic lower bound of the same kind (Bertero, De Mol & Viano 1978).

In view of numerical applications it is perhaps more interesting to guarantee the uniform stability, i.e. stability with respect to the uniform norm:

$$\| x \|_{\infty} = \sup_s \left| x(s) \right| \tag{3.18}$$

which is more stringent than L^2 - stability.

For general first kind Fredholm integral equations, compactness with respect to the uniform norm is ensured by Tikhonov's regularizing condition (Tikhonov 1963) :

$$\|Bx\|^2 = \int_{-1}^{1} \left[k(s)\left|x'(s)\right|^2 + m(s)\left|x(s)\right|^2 \right] ds \leqslant E^2 \qquad (3.19)$$

where k(s) and m(s) are strictly positive functions. However, in order to find explicit stability estimates, we have to consider the particular case where A^*A and B^*B commute and to expand x in terms of the eigenfunctions of A:

$$x(s) = \sum_{k=0}^{+\infty} x_k\, u_k(s) \quad . \qquad (3.20)$$

Let us further suppose that the eigenfunctions u_k are bounded, so that: $\|u_k\|_\infty \leqslant N_k$; then:

$$\|x\|_\infty \leqslant \sum_{k=0}^{+\infty} N_k \left|x_k\right| \leqslant \left(\sum_{k=0}^{+\infty} \frac{N_k^2}{\beta_k^2} \right)^{1/2} \|Bx\| \qquad (3.21)$$

thanks to the Schwarz inequality. Hence, in the domain of B, we may introduce the following norm:

$$|||\, x\, ||| = \sum_{k=0}^{+\infty} N_k \left| x_k \right| \qquad (3.22)$$

provided that:

$$\sum_{k=0}^{+\infty} \frac{N_k^2}{\beta_k^2} < \infty \quad . \qquad (3.23)$$

Then it is not difficult to prove the compactness, with respect to the norm (3.22), of the set of vectors satisfying the stabilizing constraint (2.3),(2.9),(3.23). It follows that the problem is stable with respect to this latter norm, and hence also to the uniform norm.

For the linear prolate spheroidal functions (2.12), we have $N_k = (\text{constant}) \cdot \sqrt{k + 1/2}$ and hence, if we take β_k proportional to k^α , condition (3.23) imposes $\alpha > 1$. The condition $\alpha > 1$ is not only sufficient but also necessary to get uniform stability. Indeed, the set of objects characterized by $\| D^{\alpha/2} x \| \leqslant E$, $\alpha \leqslant 1$, is not compact with respect to the uniform norm, since it contains unbounded functions like $x(s) = \left| \ln \left(\frac{1-s^2}{2} \right) \right|^{\nu}$ with $\nu < 1/2$ (see eq. (3.13)).

Finally, the best stability estimate $M(\varepsilon, E)$ (2.7) with $\langle \cdot \rangle = \||| \cdot |||$ is calculated as follows:

$$\sum_{k=0}^{+\infty} N_k |x_k| \leqslant \left(\| Ax \|^2 + \left(\frac{\varepsilon}{E}\right)^2 \| Bx \|^2 \right)^{1/2} \left(\sum_{k=0}^{+\infty} \frac{N_k^2}{\alpha_k^2 + \left(\frac{\varepsilon}{E}\right)^2 \beta_k^2} \right)^{1/2} \quad (3.24)$$

so that:

$$M(\varepsilon, E) \leqslant \sqrt{2}\, \varepsilon \left(\sum_{k=0}^{+\infty} \frac{N_k^2}{\alpha_k^2 + \left(\frac{\varepsilon}{E}\right)^2 \beta_k^2} \right)^{1/2} \quad (3.25)$$

which is also a stability estimate for the uniform norm.

Let us finally spend a few words about image extrapolation. If we assume that the image corresponds to an object whose energy is bounded, then the stabilizing constraint is expressed as follows (Bertero, De Mol & Viano 1978): $\beta_k = \alpha_k^{-1}$. Hence, from (3.10) with $\alpha = 1/2$, we get in the case of the norm of $L^2(-\infty, +\infty)$

$$M(\varepsilon, E) \leqslant \sqrt{\varepsilon E} \quad . \quad (3.26)$$

For the uniform norm we have similarly (Bertero, De Mol & Viano 1978):

$$M_{\infty}(\varepsilon, E) \leqslant (\text{constant}) \cdot E \left(\frac{\varepsilon}{E}\right)^{\alpha} \quad (3.27)$$

where $0 < \alpha < 1/2$. Let us remark however that these results apply to the rather peculiar case of an image measured over the whole interval $[-1, 1]$. If the image is known over a smaller interval, the restored continuity is probably weaker.

3.2. Probabilistic method. The role of the weak stability estimate $M_u(\varepsilon, E)$ (see eq. (3.1)) is played here by the least mean square error $\delta(u, \varepsilon)$ given by (see Sect. 2.2):

$$\delta(u,\varepsilon) = \delta(u,\varepsilon;L_o) = \left(\left[R_\xi - L_o R_\eta L_o^*\right] u, u\right)_X^{1/2} \qquad (3.28)$$

Let us summarize briefly the results obtained in the eigenfunction expansion case (for stability theorems valid in the general case of non-commuting operators, the interested reader is referred to (Bertero, De Mol & Viano 1978)).

Assuming the existence of A^{-1}, from eqs. (2.20) and (3.28) we have:

$$\delta(u,\varepsilon) = \varepsilon\left(\sum_{k=0}^{+\infty} \frac{\rho_k \nu_k}{\alpha_k^2 \rho_k + \varepsilon^2 \nu_k} \left|(u, u_k)_X\right|^2\right)^{1/2} \qquad (3.29)$$

and it is easy to see that $\delta(u, \varepsilon) \rightarrow 0$, when $\varepsilon \rightarrow 0$, for any $u \in X$; besides $\delta(u, \varepsilon) = O(\varepsilon)$ iff $u \in$ range $(A^* N^{1/2})$.

The equivalent of the strong stability estimate $M(\varepsilon, E)$ is the global mean square error (see eq. (2.18)) $\delta(\varepsilon) = \delta(\varepsilon, L_o)$ which is given by:

$$\delta(\varepsilon) = \varepsilon\left(\sum_{k=0}^{+\infty} \frac{\rho_k \nu_k}{\alpha_k^2 \rho_k + \varepsilon^2 \nu_k}\right)^{1/2}. \qquad (3.30)$$

$\delta(\varepsilon)$ is finite, and tends to zero with ε, when ξ has finite variance (i.e. $\sum_{k=0}^{+\infty} \rho_k < +\infty$). For object restoration and image extrapolation problems we are led to the same conclusions as in the preceding section. For instance, in the case of object restoration, when the signal-to-noise ratio behaves like $\rho_k/\nu_k = (\text{constant})\cdot k^{-2\mu}$, $(\mu > 0)$, we get the following logarithmic bound : $\delta(\varepsilon) = O(|\ln\varepsilon|^{-\mu(1-\eta)})$ for $\varepsilon \rightarrow 0$, η being an arbitrary small positive quantity (Bertero, De Mol & Viano 1978).

4. Final remarks

To take into account more realistic cases than those represented by eq. (1.2), one needs some further extensions in the following directions:

a) One has to find stability estimates in the two-dimensional case and for non-ideal systems. This can be realized within the general formalism here described but, in order to get explicit estimates, one needs more information about the behaviour of the eigenvalues.

b) One has to consider also the case of incoherent illumination, where a positivity constraint on the solution is required.

REFERENCES

ARSENIN V.Ya. and IVANOV V.V.: The solution of certain convolution type integral equations of the first kind by the regularization method. Ž. Vyčisl. Mat. i Mat. Fiz. 8, 2, 310 - 321 (1968a).

ARSENIN V.Ya. and IVANOV V.V.: The effect of regularization of order p. Ž. Vyčisl. Mat. i Mat. Fiz. 8, 3, 661 - 663 (1968b).

BERTERO M. and VIANO G.A.: On probabilistic methods for the solution of improperly posed problems. To appear in Boll. Un. Mat. Ital. (1978).

BERTERO M.,DE MOL C. and VIANO G.A.: On the problems of object restoration and image extrapolation in optics, preprint Genova(1978).

FRANKLIN J.N.: Well-posed stochastic extensions of ill-posed linear problems. J. Math. Analysis and Applications 31 , 682 - 716 (1970).

FRIEDEN B.R.: Evaluation, design and extrapolation methods for optical signals, based on use of the prolate functions. Progress in Optics, Vol. 9, 311 - 407 (1971).

FRIEDEN B.R.: Image enhancement and restoration. Topics in Applied Physics, Vol. 6, 177 - 248 (1975).

JOHN F. : A note on improper problems in partial differential equations. Comm. Pure Appl. Math. 8, 591 - 594 (1955).

JOHN F. : Continuous dependence on data for solutions of partial differential equations with a prescribed bound, Comm. Pure Appl. Math., 13, 551 - 585 (1960).

LAVRENTIEV M.M. : On the Cauchy problem for the Laplace equation. Isvestiya Akademii Nauk SSSR. Seriya Matematiceskaya 20, 819 - - 842 (1956).

LAVRENTIEV M.M. : Some improperly posed problems of mathematical physics (Springer, 1967).

MILLER K. : Three circle theorems in partial differential equations and applications to improperly posed problems, Arch. Rational Mech. Anal. 16, 126 - 154 (1964).

MILLER K. : Least squares methods for ill-posed problems with a prescribed bound, SIAM J. Math. Anal. 1, 52 - 74 (1970).

MILLER K. and VIANO G.A. : On the necessity of nearly-best-possible methods for analytic continuation of scattering data, J. Math. Phys. 14, 1037 - 1048 (1973).

MOROZOV V.A. : Choice of parameter for the solution of functional equations by the regularization method, Soviet Math. Dokl. , 8, 4, 1000 - 1003 (1967).

PAYNE L.E. : Some general remarks on improperly posed problems for partial differential equations, in Lecture Notes in Mathematics, 316, 1 - 30 (Springer, 1973).

PAYNE L.E. : Improperly posed problems in partial differential equations, Regional Conference Series in Applied Mathematics, N. 22 (SIAM, 1975).

PUCCI C. : Sui problemi di Cauchy non ben posti. Atti Accad. Naz. Lincei, Rend. Cl. Sci. Fis. Mat. Nat. (8) 18, 473 - 477 (1955).

PUCCI C. : Discussione del problema di Cauchy per le equazioni di tipo ellittico. Annali di Matematica Pura ed Applicata, Series 4, 66, 131 - 154 (1958).

TALENTI G. : Sui problemi mal posti. To appear in Boll. Un. Mat. Ital. (1978).

TIKHONOV A.N. : Solution of incorrectly formulated problems and the regularization method, Soviet Math. Dokl. , 4, 1035 - 1038 (1963).

TIKHONOV A.N. et ARSENINE V. : Méthodes de résolution de problèmes mal posés, (MIR, 1976).

VIANO G.A. : On the extrapolation of optical image data.J. Math. Phys., 17, 1160 - 1165 (1976).

DETERMINATION OF THE INDEX PROFILE OF A DIELECTRIC PLATE FROM SCATTERING DATA

A. ROGER

Laboratoire d'Optique Electromagnétique
C.N.R.S. E.R.A. n°597
faculté Saint Jérôme
13397 Marseille Cedex 4

ABSTRACT

This paper presents a numerical method which has been tried on an optical device, in order to get information on a dielectric inhomogeneous medium from scattering data. We start from an integral form of the scattering equation, and use a perturbation technique. We show that under certains conditions, it is possible to determine the permittivity profile of all physical mediums.

I - INTRODUCTION

In the last fifteen years, the use of computers has allowed one to solve many scattering problems of electromagnetism completely. Starting with the properties of the scattering object (shape, material) and of the incident (s) wave (s), one is able to compute the diffracted waves numerically, but rigorously. Of course it is impossible to obtain the same results by using analytical formulas only, excepted in a few cases where the diffracting object has special properties of symmetry (sphere, circular cylinder).

Nowadays, many programs have been checked out, for many devices, such as inhomogeneous layers, gratings, or cylinders. [1], [2], [3]. Making inverse scattering is following the inverse path : starting with the properties of the diffracted waves and finding the properties of the object itself, if it actually exists. Our aim has not been to give new theorems or algorithms, but to find a numerical method, suitable for many physical devices. As a first trial, we have studied the inhomogeneous dielectric layer.

II - THE PHYSICAL DEVICE

We consider a dielectric layer situated between the planes $z = 0$ and $z = 1$, whose refractive index is a continuous function of z in this range. The index profile is described by the relative permittivity $\varepsilon(z)$ which is independent of x and y. This medium is backed by a perfectly conducting plane located at $z = 0$. Let the time dependence be in $\exp(-i\omega t)$ and consider a plane wave incident on this medium, with the wave vector $\vec{k}$ ($|\vec{k}|=k=2\pi/\lambda$), with the angle of incidence θ, and the amplitude 1. The electric field $\vec{E}(x,z)$ is polarized in the oy direction.

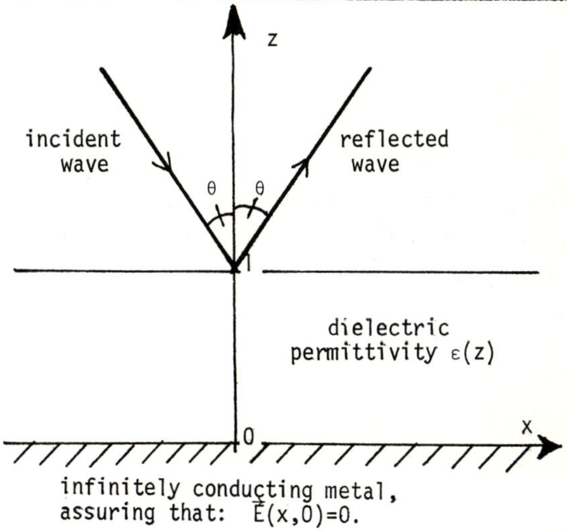

Figure 1: the physical device

The reflected wave has the amplitude $B(\theta,k)$ with $B(\theta,k) = \exp(i\Phi(\theta,k))$ and $|B(\theta,k)| = 1$. One could consider the following problem : θ being fixed, knowing $\Phi(k)$, find $\varepsilon(z)$. But this would be quite unphysical. Indeed, in the optical field, the permittivity $\varepsilon(z)$ always varies very much with k, and it should be writen $\varepsilon(z,k)$. So the only solution is to face the following problem : k being fixed, knowing $\Phi(\theta)$, find $\varepsilon(z)$.

III - BASIS EQUATION - COMPARISON WITH THE CLASSICAL CASE

We define :

$$\vec{E}(x,z) = u(z) \exp(ikx\sin\theta) \vec{e}_y \qquad (1)$$

From Maxwell equations, one arrives to the scalar wave equation :

$$\frac{d^2u}{d\,z^2} + k^2(\cos^2\theta \;-(1.-\varepsilon(z)\,)).u(z) = 0. \quad (2)$$

This is similar to the one-dimensional Schrodinger equation.

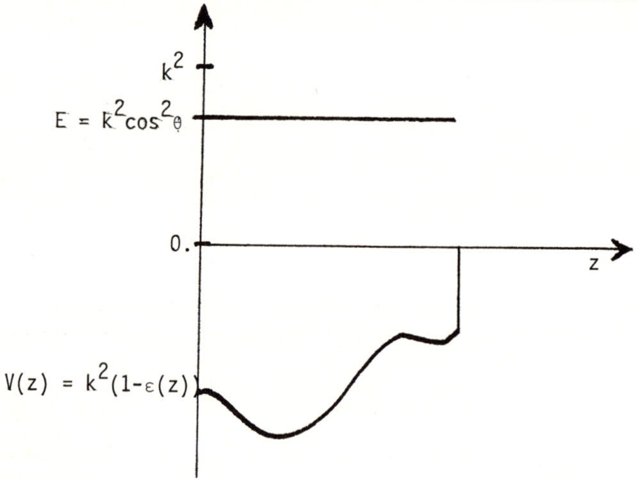

Figure 2:
comparison with
quantum mechanics

The scattering potential is

$$V(z) = k^2(\;1-\varepsilon(z)\;) \qquad (3)$$

It is always negative, and so bound states may exist. These bound states are well known : physically, they are "surface" waves propagating in the ox direction, and are used in integrated optics.

The energy is :

$E = k^2\cos^2\theta$

It is positive, and varies in the range $(\,0,k^2)$

So we are trying to reconstitute the scattering potential, knowing only the scattering amplitude in the range of energy $(0,k^2)$, and knowing nothing about the bound states. It seems to be a very unfavourable situation.

IV - INTEGRAL FORMULATION OF THE PROBLEM

It is more convenient to transform the differential equation (2). Using the boundary conditions associated with, one way find an integral expression for $B(\theta)$.

$$B(\theta) = -1 + \int_0^1 K(\theta,z).u(\theta,z).(\ 1-\ \varepsilon(z)\ ).dz \qquad (4)$$

This is only an integral form of the diffusion equation. Of course, it is not merely an integral equation, since it contains the unknown function u which implicitely depends on the potential $k^2(1-\varepsilon)$
So the relation between $\varepsilon(z)$ and $B(\theta)$is non linear.

However, in most physical cases, the permittivity $\varepsilon(z)$ is obtained by a slight perturbation from a constant permittivity ε_1.This suggests to linearize the equation (4) and to replace it by a sequence of linear equations (fig. 3)

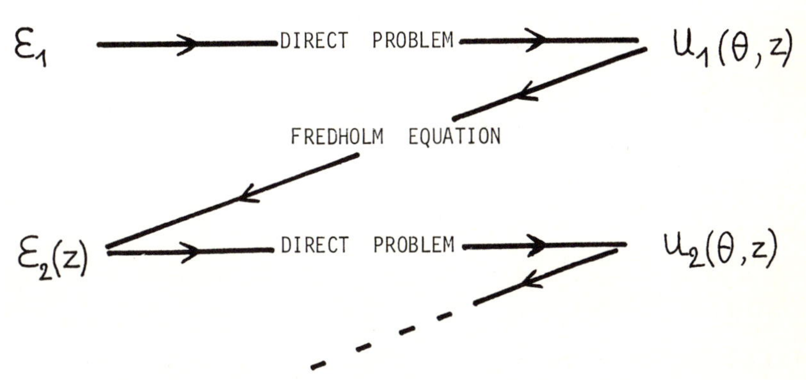

Figure 3: the iterative method.

If we know that $\varepsilon(z)$ is in the neighborhood of $\varepsilon_1$constant, we may think that the function u, solution of equation (2), will not be far from u_1, solution of (2) for $\varepsilon(z)=\varepsilon_1$. Replacing u by u_1 in equation (4) gives an actual Fredholm equation of the first kind. By solving it, one can obtain a better approximation $\varepsilon_2(z)$ of $\varepsilon(z)$, then compute a better $u_2(\theta,z)$, and so on. The only thing to do is testing this iterative method numerically. Computing $u_n(\theta,z)$ from $\varepsilon_n(z)$is very easy, and only requires a numerical algorithm for integrating differential equations.

On the contrary, solving the Fredholm equations of the first kind is an unstable problem, and requires a special study.

V - ILL POSED PROBLEM AT THE SENSE OF HADAMARD

It is well known that the inversion of a Fredholm equation of the first kind is an ill-posed problem at the sense of Hadamard. Let us briefly recall what it means. We define :

$$\int_0^1 N(s,t).f(t).dt = g(s)$$

The integral operator defined by $N(s,t)$ is called A . $(A^* A)$ is known to be a compact operator, and its eigen values λ_n tend to zero when n tend to infinite. (A^* is the adjoint of A) So it is easy to show that the operator A^{-1} is unbounded. In consequence, arbitrarily small variations on the function g (such as experimental errors) may produce arbitrarily large variations on f. (fig. 4). It is absolutely necessary to find a way to restore the stability.

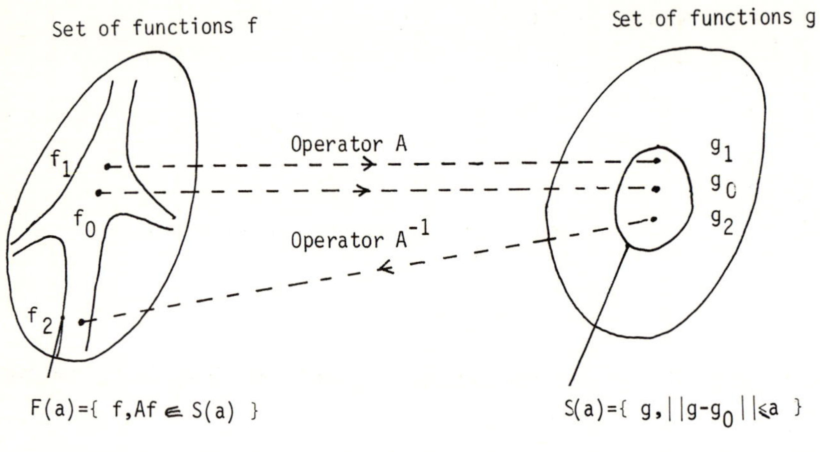

Figure 4: Scheme of the effect of the operator A.

VI - A METHOD OF REGULARIZATION DUE TO TIKHONOV

To get this difficulty over, we have found inspiration in the works of A. Tikhonov [4]. Let S(a) be the sphere of radius a, around go, and let F(a) be the set of elements f such as $Af \in S(a)$, (fig. 4). The instability is linked with the presence in F(a) of very large elements. So it seems natural to restrict the research of f inside a sphere. We must try to satisfy the two conditions :

$$||f|| \leq M$$

$$||Af-g|| \leq a$$

a being, for instance, an experimental error, and g the measured data.

Instead of dealing with these two constraints separately, we may combine them quadratically into a single constraint (losing at most a factor of $\sqrt{2}$) :

$$||Af-g||^2 + \left(\frac{a}{M}\right)^2 ||f||^2 \leq 2a^2$$

We are led to minimize the functional $P_{q_0}(f)$:

$$P_{q_0}(f) = ||Af-g||^2 + q_0 ||f||^2$$

with $q_0 = \left(\frac{a}{M}\right)^2$

The normal equation associated with this variational problem is :

$$A^* A f + q_0 f = A^* g$$

This equation is well conditionned, indeed all the eigenvalues of the operator $(A^*A + q_0)$ are greater than q_0.

VII - NUMERICAL RESULTS

The permittivity function $\varepsilon(z)$ is first given. Then the reflection amplitude $B(\theta)$ is computed. A random error is added, in order to simulate experimental errors. Then we have applied the iterative method of the figure (3), and compared the initial and final profiles.

A first result given by the numerical experiment has been that it is necessary to impose an additional constraint :

$$\left\|\frac{df}{dz}\right\| \leq M_1$$

The associated equation being :

$$\left(A^{*}A + q_0 + q_1\frac{d^2}{dz^2} \right).f = A^{*}g \qquad \text{with } q_1 = \left(\frac{a}{M_1}\right)^2$$

Of course, since we use such regularization techniques, we are forced to restrict the search of f to continuous, derivable (and slowly oscillating) functions. Numerous tests have given us a numerical rule for choosing qo and q_1 [5].

The numerical experiment has also shown that when the iterative process converges, it is actually very quick. In any case, three steps have been sufficient, and very often two are enough.

VIII - PHYSICAL RESULTS

VIII - 1 - RANGE in $\varepsilon(z)$

We have chosen a perturbation formalism : $\varepsilon(z)$ has been explicitely supposed not to be far from a constant value ε_1 . In consequence, the calculation fails when $||\varepsilon(z)-\varepsilon_1||$ typically exceeds 20% of ε_1 . For us, this is not very limitative, since the permittivity perturbations in integrated optics are always very weak.

VIII - 2 - RANGE in l

The most significant limitation concerns the thickness l of the layer and the wave vector k. We generally cannot determine the index profile when $l = 1.5\,\lambda$ ($\lambda = 2\pi/k$)

This is easily explained. When l and k increase, the associated potential given by formula (3) increases also, and bound states appear. We have nowhere taken these bound states into account, and the information they contain is completely lost for us. The numerical experiment says it is impossible to compute the permittivity profile when there is more than one or two bound states. One can only find the average value of $||\varepsilon(z)-\varepsilon_1||$. This would seem to show that, when more than two bound states are present, they contain "most of" the information on the potential. On the contrary, when there is no bound states we are able to reconstitute even complicated profiles (fig. 5).

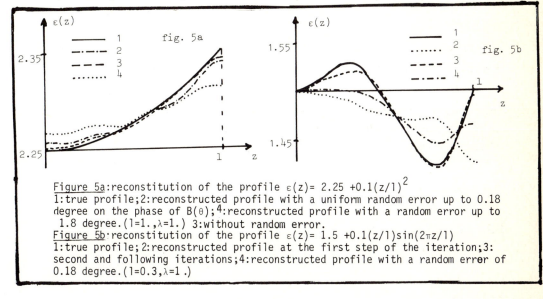

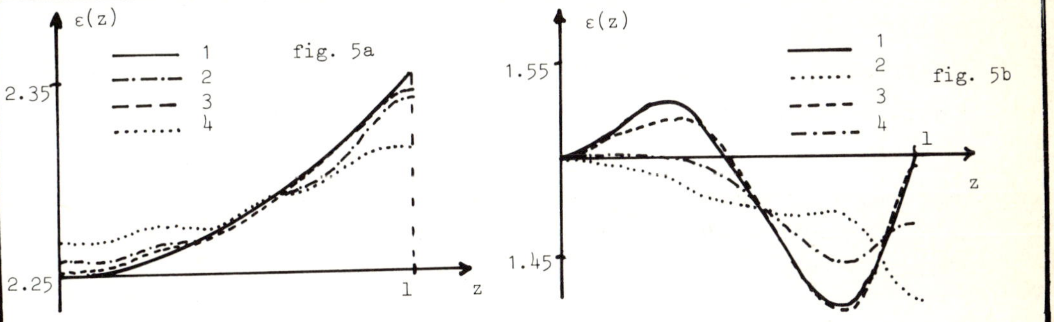

Figure 5a:reconstitution of the profile $\varepsilon(z)= 2.25 +0.1(z/l)^2$
1:true profile;2:reconstructed profile with a uniform random error up to 0.18 degree on the phase of $B(\theta)$;4:reconstructed profile with a random error up to 1.8 degree.(l=1.,λ=1.) 3:without random error.
Figure 5b:reconstitution of the profile $\varepsilon(z)= 1.5 +0.1(z/l)\sin(2\pi z/l)$
1:true profile;2:reconstructed profile at the first step of the iteration;3: second and following iterations;4:reconstructed profile with a random error of 0.18 degree.(l=0.3,λ=1 .)

IX - CONCLUSION

We have started from an integral form of the scattering equation. Using a perturbation technique has led us to solve Fredholm equations of the first kind. This has been achieved by a regularization method due to Tikhonov. Finally, we are able to reconstitute permittivity profiles in all physical cases, at least when there are not more than two bound states.

This method seems to be very adaptable for many other devices, such as gratings and cylinders. Work is in progress in this direction.

REFERENCES

[1] : Electromagnetic grating theories : limitations and successes. R. PETIT. Nouv. Rev. Optique, 1975, t. 6, n° 3, pp 129-135.

[2] : Sur la diffraction d'une onde plane par un cylindre diélectrique. P. VINCENT et R. PETIT. Optics Com. Vol. 5, n° 4 (1972).

[3] : Diffraction d'une onde électromagnétique plane par un objet cylindrique non infiniment conducteur de section arbitraire. D. MAYSTRE et P. VINCENT. Optics Com. Vol. 5, n° 5 (1972).

[4] : TIKHONOV (A) - ARSENINE (V). Méthodes de résolution de problèmes mal posés. Editions de Moscou (1976).

[5] : On a problem of inverse scattering in optics : the dielectric inhomogeneous medium. A. ROGER, D. MAYSTRE, M. CADILHAC. J. Optics (Paris), 1978, vol. 9, n° 1, pp 27-35.

INVERSION-LIKE INTEGRAL EQUATIONS

Henri Cornille

D.P.H.T.Saclay

BP n°2, 91190 Gif-sur-Yvette, France

ABSTRACT

We present the recent results obtained on the construction and the application of the Inversion-like Integral Equations (I.E) associated with linear differential or partial differential systems. In this purely algebraic method we construct a class of solutions and of potentials associated to linear differential systems without performing the analytical study of the associated scattering problem. In this approach the determination of the I.E. is always reduced to the resolution of a well defined set of non linear partial differential equation (n.ℓ.p.d.e).

I. INTRODUCTION

In the classical "Inverse Method", we start from the scattering data, deduce the kernel of the Inversion Equation (I.E) and from the solution of the I.E. we construct both the potentials and the solutions of the differential system. This connection between the scattering data and the kernel of the I.E. is generally performed by a study of the analytic properties of the direct and inverse scattering problem in the complex eigenvalue plane (Jost functions). This method is only tractable in the most simple cases : scalar second order differential equations, first order differential system with a second or third order eigenvalue problem. For higher order eigenvalue problem or for partial differential system, the extension of this analytic method seems "up to now" very difficult. On the other hand, the physical interest of defining formally and studying a scattering problem associated with any linear differential (or partial differential) system is questionable.

In fact the main actual interest in the "Inverse Method" is due to the consequences of the discovery, in a particular case, ten years ago by Gardner, Greene, Kruskal and Miura (1967), that the potentials of linear differential systems can also be the solutions of interesting physical n.ℓ.p.d.e. In this view, the most important problem seems the construction of a formalism generating a class of potentials of linear systems leaving aside the secondary problem of the connection between the kernel of the I.E. and the scattering data. In such a program, an algebraic approach, like the Zakharov-Shabat(1974)one, seems well adapted. However here we focuse our attention to the construction of the Inversion-like I.E which we think is the main tool in order to derive the solutions of the possible associated n.ℓ.p.d.e.

As we shall see, constructing an I.E. associated to a linear system is in itself a problem equivalent to finding the solutions of a well defined integrable n.ℓ.p.d.e. This is perhaps the main reason why the Inversion Formalism seems so deeply connected with the resolution of integrable n.ℓ.p.e.

Our method is the following. First we write down a representation of a set of solutions of the whole differential system as transforms of the reduced known set of solutions when the potentials are switched off (or are known). Secondly we put this representation into the differential system and we get (boundary conditions appear at this stage) that these transforms must satisfy well defined n.ℓ.p.d.e. Thirdly we guess an integral equation (I.E) such that the solutions satisfy the above n.ℓ.p.d.e. In this way we get that the kernels of the I.E. must satisfy well defined integrable linear partial differential equations (ℓ.p.d.e) and also boundary conditions in order that all the derivations be correct (we remark that the general solution of these integrable ℓ.p.d.e. is the equivalent of the general construction of the kernels of the I.E. from the scattering data in the analytic method).

II. A PEDESTRIAN ALGEBRAIC METHOD

We sketch very briefly the formal method which will be illustrated in the following sections.

i) Let us consider Δ_o a linear differential (or partial differential) operator, which contains a constant or eigenvalue term, Q a "potential" and ψ a solution of :

$$(\Delta_o - Q)\psi = 0 \qquad \text{(I)}$$

We associate another linear differential system

$$(\Delta_o - Q_o)\psi_o = 0 \quad , \qquad \text{(I.a)}$$

where Q_o and ψ_o are known.

ii) We postulate a representation of ψ in terms of ψ_o

$$\psi = \psi_o + \mathcal{L}(\psi_o \tilde{\mathcal{K}}) \quad , \qquad \text{(II)}$$

where $\mathcal{L}$ is some integral functional and $\tilde{\mathcal{K}}$ the transform of ψ .

iii) We put (II) into (I) and we find in general.

$$\text{(III)} \qquad \begin{cases} \tilde{\mathcal{K}} & \text{satisfies n.ℓ.p.d.e.} \\ \tilde{\mathcal{K}} & \text{is linked to } Q \text{ , and } Q_o \text{ .} \end{cases}$$

iv) We seek both an integral equation (that we find in general of the form)

$$\mathcal{K} = \tilde{\mathcal{F}} + \int \mathcal{F}\mathcal{K} \; , \; \tilde{\mathcal{F}} \text{ linked to } \mathcal{F} \qquad \text{(IV)}$$

and the properties of the kernel $\mathcal{F}$ such that $\mathcal{K}$, the solution of (IV), satisfies the same n.ℓ.p.d.e as in Eq.(III). In other words we try to establish that the solutions of the n.ℓ.p.d.e Eq.(III) are also solutions of a linear integral equation (IV) which means of course that in order to satisfy this property the n.ℓ.p.d.e (III) must be very particular.

In general we find that $\mathcal{F}$ satisfies an ℓ.p.d.e which is completely integrable and which is linked to the linear part of the n.ℓ.p.d.e Eq.(III).

v) From the completely integrable ℓ.p.d.e satisfied by $\mathcal{F}$ we construct a class of $\mathcal{F}$ which gives a class of $\mathcal{K}$ (using iv) and a class of ψ and Q from (III,II) where we put $\tilde{\mathcal{K}} \equiv \mathcal{K}$. Of course boundary conditions appear in the derivation of (III-IV) and $\mathcal{F}$ must satisfy well-defined conditions in order that the whole formalism be correct.

III. SCALAR SECOND ORDER CASE :

(Cornille 1973 and 1976)

i) Let us start with

$$[\mu^2(x)\ (\frac{d^2}{dx^2} - Q(x)) - \gamma]\ \psi = 0 \quad , \tag{I}$$

and associate

$$[\mu^2(x)\ (\frac{d^2}{dx^2} - Q_o(x)) - \gamma]\ \psi_o = 0 \quad , \tag{I.a}$$

where Q_o and ψ_o are known.

ii) We write a representation of ψ

$$\psi(x) = \psi_o(x) + \int_x^a \mu^{-2}(y)\, \psi_o(y)\, \tilde{K}(x,y)dy \,, \tag{II}$$

where a is a constant.

iii) We put (II) into (I) and get

$$\text{(III)} \begin{cases} \left[D_o(x)-D_o(y)+2\mu \frac{d}{dx}\left(\frac{\tilde{K}(x,x)}{\mu(x)}\right)\right] \tilde{K}(x,y)=0 \,, \quad D_o(x) = \mu^2\left(\frac{\partial^2}{\partial x^2} - Q_o\right) \\ Q(x) = Q_o - 2\mu^{-1} \frac{d}{dx} \left(\frac{\tilde{K}(x,x)}{\mu(x)}\right) \\ + \text{ boundary conditions.} \end{cases}$$

iv) We define

$$\text{(IV)} \begin{cases} K(x,y) = F(x,y) + \int_x^a F(s,y)\, K(x,s)\, \mu^{-2} ds \\ (D_o(x) - D_o(y))\, F(x,y) = 0 \end{cases}$$

and get with some algebra that K satisfies the same n.ℓ.p.d.e as above in Eq.(III) for $\tilde{K}$.(We remark that F satisfies the linear part of the n.ℓ.p.d.e. Eq.(III)). In this derivation well-defined boundary conditions appear.

v) We consider the eigenfunctions $D_o\, \phi_n^o = \delta_n^o\, \phi_n^o$ and construct the kernel F satisfying Eq.(IV).

$$F = \sum_{\text{discrete terms}} \phi_n^o(x)\, C_n\, \phi_n^o(y) \text{ or } \int_{\text{continuum}} .$$

From a given F we construct K by Eq.(IV), put $\tilde{K} \equiv K$ in (II-III) and so get a potential Q and a solution ψ of (I) . We consider $a = 0$ or $+\infty$, and we must clean all the derivations taking into account the boundary conditions. For instance if $a = +\infty$ and $\mu = 1$, then the general solution for F is a linear combination of a x+y dependent function and of a x-y dependent one. However, due to the boundary condition at infinity, the last one must be put to zero.

IV. LINEAR FIRST ORDER DIFFERENTIAL SYSTEMS

(Cornille 1977-a)

i) Let us start with a $n\times n$ differential system :

$$\text{(I)} \qquad [\Delta_o(x) - ik\Lambda + Q(x)]\ \psi = 0 \quad ,$$

where ψ is a column vector, Δ_o and Λ being diagonal, $\Delta_o(x) = (\delta_{ij}\ \mu_i(x)\ \frac{\partial}{\partial x})$, $\mu_i(x) > 0$, $\Lambda = (\delta_{ij}\lambda_i)$, λ_i fixed numbers, $Q = \begin{pmatrix} q_1^1 & \cdots & q_1^n \\ \hline q_n^1 & & q_n^n \end{pmatrix}$. We associate the system

$$\text{(I.a)} \qquad (\Delta_o(x) - ik\Lambda)\ \ \psi^o = 0 \ .$$

If we define $[\mu_j(x)\ \frac{\partial}{\partial x} - i\lambda_j k]\ v_j(x) = 0$ then the column vectors $\psi_j^o = (\delta_{ij}v_j)$ satisfy $(\Delta_o - ik\Lambda)(\psi_1^o, \psi_2^o, \ldots, \psi_n^o) = 0$.

ii) Let us postulate the following representation

$$\text{(II)} \qquad \psi_j = \left(\delta_{ij}v_j(x) + \int_x^\infty (\mu_j(y))^{-1}\ v_j(y)\ K_i^j(x,y)\right) dy$$

iii) We put (II) into (I) and get both the (n.ℓ.p.d.e) which must be satisfied by the $\{K_j^i\}$ and the links with the potentials $\{q_j^i\}$

$$\text{(III)} \quad \begin{cases} (\mu_j(x)\ \frac{\partial}{\partial x} + \frac{\lambda_j}{\lambda_i}\mu_i(y)\frac{\partial}{\partial y})\tilde{K}_j^i(x,y) = \sum\limits_{m\neq j} K_m^i\ \hat{\tilde{K}}_j^m\ (\frac{\lambda_i}{\lambda_m} - \frac{\mu_j(x)}{\mu_m(x)}) \\ q_j^i = (\ \frac{\lambda_j}{\lambda_i} - \frac{\mu_j(x)}{\mu_i(x)}\ )\ \hat{\tilde{K}}_j^i\ ,\ \hat{\tilde{K}}_j^i = \tilde{K}_j^i(x,x)\ . \end{cases}$$

iv) We consider the linear integral system of equations

$$(\mathrm{IV})\quad \begin{cases} K^i_j(x,y) = F^i_j(x,y) + \sum\limits_{m\neq i} \int_x^\infty \mu_m^{-1}(s)\, F^i_m(s,y)\, K^m_j(x,s)\, ds \\ \left(\mu_j(x)\dfrac{\partial}{\partial x} + \dfrac{\lambda_j}{\lambda_i}\,\mu_i(y)\dfrac{\partial}{\partial y}\right) F^i_j = 0\ . \end{cases}$$

We remark that the kernels F^i_j satisfy the linear part of the n.ℓ.p.d.e. Eq.(III). With some algebra we get that if well-defined boundary conditions are satisfied, the solutions K^i_j of Eq.(IV) satisfy the n.ℓ.p.d.e. Eq.(III).

(v) In order to construct the kernels F^i_j, we consider for simplicity only discrete terms and assume $\int^x \mu_j^{-1}(u)\, du \underset{x\to\infty}{=} +\infty$ for any μ_j function. We get two cases following the sign of $\lambda_i\,\lambda_j$

<u>$\lambda_i\lambda_j < 0$</u> : $F^i_j = \sum \alpha^i_{j,m}\ \exp -\nu^i_{j,m}\left[\int^x \mu_j^{-1}\,du - \lambda_i\lambda_j^{-1}\int^y \mu_i^{-1}du\right], \nu^i_{j,m} > 0$

and $F^i_j \to 0$ either when $x\to\infty$ or $y\to\infty$.

<u>$\lambda_i\lambda_j > 0$</u> : $F^i_j = \sum \alpha^i_{j,m}\ \exp \nu^i_{j,m}\left[\int^x \mu_j^{-1}\,du - \lambda_i\lambda_j^{-1}\int^y \mu_i^{-1}du\right], \nu^i_{j,m} > 0$

and $F^i_j \to 0$ when $y\to\infty$, while $F^i_j \to \infty$ (has a bad behaviour) when $x \to \infty$. These kernels lead to difficulties into the formalism, we must have $F^i_i = 0$, $F^i_j = 0$ if $\lambda_j = \lambda_i$ and $\mu_j = \mu_i$. However these conditions are not sufficient in order to satisfy the boundary conditions. We get for the F^i_j kernels with $\lambda_i\lambda_j > 0$ the following rules : if F^i_j, F^j_i are present, one of them must be zero and more generally if $F^i_j F^j_k F^k_\ell \ldots F^n_i$ are present, one of them must be zero. For the remaining non zero F^i_j kernels, their parameters $\nu^i_{j,m}$ must be constrained in order to avoid a breakdown of the formalism.

In conclusion, for the validity of this inversion-like procedure, it appears convenient to associate to each kernel F^i_j the sign of the product $\lambda_i\lambda_j$. If this sign is negative, then F^i_j can be choosen square integrable whereas this is not possible if this sign is positive. Then either we put to zero these badly behaving kernels (reducing the class of reconstructed potentials) or if we introduce such kernels we have to define the rules in order to have a correct formalism.

Once the boundary conditions are satisfied we can introduce a parameter t and assume, like in the Zakharov-Shabat (1974) formalism, that the F^i_j satisfy well-defined ℓ.p.d.e. However the Zakharov-Shabat (1974) formalism ignores these conditions on the kernels of the I.E. generating the potentials so that it is necessary to study how the associated n.ℓ.p.d.e. can still be deduced when the correct boundary conditions on the F^i_j are introduced. In this way we can construct explicitly a class of solutions of the n.ℓ.p.d.e. whose linear parts correspond to the operator const. $\frac{\partial}{\partial t} + \frac{\partial}{\partial x}$ and const $\frac{\partial}{\partial t} + \frac{\partial^2}{\partial x^2}$. These n.$\ell$.p.d.e. which are written down in the Ablowitz-Haberman paper (1975-a) are such that the non linear parts can include triads, multitriads, quartets, multi-quartets, cubic, self-model, self-self interactions.

At the end, let us emphasize that the I.E.(IV) and the corresponding reconstructed class of potentials *depend mainly upon the initial set of solutions (II)* of the system (I) that we have choosen. A priori it can happen that there exists *another set of solutions (II) linked to (I)* such that an algebraic approach like (III-IV) finally leads to an enlarged class of reconstructed potentials where perhaps these difficulties with the $\lambda_i\lambda_j > 0$ kernels F^i_j could disappear.

V. LINEAR FIRST ORDER PARTIAL DIFFERENTIAL SYSTEMS

i) Let us start with a $n\times n$ linear partial differential system

$$\text{(I)} \qquad [\Delta_o(x_1,x_2, \ldots x_n) + ik\Lambda+Q]\, \psi = 0 \ ,$$

where ψ is a column vector, Δ_o and Λ are diagonal, $\Delta_o = (\delta_{ij}\frac{\partial}{\partial x_i})$, $\Lambda = (\delta_{ij}\lambda_i)$, λ_i fixed numbers and $Q = \begin{pmatrix} q_1^1 \ldots q_1^n \\ \ldots\ldots\ldots \\ q_n^1 \ldots q_n^n \end{pmatrix}$. We associate the system

$$\text{(I.a)} \qquad [\Delta_o(x_1,x_2,\ldots x_n) + ik\Lambda]\, \psi^o = 0 \ .$$

If we define $(\frac{\partial}{\partial x} + ik\lambda_i)\, u_i = 0$ then the column vectors $\psi_j^o = (\delta_{ij}u_i)$ satisfy

$$(\Delta_o - ik\Lambda)\, (\psi_1^o,\psi_2^o,\ldots \psi_n^o) = 0 \ .$$

ii) Let us postulate the following representation :

$$\text{(II)} \qquad \psi_j = (\delta_{ij}u_j(x) + \int_x^\infty u_j(y)\, \tilde{K}_i^j(x_1, \ldots x_n;y)\, dy \ .$$

iii) We put (II) into (I) and get both the n.ℓ.p.d.e. which must be satisfied by the set $\{\tilde{K}_j^i\}$ and the connection with the potentials q_j^i

$$\text{(III)} \qquad \begin{cases} 0_{j\,x_j}^i\ \tilde{K}_j^i = \sum\limits_{m\neq j} \tilde{K}_m^i\ \hat{\tilde{K}}_j^m\ \dfrac{\lambda_j}{\lambda_m} \\[2ex] q_j^i = \lambda_j\lambda_i^{-1}\ \hat{\tilde{K}}_j^i \\[2ex] \dfrac{\partial}{\partial x_j} + \lambda_j\lambda_i^{-1}\dfrac{\partial}{\partial y} = 0_{jx_j}^i \ , \quad \hat{K}_j^m(x_1,\ldots x_n;y=x_m) = \hat{\tilde{K}}_j^m \\[2ex] + \text{boundary conditions.} \end{cases}$$

iv) We consider the linear integral system of equations

$$
\text{(IV)}\quad
\begin{cases}
K^i_j(x_1,\dots x_n;y) = \tilde{F}^i_j + \sum_m \int_{x_m}^{\infty} F^i_m(s;x_1,\dots x_n;y)K^m_j(x_1,\dots x_n;s)ds \\
\tilde{F}^i_j = F^i_j(s=x_j;x_1,\dots x_n;y) \\
\lambda_1^{-1}\frac{\partial}{\partial x_1} F^i_j = \lambda_2^{-1}\frac{\partial}{\partial x_2} F^i_j = \dots \lambda_n^{-1}\frac{\partial}{\partial x_n} F^i_j = -(\lambda_j^{-1}\frac{\partial}{\partial s} + \lambda_i^{-1}\frac{\partial}{\partial y}) \times \\
\qquad \times F^i_j(s;x_1,\dots x_n;y) \quad .
\end{cases}
$$

With some algebra we get that the solutions $\{K^i_j\}$ satisfy the same n.ℓ.p.d.e. as in (III) if the kernels F^i_j satisfy well-defined boundary conditions. In the discussion of the classes of potentials reconstructed from (IV) we find two cases depending upon F^i_j is independent or not of the coordinates $x_1,x_2, \dots x_n$.

V.a. $F^i_j = F^i_j(s;y)$ are independent of $x_1,x_2,\dots x_n$

(Cornille 1977-b)

Taking into account $F^i_{j,x_m} = 0$ in (IV) we get $(\lambda_j^{-1}\frac{\partial}{\partial s} + \lambda_i^{-1}\frac{\partial}{\partial y})F^i_j=0$ or equivalently F^i_j depends upon only one variable $F^i_j(u^i_j = \lambda_j s-\lambda_i y)$ and $\tilde{F}^i_j = F^i_j(x_j\lambda_j-\lambda_i y)$. Consequently we get :

1. The degenerate kernels are of the pure exponential type and the solutions $\hat{K}^i_j$ cannot be confined in the R^n space. Concerning the boundary conditions at infinity we have the same kind of difficulty as in the one dimensional case for the kernels F^i_j with $\lambda_i\lambda_j > 0$. For these kernels we get the same rules : F^i_j, F^j_i cannot be both present ...

$\underline{2}$. The potentials $\hat{K}^i_j$ depend in fact upon n-1 independent variables. Let us define $x_{ij} = \lambda_j x_j - \lambda_i x_i$. For n=2 we have only one variable x_{12}, for n=3, two variables x_{12}, x_{13} and more generally $x_{12}, x_{13}, \ldots x_{1n}$.

$\underline{3}$. For $k \neq j$ we have $F^i_{j,x_k} = \tilde{F}^i_{j,x_k} = 0$ and consequently the (K^i_j) must satisfy extra n.ℓ.p.d.e. $K^i_{j,x_k} + K^i_k \hat{K}^k_j = 0$ for $n \geq 2$.

In conclusion this case is very particular and the potentials reconstructed from (IV) represent only a sub-class of the potentials associated with (I). Furthermore, if there exist non linear evolution equations with solutions confined in R^n, they cannot correspond to this I.E.

V.b. F^i_j depend upon $x_1, \ldots x_n$

(Cornille 1977-c)

Let us first remark that if $x_1 = x_2 = \ldots x_n$, then in order to satisfy (IV), necessarily $\lambda_1 = \lambda_2 = \ldots \lambda_n$ and from the relation (III) in section IV we get $q^i_j = 0$. Consequently this is not possible in the one dimensional case and the existence of such F^i_j is a property particular to the multidimensional case.

Taking into account the set of ℓ.p.d.e. satisfied by F^i_j in the relation (IV) we remark that we have n+2 variables $x_1, x_2, \ldots x_n$, s,y and only n relations. It follows that the F^i_j *depend upon two independent variables*

$$F^i_j = F^i_j(u^i_j ; v^i_j), u^i_j = \varepsilon^i_j(\lambda_j(x_j - s) + \sum_{m \neq j} \lambda_m x_m), v^i_j = \eta^i_j\,[\lambda_i(x_i - y) + \sum_{m \neq i} \lambda_m x_m] ,$$

$(\varepsilon^i_j)^2 = (\eta^i_j)^2 = 1$. Consequently we get :

1. The degenerate kernels $F^i_j = g^i_j(u^i_j)h^i_j(v^i_j)$ depend upon two arbitrary functions (such that of course they lead to a correct formalism concerning the boundary conditions) which are not necessarily of the pure exponential type. We can, for instance, choose for g^i_j , functions going to zero when $|u^i_j| \to \infty$

$$\text{(V)} \qquad g^i_j = (u^i_j)^{m_o} \; \exp-(u^i_j)^{2m} \;, \; m \quad \text{integer}$$

and functions of the same type for h^i_j .

2. We consider kernels of the Eq.(V) type in order to study the possibility of existing confined solutions $\hat{K}^i_j$ of Eq.(IV). In order to have a crude insight of what can happen, let us first remark that $\hat{K}^i_j = \tilde{F}^i_j\ (y=x_i)$+ other terms and investigate for the moment the first term which is :

$$g^i_j\left(\sum_{m\neq j} \lambda_m x_m\right) \quad h^i_j\left(\sum_{m\neq i} \lambda_m x_m\right) \quad .$$

In a two dimensional space we have $g^i_j(x_i)h^i_j(x_j) \to 0$ when $\sqrt{x_1^2+x_2^2} \to \infty$. In a three dimensional space we must consider $g^i_j(\lambda_i x_i+\lambda_k x_k)h^i_j(\lambda_j x_j+\lambda_k x_k)$ and we see that even for kernels of the Eq.(V) type, there exist asymptotic directions where the solutions are not confined. However if one variable, let us say $|x_k|$ is finite, then the solutions are still confined in the remaining x_i , x_j plane. For n > 3 we again find for this approximation that the solutions are not confined. Considering particular cases of degenerate kernels where the solutions $\hat{K}^i_j$ can be written in closed form, lead to the some results. As an example let us consider n=2, $F^1_1=F^2_2=0$, $F^1_2=g^1_2h^1_2$, $F^2_1=g^2_1h^2_1$, then we get :

$$
\text{(VI)} \quad \begin{cases} D\,\hat{K}^i_j = g^i_j(\lambda_i x_i) h^i_j(\lambda_j x_j), \quad i=1\ j=2 \text{ and } i=2\ j=1, D=1-A^1_{12}\,A^2_{21} \\ A^i_{ij} = \int_0^\infty g^i_j\,(-\lambda_j u+\lambda_i x_i)\; h^j_i(-\lambda_j u+\lambda_i x_i)\,du\ . \end{cases}
$$

We see that for kernels of the Eq.(V) type, $D\hat{K}^i_j \to 0$ asymptotically in the x_1, x_2 plane whereas D is bounded. It follows that for $D \neq 0$, the potentials $\hat{K}^1_2$ and $\hat{K}^2_1$ are confined in the x_1, x_2 plane.

3. For $n=2$ we find that the solution have not to satisfy extra n.ℓ.p.d.e. In other words the potentials reconstructed from our I.E.(IV) are not restricted.

For $n \geq 3$, due to the fact that the operator $(\lambda_k^{-1} \frac{\partial}{\partial x_k} - \lambda_q^{-1} \frac{\partial}{\partial x_q})$ when applied to F^i_j as well as $\tilde{F}^i_j$ give zero for $k \neq j$, $q \neq j$, we get that the potentials $\hat{K}^i_j$, solutions of Eq.(IV) must satisfy extra n.ℓ.p.d.e. For instance for $n=3$ we get :

$$
\text{(VII)} \qquad \left(\frac{\lambda_k}{\lambda_j} \frac{\partial}{\partial x_j} + \frac{\lambda_k}{\lambda_i} \frac{\partial}{\partial x_i} - \frac{\partial}{\partial x_k} \right) \hat{K}^i_j = 2\hat{K}^i_k\,\hat{K}^k_j \qquad ; i \neq j \neq k \neq i.
$$

If for $n=3$ we consider one x_j as a time and the other two as coordinates, (VII) represents a three-wave evolution equation in a two-dimensional space. If further we restrict to finite time values, there exists an infinite number of confined solutions (coming from kernels g^i_j and h^i_j of the type (V)) in the two spatial dimensional x_i, x_k space.

4. Note added after the meeting (Cornille 1978-a)

Let us consider Eq.(IV) for $n=2$, $\lambda_1 = \varepsilon^1_2 = \eta^2_1 = 1$, $\lambda_2 = \varepsilon^2_1 = \eta^1_2 = -1$, $F^1_1 = F^2_2 = 0$, $F^i_j = F^i_j(s-x_j+x_i; y-x_i+x_j)$, $F^2_1(u,v) = \eta(F^1_2(v,u))^*$ and further let us assume for F^1_2 the following ℓ.p.d.e. when a time parameter is introduced

$$\text{(VIII)} \qquad \left(\frac{i\partial}{\partial t} + \alpha_1 \frac{\partial^2}{\partial s^2} + \alpha_2 \frac{\partial^2}{\partial y^2} \right) F_2^1 = 0 \quad .$$

Then one can show that the solution $\hat{K}_2^1$ of Eq.(IV) satisfies

$$\text{(IX)} \quad \left(\frac{i\partial}{\partial t} + \alpha_1 \frac{\partial^2}{\partial x_1^2} + \alpha_2 \frac{\partial^2}{\partial x_2^2} \right) \hat{K}_2^1 + 2\hat{K}_2^1 (\alpha_1 \hat{K}^2_{2,x_1} + \alpha_2 \hat{K}^1_{1,x_2}) = 0, \quad \hat{K}^i_{i,x_i} + \eta |\hat{K}_2^1|^2 = 0$$

which is the generalized non linear Schrödinger equation in two spatial dimensions and which reduces to the usual cubic term only if $\hat{K}^i_{i,x_j}$ and $\hat{K}^i_{i,x_i}$ are proportional. This equation has been derived recently from two different approaches. Firstly by Ablowitz and Haberman (1975-b) using the compatibility conditions of two linear partial differential systems and by Morris (1977) from the "prolongation structure" of Wahlquist and Estabrook (1975). Considering for F_2^1 degenerate kernels $F_2^1 = g_2^1(s-x_2+x_1;t)\, h_2^1(y-x_1+x_2;t)$ satisfying the heat Eq.(VIII), one can show for any finite time, that there exists an infinite number of solutions (given by (VI)) which are confined in the x_1, x_2 planes. On the contrary when the non linear part of Eq.(IX) reduces to the usual cubic term, then we find that the solutions corresponding to degenerate kernels are not confined.

VI. CONCLUSION

The Inversion-like procedure seems very useful for the study of two problems :

i) The generation of a class of potentials associated to linear differential systems in one or more than one dimension. We have considered some simple cases, but many other extensions can be done.

Let us recall that the determination of the I.E. is always reduced to the resolution of a set of n.ℓ.p.d.e. for the transforms of a set of solutions of the linear differential system. Consequently this set of n.ℓ.p.d.e. (and the I.E) depends mainly upon the set of solutions that we consider.

The main result is that in more than one dimension, in general the kernels of the I.E. do not depend upon only one variable like in the one dimensional case. Consequently the degenerate kernels of the I.E are not necessarily of the pure exponential type, and for instance, there exist, in two-dimensional space, confined reconstructed potentials.

Further in the multidimensional case, for the same set of n.ℓ.p.d.e. satisfied by the transforms $K^i_{\cdot j}$ we have found at least two different I.E. (one generalizing the other). As a signature of the fact that an I.E. is not the more general one we have found that the reconstructed $K^i_{\cdot j}$ (or sometimes even the potentials $\hat{K}^i_{\cdot j}$) satisfy extra n.ℓ.p.d.e. and that there do not exist confined reconstructed potentials. This is why we think that the I.E. presented here is not the more general one in more than two dimensions. The structure of the multidimensional potential space reconstructed from the Inversion formalism appears very rich and complex.

ii) The study of the solutions of the classical non linear evolution equations associated to linear system in one or more than one dimension. In two spatial dimensions, we have now two examples : the three waves non linear equation and the generalization of the non linear Schrödinger equation for which we know that there exist confined solutions at finite time. Let us first remark for the solutions corresponding to the most simple degenerate kernels (what is called solitons in the one dimensional case) that whereas we have only one functional in one dimension we have found an infinite number of functionals in two spatial dimensions. Secondly, in order to have confined solutions at finite time, whereas in the first example (non linear three waves) we have only to change the linear part of the equation going from one to two spatial dimensions, on the contrary in the second case (non linear Schrödinger equation) we have to modify both the linear and the non linear parts. It remains of course, to investigate the extensions of the other classical evolution equations. If we try to understand, through the Inversion-like formalism, the existence of integrable non linear differential systems, it seems convenient to enlarge both the set of variables and the set of functions : instead of the set $\{\hat{K}^i_j\}$ linked to the potentials we consider the whole set of transforms $\{K^i_j(x_1,\ldots x_n;y)\}$. We observe in all solvable cases that these $\{K^i_j\}$ are the common solutions of two different n.ℓ.p.d.e. The first one mixes the $\{K^i_j\}$ and the $\{\hat{K}^i_j\}$ but cannot be reduced to an n.ℓ.p.d.e. for the $\{\hat{K}^i_j\}$ alone. (It is this n.ℓ.p.d.e. that we have to solve in order to establish the I.E.).

The second one mixes also the $\{K^i_j\}$ and the $\{\hat{K}^i_j\}$ but contains a sub-set of n.ℓ.p.d.e. which give the classical evolution equations for the potentials. So at least one question arises : what is the meaning of the n.ℓ.p.d.e. satisfied by those K^i_j for which their restrictions $\hat{K}^i_j$ is not connected to the potentials?

REFERENCES

ABLOWITZ M. and HABERMAN R. (1975-a) : J.M.P. 16, 2301 ;
(1975-b) : PRL 18, 1185.

CORNILLE H. (1973) : Nuovo Cimento 141, 14A ; (1976), J.M.P 17, 2143;
(1977-a) : J.M.P. 18, 1855 and Rockefeller preprint C.O.O.2232B-124 ;
(1977-b) : Saclay preprint DPh-T/77/45 (to appear in J.M.P.);
(1977-c) : Saclay preprint DPh-T/77/108 ; (1978-a) to appear.

GARDNER C.S., GREENE J.M., KRUSKAL M.D., MIURA R.M. (1967) : P.R.L. 19, 1095.

MORRIS H.C. (1977) : J.M.P. 18, 285.

WAHLQUIST H.D. and ESTABROOK F.B. (1975) : J.M.P. 16, 1.

ZAKHAROV V.E. and SHABAT A.B. (1974) : Func. Anal. Appl. 8, 226.

INVERSE METHOD FOR OFF-SHELL CONTINUATION OF THE SCATTERING AMPLITUDE IN QUANTUM MECHANICS

Bengt Karlsson

Institute of Theoretical Physics
Fack - 40220 Göteborg 5 - Sweden

Abstract

The talk consists of two separate parts. In the first part we show that the multiple-scattering formulation of the quantum mechanical three-body problem requires as input the half-off-shell two-body scattering amplitude. We indicate how this input information can be obtained from the physical (on-shell) two-body scattering amplitude using the methods of the inverse problem of scattering.

In the second part we present a non-conventional (dispersion theory) derivation of the basic equations of the inverse problem of scattering (for fixed 1); we obtain in addition to the well-known Marchenco and Gel'fand-Levitan equations an infinite set of alternative equations.

We finally present a momentum space formulation of the inverse problem which is particularly suited for the application mentionned in the first part of the talk.

1. INTRODUCTION

With the development of the various multiple-scattering approaches, notably that of Faddeev, to the quantum mechanical three and N-body problem, the two-body potential has been replaced by the two-body scattering amplitudes (or in the N-body case the general subsystem amplitudes) as the relevant dynamical input. It is clearly not only the "physical" two-body amplitude (i.e. that measurable in a two-body scattering process) that is needed but also the scattering amplitude for a two-body collision in the neighborhood of the third particle, the so called off-the-energy-shell scattering amplitude. If a two-body potential is given, this more general off-shell amplitude can readily be obtained from the complete solution to the two-body problem. On the other hand, adopting the "inverse problem of scattering" point of view, the relevant inverse problem to pose is no longer how to determine the two-body potential from scattering data but rather how to continue the scattering data off-shell.

The aim of this talk is to show more in detail

1) What two-body information is required in the three-body problem, and
2) How inverse methods can be used in the construction of this information from two-body scattering data.

Since we hope that the inverse methods that we will discuss have some general interest independent of the particular application we have in mind here, the presentation will be such that the later part of the talk (sections 3.3-6) can be comprehended independently of the former.

2. TWO-AND THREE-BODY QUANTUM MECHANICS

The practical problem of two-body quantum mechanics is the solution of the eigenvalue problem of the two-body hamiltonian,

$$\left(\frac{q^2}{2\mu} + V\right)|\psi\rangle = e|\psi\rangle \qquad (2.1)$$

Where $\frac{q^2}{2\mu}$ is the kinetic energy of the relative motion of the two particles, V is the potential that represents the interaction between the particles, and e is the energy of the system in the center-of-mass frame. Two cases which correspond to different boundary conditions at infinity are to be distinguished,

1) The bound state case, $e = -e_B < 0$

$$|\varphi\rangle = -\left(\frac{q^2}{2\mu} + e_B\right)^{-1} V|\varphi\rangle \qquad (2.2)$$

2) The scattering case, $e = \frac{k^2}{2\mu} > 0$

$$|\psi^+_{\vec{k}}> = |\vec{k}> - (\frac{q^2}{2\mu} - e - i\epsilon)^{-1} V |\psi^+_{\vec{k}}> \qquad (2.3)$$

In "coordinate" or "position" space, Eqn (2.3) takes the familiar form

$$\psi^+_{\vec{k}} (\vec{r}) = e^{i\vec{k}.\vec{r}} - \frac{\mu}{2\pi} \int d\vec{r}' \frac{e^{ik|\vec{r}-\vec{r}'|}}{|\vec{r}-\vec{r}'|} V(\vec{r}') \psi^+_{\vec{k}} (\vec{r}')$$

$$\underset{r \to \infty}{\rightarrow} e^{i\vec{k}.\vec{r}} - \mu(2\pi)^2 t(\vec{k}',\vec{k};e+i\epsilon) \frac{e^{ikr}}{r} , \vec{k}' = k\hat{r} \qquad (2.4)$$

where $t(\vec{k}',\vec{k};e+i\epsilon) = <\vec{k}'|V|\psi^+_{\vec{k}}>$ is the usual expression for the scattering amplitude.

In 'momentum" space (the Fourier transform of coordinate space) the same equation takes the form

$$\psi^+_{\vec{k}} (\vec{q}) = \delta(\vec{q}-\vec{k}) - \frac{1}{\frac{q^2}{2\mu} - e - i\epsilon} \int <\vec{q}|V|\vec{q}'> d\vec{q}' \psi^+_{\vec{k}} (\vec{q}')$$

$$\equiv \delta(\vec{q}-\vec{k}) - \frac{1}{\frac{q^2}{2\mu} - e - i\epsilon} t(\vec{q},\vec{k};e+i\epsilon) \qquad (2.5)$$

where we have <u>defined</u>

$$t(\vec{q},\vec{k};e+io) = <\vec{q}|V|\psi^+_{\vec{k}}> \qquad (2.6)$$

This entity differs from the scattering amplitude introduced above because here in general $\frac{q^2}{2\mu} \neq e$, i.e. the momentum $\vec{q}$ is not "on-the-energy-shell". Therefore, $t(\vec{q},\vec{k};e+io)$ is known as the half-off-shell scattering amplitude. Clearly, the momentum space wave function and the half-off-shell amplitude are entirely equivalent entities.

In three-body quantum mechanics, the kinetic part of the hamiltonian

$$H_0 = \frac{1}{2\mu_\beta} q_\beta^2 + \frac{1}{2n_\beta} p_\beta^2 \quad (\beta = 1,2, \text{ or } 3) \qquad (2.7)$$

has two terms, $\frac{1}{2\mu_\beta} q_\beta^2$ representing the relative motion of the β-pair (if β=1, the β-pair is the two particles 2 and 3) and $\frac{1}{2n_\beta} p_\beta^2$ representing the relative motion of the third particle and the center-of-mass of the β-pair (we consider the system in the overall center-of-mass frame). If V_β is the potential between the particles of the β-pair, the three-body Schrodinger equation is

$$(H_0 + V_1 + V_2 + V_3)|\psi\rangle = E|\psi\rangle \tag{2.8}$$

where E is the total energy of the system.

The Faddeev reformulation of this problem starts with a splitting of the wave function into three pieces, $|\psi\rangle = |\psi_1\rangle + |\psi_2\rangle + |\psi_3\rangle$, such that

$$\begin{aligned}(H_0 + V_1 - E)|\psi_1\rangle &= - V_1(|\psi_2\rangle + |\psi_3\rangle)\\ (H_0 + V_2 - E)|\psi_2\rangle &= - V_2(|\psi_3\rangle + |\psi_1\rangle)\\ (H_0 + V_3 - E)|\psi_3\rangle &= - V_3(|\psi_1\rangle + |\psi_2\rangle)\end{aligned} \tag{2.9}$$

These equations evidently reduce to the Schrodinger equation upon summation.

For any choice of physical three-body situation, appropriate boundary conditions have to be added to Eqn (2.9). For the interesting case of the scattering of an initial two-body pair, say α, with the third particle, the set of differential equations is then converted into the following set of integral equations,

$$|\psi^+_\beta\rangle = |\vec{p}^{(o)}_\alpha \varphi_\alpha\rangle\, \delta_{\beta\alpha} - (H_0 + V_\beta - E - i\epsilon)^{-1} V_\beta \sum_{\gamma\neq\beta} |\psi^+_\gamma\rangle \tag{2.10}$$

where $|\vec{p}^{(o)}_\alpha \varphi_\alpha\rangle$ represents the initial state. The kernel of this equation is often written in terms of the two-body transition operator $t(E+i\epsilon)$, defined by the relation

$$(H_0 + V_\beta - E - i\epsilon)^{-1} V_\beta = (H_0 - E - i\epsilon)^{-1}\, t_\beta(E+i\epsilon) \tag{2.11}$$

In order to see more in detail what goes into the construction of the kernel of the Faddeev equations (2.10) we will use a basis in which the resolvent operator is diagonal. Assuming that the two-body bound states $|\varphi\rangle$ and scattering states $|\psi^+\rangle$, form a complete basis set, the kernel takes the form ($e = E - \frac{1}{2n} p^2$)

$$\left(\frac{1}{2\mu} q^2 + V - e - i\epsilon\right)^{-1} V = \frac{|\varphi\rangle\langle\varphi| V}{- e_B - e} + \int \frac{|\psi^+_{\vec{k}}\rangle d\vec{k} \langle\psi^+_{\vec{k}}| V}{\frac{k^2}{2\mu} - e - i\epsilon} \tag{2.12}$$

The appearence of the potential V on the right hand side of this equation is not significant ; indeed, using the two-body relations (2.1) and (2.6) one sees that

$$\begin{aligned}\langle\varphi|V|\vec{q}\rangle &= \langle\varphi|\vec{q}\rangle\left(- \frac{1}{2\mu} q^2 - e_B\right)\\ \langle\psi^+_{\vec{k}}|V|\vec{q}\rangle &= \langle\vec{q}|V|\psi^+_{\vec{k}}\rangle^* = t(\vec{q},\vec{k}\ ;\ e+i\epsilon)^*\end{aligned} \tag{2.13}$$

Recalling the relation (2.5) between the off-shell scattering amplitude and the momentum space wave function we therefore conclude that the relevant input to the three-body Faddeev equations is the two-body bound state and scattering wave functions.

In a two-body experiment, the scattering wave functions can only be obtained asymptotically in coordinate space or at the on-shell pole in momentum space. One therefore has to rely on theoretical models for the construction of the full input to the three-body equations, i.e. for off-shell continuation. Common practice is of course to choose a potential that reproduces the two-body data as well as possible, and then solve the two-body problem for the off-shell (or non-asymptotic) wave functions. Here, we want to see to what extent inverse methods can be used to supply the same information.

3 . INVERSE METHODS. PRELIMINARIES

The discussion of the preceeding section resulted in the following formulation of the inverse two-body problem : given the Schrodinger equation and conventional scattering data (fixed ℓ, all energies), find the wave function, preferably in momentum space, without knowing the details of the interaction potential. This is only a slight reformulation of the classical inverse problem of scattering (in which one asks for the potential) and this problem will now be studied at some length. In this section some general arguments will be given. More detailed results (for local potentials) will be given in the next sections.

3.1 Unitarity constraints

Assume first that the only constraints on the potential are that the solutions to the Schrodinger equation form a complete orthogonal set, i.e. (in momentum space and with no bound states)

$$\int_0^\infty \psi_\ell^+(k,q)^* \; \psi_\ell^+(k',q)dq = \delta(k-k') \qquad (3.1)$$
$$\int_0^\infty \psi_\ell^+(k,q)^* \; \psi_\ell^+(k,q')dk = \delta(q-q')$$

This means that the Møller wave operator is unitary (isometric if there are bound states),

$$\Omega\,\Omega^+ = \Omega^+\Omega = 1 \qquad (3.2)$$

where $(k = \sqrt{2\mu e})$

$$\langle q|\Omega|k\rangle \equiv \psi_\ell^+(k,q) = \delta(k-q) - \frac{2\mu}{q^2-k^2-i\epsilon}\, t_\ell(q,k;e+i\epsilon) \qquad (3.3)$$

The constraint (3.1) can be shown to force the phase of the wave function <u>and</u> of the <u>off-shell</u> scattering amplitude to be the physical phase shift $\delta_\ell(k)$ independently of q, i.e.

$$\psi_\ell(k,q) = e^{i\delta_\ell(k)}\, R_\ell(q,k) \qquad (3.4)$$
$$t_\ell(q,k;e+i\epsilon) = e^{i\delta_\ell(k)}\, |t_\ell(q,k;e+i\epsilon)|$$

(Recall that on shell, $t_\ell(k,k\ ;\ e+i\epsilon) \propto e^{i\delta_\ell(k)} \sin \delta_\ell(k)$).

However, Eqn (3.1) puts further constraints on $R_\ell(q,k)$ and $|t_\ell(q,k;e+i\epsilon)|$, first investigated by Baranger et al., essentially that the antisymmetric part of $|t_\ell(q,k;e+i\epsilon)|$ can be reconstructed from its symmetric part. Since the physical, experimentally accessible, part of $|t_\ell(q,k\ ;\ e+i\epsilon)|$ is (at most) the diagonal of the symmetric part, these unitarity constraints alone do not suffice to solve the inverse problem posed above.

3.2 Separable potentials

In order to proceed with the inverse problem, further assumptions have to be made about the potential. Before we turn to the case of local potentials, it should be recalled that separable potentials, i.e. $\mathcal{V}(r,r') = \sum_n \lambda_n v_n(r)\ v_n(r')$, play an important role in three body quantum mechanics (if the number of separable terms is small) since for such potentials the numerical complexity in solving the Faddeev equations is reduced by an order of magnitude (as compared to local potentials). We will not discuss this interesting case here but again refer to the literature, in particular to the review by Chadan on the solution to the inverse problem for separable potentials.

3.3 Local potentials

From here on we will assume as starting point the Schrodinger equation with a local potential. In this case the classical methods of the quantum mechanical inverse problem are available for the reconstruction of the potential $\mathcal{V}(r)$ from the scattering data (e.q. $S(k) = \exp(2i\delta(k))$) and once $\mathcal{V}(r)$ is known, the direct two-body problem can be solved for the wave functions (recall that the inverse problem we have in mind is the determination of (momentum space) wave functions from the scattering data). However, any one familiar with the Marchenko and Gel'fand-Levitan techniques realizes that this is an unnecessarily complicated proceedure : these techniques directly provide the wave functions as intermediate steps in the construction of the potential. Therefore, our inverse problem is in principle already solved by these classical methods.

On the other hand, it has repeatedly been found that these techniques are very inconvenient from a practical point of view, since at one point a Fourier transform of experimental data has to be carried out (one starts with a momentum space function $S(k)$ and wants a coordinate space function $\mathcal{V}(r)$). In the particular application of interest here we do not care in particular for coordinate space wave functions ; rather, the momentum space wave functions, i.e. the off-shell scattering amplitudes, are what we want. But in such a case there should be no need to carry out the troublesome Fourier transform of experimental data : we start in momentum space, end up in momentum space and ought to be able to avoid coordinate space entities also in the intermediate steps (the corresponding version of the classical inverse problem would be to ask for the potential not in coordinate but in momentum space).

In order to carry out this reformulation of the classical inverse methods one could of course take the conventional formulations as starting points. This will not be done here, rather, we take the opportunity to demonstrate how analyticity properties derivable from a study of the direct problem of scattering provide an excellent starting point for the construction of inverse methods.

4 . INVERSE METHODS. A MATTER OF ANALYTICITY

In this section we are going to demonstrate that the classical results of the inverse problem of scattering can all be obtained in a very straightforward way by exploiting analyticity properties of the solutions to the Schrodinger equation obtained in a study of the direct problem of scattering. For simplicity of presentation, the discussion of this section will be restricted to the $\ell=0$ (s-wave) case without bound states. A more general case will be discussed in Appendix II.

Consider the differential equation

$$[-\frac{d^2}{dr^2} + \mathcal{V}(r)]\ \psi(k,r) = k^2\psi(k,r) \qquad (4.1)$$

and assume that $\int_0^\infty |\mathcal{V}(r)|\ r^\alpha\ dr < \infty,\quad \alpha = 1,2.$ The following properties of the solutions to this equation are taken from Newton's book on scattering theory. The "physical" solution to Eqn (4.1) is determined by the conditions that

$$\begin{aligned} r \to 0 \quad &: \quad \psi^+(k,r) \to 0 \\ r \to \infty \quad &: \quad \psi^+(k,r) \to \frac{1}{2i}\,(S(k)\ e^{ikr} - e^{-ikr}) \end{aligned} \qquad (4.2)$$

(The condition in the last case is that the coefficient of e^{-ikr} is $i/2$. The coefficient $S(k) = \exp(2i\delta(k))$, δ real, is determined by the detailed behaviour of the potential).

In the inverse problem, $S(k)$ is assumed to be known for all k, and the task is to determine $\psi(k,r)$ (without knowing $\mathcal{V}(r)$, except for the general conditions just below eqn (4.1)).

The following results from the study of the solutions of Eqn (4.1) will turn out useful for the solution of this inverse problem.

The "regular" solution $\varphi(k,r)$ of Eqn (4.1) is defined by the conditions

$$r \to 0 \quad \begin{cases} \varphi(k,r) \to 0 \\ \frac{d}{dr} \varphi(k,r) \to 0 \end{cases} \tag{4.3}$$

$\varphi(k,r)$ has the important property of being an entire function of k^2 for any fixed r, and behaves at $|k| \to \infty$ according to

$$\varphi(k,r) - \frac{\sin kr}{k} = o(e^{|\mathrm{Im}\, k|r}), \quad |k| \to \infty \tag{4.4}$$

The Jost solutions $f_\pm(k,r)$ are defined by the conditions

$$r \to \infty : f_\pm(k,r) \to e^{\pm ikr} \tag{4.5}$$

$f_+(k,r)$ is for fixed r a regular analytic function in $\mathrm{Im}\, k \geq 0$, continuous with a continuous k-derivative down to $\mathrm{Im}\, k = 0$, and

$$f_+(k,r) - e^{ikr} = o(e^{-|\mathrm{Im}\, k|r}), \quad |k| \to \infty, \quad \mathrm{Im}\, k \geq 0. \tag{4.6}$$

$f_-(k,r)$ has similar properties in the lower half of the complex k-plane.

The Jost functions $\mathcal{J}_\pm(k) = f_\pm(k,0)$ have the same analyticity properties as the Jost solutions and in particular we have that $\mathcal{J}_+(k)$ is regular and analytic in $\mathrm{Im}\, k > 0$ and behaves according to

$$\mathcal{J}_+(k) - 1 = o(1), \quad |k| \to \infty, \quad \mathrm{Im}\, k \geq 0 \tag{4.7}$$

at infinity. Furthermore, it is known that $\mathcal{J}_+(k) \neq 0$ in $\mathrm{Im}\, k \geq 0$ when there are no bound states present.

The solutions to the Schrodinger equation (4.1) defined above cannot all be linearly independent. It is easily verified using the information compiled above that

$$\begin{aligned} \psi^+(k,r) &= \frac{k\, \varphi(k,r)}{\mathcal{J}_+(k)} \\ &= \frac{1}{2i} \left(S(k)\, f_+(k,r) - f_-(k,r)\right) \end{aligned} \tag{4.8}$$

and that $S(k) = \mathcal{J}_-(k)/\mathcal{J}_+(k)$. For real k we also have $f_-(k,r) = f_+(-k,r) = f_+(k,r)^*$ so that in particular the two Jost functions $\mathcal{J}_\pm(k)$ have the representations

$$\mathcal{J}_\pm(k) = |\mathcal{J}_+(k)|\, e^{\mp i\delta(k)} \tag{4.9}$$

From the first of the relations (4.8) and the properties of $\varphi(k,r)$ and $\mathcal{J}_+(k)$ we can also conclude that in the absence of bound states (i.e. when $\mathcal{J}_+(k) \neq 0$, $\mathrm{Im}\, k \geq 0$),

$\psi^+(k,r)$ is also a regular analytic function in $\text{Im}\,k > 0$, continuous with a continuous k-derivative down to $\text{Im}\,k = 0$, with the behaviour at infinity

$$\psi^+(k,r) - \sin kr = o(e^{|\text{Im}\,k|r}), \quad |k| \to \infty, \quad \text{Im}\,k \geq 0 \qquad (4.10)$$

As a first example of what can be done with all this information on the analyticity properties (in the k-plane) of the solutions to the Schrodinger equation we derive an explicit expression for $|\mathcal{J}_+(k)|$ in terms of $\delta(k)$.

Consider the Cauchy integral $(\mathcal{J}_+(k) \equiv \mathcal{J}(k))$

$$\ln \mathcal{J}(k) = \frac{1}{2\pi i} \int_C \frac{\ln \mathcal{J}(k')}{k'-k-i\epsilon}\, dk' \qquad (4.11)$$

where k is real and C is a contour running along the real axis from $k' = -\Lambda$ to $k' = \Lambda$, and then back to $k' = -\Lambda$ along a semi circle in the upper half k'-plane. Since $\mathcal{J}(k) \to 1$, $|k| \to \infty$, $\text{Im}\,k \geq 0$, there will be no contribution from the semi circle part of the contour when $\Lambda \to \infty$. Using the relation

$$\frac{1}{k'-k-i\epsilon} = \mathcal{P}\frac{1}{k'-k} + i\pi\,\delta(k'-k) \qquad (4.12)$$

we therefore get

$$\ln \mathcal{J}(k) = \frac{1}{i\pi} \int_{-\infty}^{\infty} \frac{\ln \mathcal{J}(k')}{k'-k}\, dk' \qquad (4.13)$$

Recalling that $\text{Re} \ln \mathcal{J}(k) = \ln |\mathcal{J}(k)|$ and that $\text{Im} \ln \mathcal{J}(h) = -\delta(k)$ we finally get, upon taking the real part of (4.13),

$$|\mathcal{J}(k)| = \exp\{-\frac{1}{\pi} \int_{-\infty}^{\infty} \mathcal{P}\frac{\delta(k')}{k'-k}\, dk'\} \qquad (4.14)$$

which is the advertised result.

The physical interest in this result lies in the fact that we have been able to determine the normalization of $\psi^+(k,r)$ at $r = 0$ from the behaviour of $\psi^+(k,r)$ at $r \to \infty$ (see equations (4.8) and (4.3)).

We are now prepared to attact the more difficult problem of the determination of $\psi^+(k,r)$ at any arbitrary value of r from its asymptotic behaviour (i.e. from $S(k)$).

For this purpose we consider the Cauchy integral

$$f_+(k,r)\, e^{-ikr} - 1 = \frac{1}{2\pi i} \int_C \frac{f_+(k',r)\, e^{-ik'r} - 1}{k'-k-i\epsilon}\, dk' \qquad (4.15)$$

where C is the same contour as in the previous example and where $\mathrm{Im}\, k \geq 0$. Recalling Eqn (4.6) we first note that there is no contribution from the semi circle part of C when $\Lambda \to \infty$.

Let us now try to close the remaining real axis part of C with a semi-circle in the lower half of the k'-plane or, equivalently, make the change of variables $k' \to -k'$ and then close the contour again in the upper half plane. Using the relation (4.8) and the fact that $f_-(k,r) = f_+(-k,r)$, we first get

$$f_+(k,r)e^{-ikr} - 1 = \frac{1}{2\pi i} \int_{-\infty}^{\infty} \frac{[S(k')f_+(k',r) - 2i\psi^+(k',r)]e^{ik'r} - 1}{-(k'+k+i\epsilon)} dk'$$

$$= \frac{1}{2\pi i} \int_{-\infty}^{\infty} \frac{(S(k')-1)\, f_+(k',r)e^{ik'r}}{-(k+k'+i\epsilon)} dk'$$

$$+ \frac{1}{2\pi i} \int_{-\infty}^{\infty} \frac{[f_+(k',r) - 2i\,\psi^+(k',r)]\, e^{ik'r} - 1}{-(k'+k+i\epsilon)} dk'. \quad (4.16)$$

In the last integral, the contour can be replaced by C with no change in the value of the integral (From (4.10) one sees that $2i\,\psi^+(k',r)\, e^{ik'r} \to -1$, $|k'| \to \infty$, $\mathrm{Im}\, k' \geq 0$). If there are no bound states, the integrand is regular analytic for $\mathrm{Im}\, k > 0$ and the integral vanishes. Therefore, we are left with the result

$$f_+(k,r) = e^{ikr} - \frac{1}{2\pi i} \int_{-\infty}^{\infty} \frac{S(k')-1}{k'+k+i\epsilon}\, e^{i(k+k')r}\, f_+(k',r)\, dk' \quad (4.17)$$

which is an integral equation for the determination of $f_+(k,r)$ from $S(k)$ at any fixed value of r.

Equation (4.17) is our solution to the inverse problem of scattering.

Before we proceed with the momentum space reformulation of this result we would like to make a couple of comments

(i) The Marchenko equation is closely related to Eqn (4.17) as can easily be verified using the representation

$$f_+(k,r) = e^{ikr} + \int_r^{\infty} A(r,r')\, e^{ikr'} dr' \quad (4.18)$$

for $f_+(k,r)$.

(ii) Using similar techniques one can derive an equation for $\varphi(k,r)$ in terms of $|\mathcal{J}(k)|$ (see Appendix I).

$$\varphi(k,r) = \frac{\sin kr}{k} + \frac{1}{\pi k} \int_{-\infty}^{\infty} k' \left[\frac{1}{|\mathcal{J}(k')|^2} - 1\right] \frac{\sin (k+k')r}{k+k'} \varphi(k',r)\, dk' \quad (4.19)$$

This equation is obviously related to the Gel'fand-Levitan equation.

(iii) The techniques used here can be generalized to include bound states and to the treatment of the $\ell \neq 0$ case. This generalization is carried out in Appendix II.

(iv) The equation (4.17) is not the unique result of the techniques used here. For instance, $S(k') - 1$ could be replaced by any function $S(k') - \widetilde{S}(k')$ such that $\widetilde{S}(k')$ has no singularities in the upper half plane and $\widetilde{S}(k') \leq$ const, $|K'| \to \infty$, Im $k' \geq 0$. Below we will use this freedom to replace $S(k') - 1$ by

$$S(k') - S^*(k) = 2i \sin (\delta(k') + \delta(k))\, e^{i(\delta(k)-\delta(k'))} \quad (4.20)$$

which makes the kernel of the integral equation for $f_+(k,r)$ non-singular. Another interesting choice is $S(k') - \frac{k'+k+i\mu\, S^*(k)}{k'+k+i\mu}$ where $\mu > 0$ is arbitrary. In addition to cancelling the pole at $k' = -k$, this expression reduces to $S(k') - 1$ for $k' \to \pm \infty$.

(v) For the class of Yukawa-like potentials, eq. (4.17) is closely related to Martin's solution of the inverse problem of scattering. This case will be discussed in Appendix III.

To summarize, the classical inverse problem of scattering has been solved in a straightforward way by exploiting the analyticity structure (in the energy parameter) of the solutions to the Schrodinger equation. This approach provides a whole class of equations that all solve the inverse problem and therefore leads to the interesting question : given a particular set of incomplete scattering data, which equation is the more suitable ?

5 . INVERSE METHOD. MOMENTUM SPACE FORMULATION.

As has already been pointed out, for the practical solution of an inverse problem where $S(k)$ has been deduces from experimental data, it is important to avoid Fourier transforms or similar transitions from the momentum space entity $S(k)$ into coordinate space entities (like $f_+(k,r)$). One might therefore attempt to convert eq. (4.17) into a momentum space equation. The straightforward proceedure would be to define $f(k,q) = \int_0^{\infty} f_+(k,r) \sin qr\, dr$, and one finds

$$f(k,q) - \dots - \frac{1}{2\pi i} \int_{-\infty}^{\infty} (S(k')-1)\, \frac{Q(k,q;k',q')}{k+k'+i\epsilon}\, f(k',q')\, dk'\, dq' \quad (5.1)$$

where $Q(k,q;k',q')$ is a singular kernel function. From practical point of view this equation has the drawback of being an integral equation in two variables (in addition to having a singular kernel).

A much more elegant momentum space formulation can be obtained in the following way (the method works also in the $\ell = 0$ case with bound states). Define a "sideways" Laplace transform of $f_+(k,r)\ e^{-ikr}$ through

$$\hat{f}(k,q) = \frac{2}{\pi} \int_0^\infty f_+(k,r)\ e^{-ikr}\ e^{2iqr}\ dr \qquad (5.2)$$

and let $\hat{g}(k,q) = \mathrm{Im}(\frac{1}{2i}\ e^{i\delta(k)}\ \hat{f}(k,q))$. It is easy to verify that the momentum space wave function (and the off-shell scattering amplitude) is related to this transform throught

$$\psi^+(k,q) = e^{i\delta(k)}\ [\hat{g}(k;\tfrac{1}{2}(k+q)) - \hat{g}(k;\tfrac{1}{2}(k-q))] \qquad (5.3)$$

The important property of $\hat{f}(k,q)$ is that the integral equation (4.17) takes the simple form

$$\hat{f}(k,q) = \frac{1}{\pi}\frac{i}{q+i\epsilon} - \frac{1}{2\pi i}\int_{-\infty}^{\infty} dk'\ \frac{S(k')-1}{k+k'+i\epsilon}\ \hat{f}(k',k'+q) \qquad (5.4)$$

when expressed in terms of $\hat{f}(k,q)$, and this is a one variable integral equation. Using the freedom elaborated upon in comment (iv) at the end of section 4, this equation can alternatively be written as a non-singular integral equation. With the particular choice (4.20), equation (5.4) reduces to a real, non-singular integral equation for the $\hat{g}(k,q)$ that appears in (5.3).

Equation (5.4) can easily be solved by iteration, and we expect convergence as long as there are no bound states.

A more general procedure with better convergence properties would be to use some adaptation of the Fredholm method. This method is applicable to Eqn (4.17) but not directly to Eqn (5.4) because of the peculiar dependence of $\hat{f}(k',k'+q)$ on k'. We have been able to by-pass this obstacle in the following way.

Consider Eqn (4.17) with a parameter η added,

$$f_+(k,r) = e^{ikr} + \eta \int_{-\infty}^{\infty} K(k,k';r)\ f_+(k',r)\ dk' \qquad (5.5)$$

where

$$K(k,k';r) = -\frac{1}{2\pi i}\ (S(k')-1)\ \frac{1}{k+k'+i\epsilon}\ e^{i(k+k')r} \qquad (5.6)$$

The solution to this equation can be written on the form

$$f_+(k,r) = e^{ikr} + \eta \frac{1}{\Delta(r)} \int_{-\infty}^{\infty} Y(k,k';r)\, e^{ik'r}\, dk' \qquad (5.7)$$

where

$$\Delta(r) = \det\,(1-\eta K) \equiv \exp(\mathrm{Tr}\,\ell n\,(1-\eta K)) \qquad (5.8)$$

is a real function. In the Fredholm theory, $\Delta(r)$ and $Y(k,k';r)$ can be represented as everywhere convergent power series expansions in η

$$\Delta(r) = \sum_{n=0}^{\infty} \eta^n \Delta_n(r)\ ;\ \Delta_0(r) = 1$$

$$Y(k,k';r) = \sum_{n=0}^{\infty} \eta^n Y_n(k,k';r)\ ;\ Y_0(k,k',r) = K(k,k';r) \qquad (5.9)$$

Inserting these expansions in eqs (5.5) and (5.7) and equating powers of η leads to the recursion relation

$$Y_n(k,k';r) = \Delta_n(r)\,K(k,k';r) + \int_{-\infty}^{\infty} K(k,k'';r)\,Y_{n-1}(k'',k';r)dk'' \qquad (5.10)$$

One more recursion relation will of course be needed. The standard Fredholm choice has not turned out convenient for the subsequent manipulations but an alternative recursion relation can be obtained by differentiating the expression (5.8) for $\Delta(r)$ with respect to r. One gets

$$\frac{d}{dr}\Delta(r) = -\,\eta\,\Delta(r)\,\mathrm{Tr}((1-\eta K)^{-1}\frac{dK}{dr})$$

$$= \eta\,\Delta(r)\,\frac{1}{2\pi}\int_{-\infty}^{\infty} dk'\,(S(k')-1)\,e^{ik'r}\,f_+(k',r) \qquad (5.11)$$

and the recursion relation

$$\frac{d}{dr}\,\Delta_n(r) = \Delta_{n-1}(r)\,\frac{1}{2\pi}\int_{-\infty}^{\infty} dk\,(S(k)-1)\,e^{2ikr}$$

$$+\frac{1}{2\pi}\int_{-\infty}^{\infty} dk\,(S(k)-1)\,e^{ikr}\int_{-\infty}^{\infty} dk'\,Y_{n-2}(k,k';r)e^{ik'r} \qquad (5.12)$$

(For n=1, the last term is absent). The use of this recursion relation is probably the most non-trivial part of the present method.

The interesting point is now that the two recursion relations (5.10) and (5.12) can conveniently be transformed into the sideways Laplace transform space introduced earlier. Indeed, if we introduce

$$\hat{Y}_n(k,q) = \frac{2}{\pi}\int_0^\infty ds\; e^{i(2q-k)r} \int_{-\infty}^{\infty} Y_n(k,k';r)e^{ik'r}\,dk'$$

$$\hat{\Delta}_n(q) = \frac{2}{\pi}\int_0^\infty dr\;\Delta_n(r)\;e^{2iqr} = -\frac{2}{\pi}\int_0^\infty dr\;\frac{e^{2iqr}-1}{2iq}\,\frac{d}{dr}\Delta_n(r) \tag{5.13}$$

(note that $\Delta_n(r) \to 0$, $r \to \infty$, for $n \neq 0$), the relations (5.10) and (5.12) take the simple forms

$$\hat{Y}_n(k,q) = -\frac{1}{2\pi i}\int_{-\infty}^{\infty} dk'\;\frac{S(k')-1}{k+k'+i\epsilon}\;\{\hat{\Delta}_n(k'+q) + \hat{Y}_{n-1}(k',k'\} +$$

$$\hat{\Delta}_n(q) = \tag{5.14}$$

$$= -\frac{1}{2\pi i}\int_{-\infty}^{\infty} dk'(S(k')-1)\;\frac{1}{2q}\;(\hat{Y}_{n-2}(k',k'+q) - \hat{Y}_{n-2}(k',k'+q) +$$

$$+ \hat{\Delta}_{n-1}(q+k') - \hat{\Delta}_{n-1}(k'))$$

where

$$\hat{\Delta}_0(q) = \frac{i}{\pi(q+i\epsilon)}$$

$$\hat{Y}_0(k,q) = -\frac{1}{2\pi^2}\int_{-\infty}^{\infty} \frac{S(k')-1}{(k+k'+i\epsilon)}\;\frac{dk'}{(k'+q+i\epsilon)} \tag{5.15}$$

It finally remains to relate $\hat{Y}(k,q) = \sum_{n=0}^{\infty} \hat{Y}_n(k,q)$ and $\hat{\Delta}(q) = \sum_{n=0}^{\infty} \hat{\Delta}_n(q)$ to the momentum space wave function, the main obstacle being $\Delta(r)$ in the denominator of Eq.(5.7). Let

$$\Delta^{-1}(k,k') = \frac{2}{\pi}\int_0^\infty \sin kr\,\sin k'r\,\frac{dr}{\Delta(r)} \tag{5.16}$$

Then $\Delta^{-1}(k,k')$ is the inverse of the matrix

$$\Delta(k,k') = \frac{2}{\pi}\int_0^\infty \sin kr\,\sin k'r\;\Delta(r)\,dr$$

$$= \mathrm{Im}\,[\frac{1}{2i}\,(\hat{\Delta}(\tfrac{1}{2}(k+k')) - \hat{\Delta}(\tfrac{1}{2}(k-k')))] \tag{5.17}$$

and

$$\frac{1}{\Delta(r)}\,\delta(r-r') = \frac{2}{\pi}\int_0^\infty dk\int_0^\infty dk'\;\sin kr\,\sin k'r'\;\Delta^{-1}(k,k')$$

$$= -\frac{1}{2\pi}\int_{-\infty}^{\infty} dk\int_{-\infty}^{\infty} dk'\;\Delta^{-1}(k,k')\;e^{ikr}\;e^{ik'r'} \tag{5.18}$$

Equation (5.7) therefore corresponds to the transformed equation

$$\hat{f}(k,q) = \frac{i}{\pi(q+i\epsilon)} + \frac{2}{\pi}\int_0^\infty dr\, e^{-ikr} \int_{-\infty}^{\infty} dk'\, Y(k,k';r) e^{ik'r} \int_{-\infty}^{\infty} dr'\, e^{2iqr'} \frac{\delta(r-r')}{\Delta(r)}$$

$$= \frac{i}{\pi(q+i\epsilon)} + 2\int_{-\infty}^{\infty} d\rho\, \hat{Y}(k',\rho)\, \Delta^{-1}(2\rho,2q) \qquad (5.19)$$

We can therefore conclude that this last method from practical point of view is no more difficult to use than the simple iterative method. The main (but inessential) complication is the need to invert the (real) matrix $\Delta(k,k')$ in order to evaluate $\hat{f}(k,q)$ according to eq. (5.19).

In addition to the general convergence properties of the Fredholm method, it has the interesting property that for the simple model S-matrix corresponding to $\mathcal{J}_+(k) = \frac{k+i\alpha}{k+i\beta}$ (the simplest Bargmann case), the η-series terminates, with only Δ_0, Δ_1 and Y_0 contributing.

6 . CONCLUSIONS

We have in the previous sections tried to emphasize three points

(i) Three-body multiple scattering equations require as two-body input not potentials but two-body wave functions

(ii) For local potentials in the class $\int_0^\infty |\mathcal{V}(r)| r^\alpha dr < \infty$, $\alpha = 1,2$, analyticity is an excellent tool for solving the inverse problem of the Schrodinger equation

(iii) Since scattering data are given in momentum space, the inverse method should also be formulated in momentum space. Practical momentum space methods are available in the ℓ=0 case.

APPENDIX I

In this appendix we are going to derive the integral equation (4.19). Consider the Cauchy integral

$$\frac{f_+(k,r)e^{-ikr}}{\mathcal{J}_+(k)} - 1 = \frac{1}{2\pi i} \int_C \frac{f_+(k',r)}{\mathcal{J}_+(k')} e^{-ik'r} - 1) \frac{dk'}{k'-k-i\epsilon} \qquad \text{(AI.1)}$$

Where C is the integration contour running from $k' = -\Lambda$ to $k' = \Lambda$ along the real axis, and back to $k' = -\Lambda$ along a semicircle in the upper half k'-plane.

There is no contribution to the integral from the semicircle part of C when $\Lambda \to \infty$. Following the proceedure used in section 4, we now change variables, $k' \to -k'$, and use the relation (4.8) in the form

$$\frac{f_+(-k,r)}{\mathcal{J}_+(-k)} = \frac{f_-(k,r)}{\mathcal{J}_-(k)} = \frac{f_+(k,r)}{\mathcal{J}_+(k)} - 2ik \frac{\varphi(k,r)}{|\mathcal{J}(k)|^2} \qquad \text{(AI.2)}$$

to obtain for the right hand side of (AI.1)

$$\frac{1}{2\pi i} \int_{-\infty}^{\infty} [(\frac{f_+(k',r)}{\mathcal{J}_+(k')} - 2ik' \frac{\varphi(k',r)}{|\mathcal{J}(k')|^2}) e^{ik'r} - 1)] \frac{dk'}{-(k+k'+i\epsilon)}$$

$$= \frac{1}{\pi} \int_{-\infty}^{\infty} (\frac{1}{|\mathcal{J}(k')|^2} - 1) e^{ik'r} \varphi(k',r) \frac{k'dk'}{k+k'+i\epsilon} \qquad \text{(AI.3)}$$

$$+ \frac{1}{2\pi i} \int_{-\infty}^{\infty} [(\frac{f_+(k',r)}{\mathcal{J}_+(k')} - 2ik' \varphi(k',r)) e^{ik'r} - 1] \frac{dk'}{-(k+k'+i\epsilon)}$$

In the very last integral, the contour can be changed to C, and we can then conclude that the value of the integral is zero. Making use of (4.12) we are therefore left with

$$\frac{f_+(k,r)}{\mathcal{J}_+(k)} - i\, k\varphi(k,r) \, (\frac{1}{|\mathcal{J}_+(k)|^2} - 1) =$$

$$= e^{ikr} + \frac{1}{\pi} \mathcal{P}\int_{-\infty}^{\infty} (\frac{1}{|\mathcal{J}(k')|^2} - 1) e^{i(k+k')r} \varphi(k',r) \frac{k'dk'}{k+k'} \qquad \text{(AI.4)}$$

Equation (4.19) is finally obtained by taking imaginary part of (AI.4).

APPENDIX II

In this appendix we are going to demonstrate that the simple method applied in section 4 for finding solutions to the inverse problem can be considerably generalized. Consider the Schrodinger equation

$$[-\frac{d^2}{dr^2} + \frac{\ell(\ell+1)}{r^2} + \mathcal{V}(r)]\ \psi_\ell(k,r) = k^2\ \psi_\ell(k,r) \qquad \text{(AII.1)}$$

where we take ℓ to be any non-negative integer. We will still assume that $\int_0^\infty |\mathcal{V}(r)|r^\alpha\, dr < \infty$, $\alpha = 1,2$. From Newton's book we collect the following properties of the solution to (AII.1).

The regular solution $\varphi_\ell(k,r)$ to (AII.I) is defined by the condition

$$r \to 0\ :\ \varphi_\ell(k,r) \to r^{-\ell-1} \qquad \text{(AII.2)}$$

For fixed ℓ and r, $\varphi_\ell(k,r)$ is an entire function of k^2 that behaves at infinity according to

$$\varphi_\ell(k,r) - (2\ell+1)!!\ k^{-\ell-1}\ \sin(kr-\tfrac{1}{2}\ \ell\pi) = o(|k|^{-\ell-1}\ e^{|\mathrm{Im}k|r}),\ |k| \to \infty \qquad \text{(AII.3)}$$

The Jost solutions $f_{\ell\pm}(k,r)$ are defined by the conditions

$$r \to \infty : f_{\ell\pm}(k,r) \to e^{\pm ikr} \qquad \text{(AII.4)}$$

$f_{\ell+}(k,r)$ is a regular analytic function in $\mathrm{Im}\ k > 0$, continuous with a continuous k-derivative down to $\mathrm{Im}\ k = 0$, except at $k = 0$ where there is a pole $k^{-\ell}$. The behaviour at infinity is

$$f_{\ell+}(k,r) - e^{ikr} = o(e^{-|\mathrm{Im}k|r}),\ |k| \to \infty,\ \mathrm{Im}\ k \geq 0 \qquad \text{(AII.5)}$$

The Jost functions are defined as

$$\mathcal{J}_{\ell+}(k) = k^\ell\ e^{\frac{-i\pi\ell}{2}}\ [(2\ell-1)!!]^{-1}\ \lim_{r\to 0} r^\ell\ f_{\ell+}(k,r) \qquad \text{(AII.6)}$$

and similarly for $\mathcal{J}_{\ell-}(k)$.

The physical solution $\psi_\ell^+(k,r)$ is related to these solutions through

$$\psi_\ell^+(k,r) = \frac{k^{\ell+1}}{(2\ell+1)!!}\ \frac{\varphi_\ell(k,r)}{\mathcal{J}_{\ell+}(k)}$$
$$= \frac{1}{2i}\ e^{\frac{i\pi\ell}{2}}\ [(-1)^\ell\ S_\ell(k)\ f_{\ell+}(k,r) - f_{\ell-}(k,r)] \qquad \text{(AII.7)}$$

In addition to Equation (AII.1) we will consider the same equation with $\mathcal{V}(r) \to \hat{\mathcal{V}}(r)$, and all entities related to $\hat{\mathcal{V}}(r)$ will carry a "hat" (this is <u>not</u> the same "hat" as in section 5).

Consider now the Cauchy integral

$$\frac{1}{i\pi} \int_C dk' \, f_{\ell+}(k',r) \int_r^\infty \hat{\psi}_\ell^+(k',r') \, \hat{f}_{\ell+}(k,r') \, dr' \qquad \text{(AII.8)}$$

where C is the same contour as in section 4 and appendix I. Using the Schrodinger equation it is easily verified that

$$\int_r^\infty \hat{\psi}_\ell^+(k',r') \, \hat{f}_{\ell+}(k,r') \, dr' = \frac{1}{k'^2-(k+i\epsilon)^2} \left[\frac{d}{dr} \hat{\psi}_\ell^+(k',r) \, \hat{f}_{\ell+}(k,r) - \hat{\psi}_\ell^+(k',r) \frac{d}{dr} \hat{f}_{\ell+}(k,r)\right] \qquad \text{(AII.9)}$$

This expression is a regular analytic function in $\mathrm{Im}\, k' > 0$, except for the pole at $k' = k+i\epsilon$ and for possible bound state poles of $\hat{\psi}_\ell^+(k',r)$. Moreover, the behaviour of (AII.9) at $k' = 0$ is such as to cancel the pole of $f_+(k',r)$ (in AII.8) at this point.

The contribution from the pole at $k' = k+i\epsilon$ to (AII.8) is simply $e^{\frac{i\pi\ell}{2}} f_{\ell+}(k,r)$ (the parenthesis in (AII9) reduces to a Wronskian at this point).

Let us for simplicity of notation assume that $\mathcal{V}(r)$ and $\hat{\mathcal{V}}(r)$ both have exactly one bound state in the partial wave of interest. In such a case there is a contribution to (AII.8) from the pole of $\hat{\psi}_\ell^+(k,r)$ at $k = \hat{k}_0 = i\,|\hat{k}_0|$

$$2f_{\ell+}(\hat{k}_0,r) \int_r^\infty \frac{\hat{k}_0^{\ell+1}}{(2\ell+1)!!} \frac{\hat{\varphi}_\ell(\hat{k}_0,r')}{c(\hat{k}_0)} \hat{f}_{\ell+}(k,r') \, dr' \qquad \text{(AII.10)}$$

where $\hat{\mathcal{F}}_\ell(\hat{k}_0) = 0$ and $c(\hat{k}_0) \equiv \frac{d}{dk} \hat{\mathcal{F}}_\ell(k)\Big|_{k=\hat{k}_0}$.

We now proceed to investigate the contributions to (AII.8) coming from the integration along the real axis

$$\frac{1}{i\pi} \int_{-\Lambda}^{\Lambda} dk' \, f_{\ell+}(k',r) \int_r^\infty \frac{1}{2i} \, e^{\frac{i\pi\ell}{2}} \left[(-1)^\ell \hat{S}_\ell(k') \, \hat{f}_{\ell+}(k',r') - \hat{f}_{\ell-}(k',r')\right] \hat{f}_{\ell+}(k',r') dr' \qquad \text{(AII.11)}$$

In the $f_{\ell+} \hat{f}_{\ell-}$ part of this expression we make the change of variables $k' \to -k'$, and replace $f_{\ell+}(-k',r) = f_{\ell-}(k',r)$ by an expression following from the second of the relations (AII.7)

$$\frac{1}{i\pi}\int_{-\Lambda}^{\Lambda} dk'\ f_{\ell+}(k',r)\ \frac{1}{2i}\ e^{\frac{i\pi\ell}{2}}\ (-1)^{\ell}\ [\hat{S}_{\ell}(k') - S_{\ell}(k')]\int_{r}^{\infty} \hat{f}_{\ell+}(k',r')\ \hat{f}_{\ell+}(k,r')\ dr'$$

$$+\frac{1}{i\pi}\int_{-\Lambda}^{\Lambda} dk'\ \psi_{\ell}^{+}(k',r)\int_{r}^{\infty} \hat{f}_{\ell+}(k',r')\ \hat{f}_{\ell+}(k,r')\ dr' \qquad \text{(AII.12)}$$

If the contour of integration in this last integral had been C, then its value would have been (bound state pole at $k = k_0 = i|k_0|$)

$$2\ \frac{k_0^{\ell+1}}{(2\ell+1)!!}\ \frac{\varphi_{\ell}(k_0,r)}{c\,(k_0)}\int_{r}^{\infty} \hat{f}_{\ell+}(k_0,r')\ \hat{f}_{\ell+}(k,r')\ dr' \qquad \text{(AII.13)}$$

Since it is not, we have to subtract from (AII.13) the contribution from the semi-circle part of C. Combining this with the semicircle contribution from (AII.8) we get

$$\frac{1}{i\pi}\int\Big(f_{\ell+}(k',r)\int_{r}^{\infty} \hat{\psi}_{\ell}^{+}(k',r')\ \hat{f}_{\ell+}(k,r')dr' - \psi_{\ell}^{+}(k',r)\int_{r}^{\infty} \hat{f}_{\ell+}(k',r')\ \hat{f}_{\ell+}(k,r')dr'\Big)dk' \qquad \text{(AII.14)}$$

In the limit $\Lambda \to \infty$, the value of AII.14 is easily verified to be $e^{\frac{i\pi\ell}{2}}\ \hat{f}_{\ell+}(k,r)$.

Collecting results, we therefore have

$$f_{\ell+}(k,r) = \hat{f}_{\ell+}(k,r) + \frac{1}{2\pi}\int_{-\infty}^{\infty} dk'\ f_{\ell+}(k',r)(-1)^{\ell}[S_{\ell}(k')-\hat{S}_{\ell}(k')]$$
$$\int_{r}^{\infty} \hat{f}_{\ell+}(k',r')\ \hat{f}_{\ell+}(k,r')\ dr'$$
$$+\ e^{\frac{-i\pi\ell}{2}}\ \{-(\text{AII.10}) + (\text{AII.13})\} \qquad \text{(AII.15)}$$

Using the fact that $\hat{\varphi}_{\ell}(\hat{k}_0,r') \propto \hat{f}_{\ell+}(\hat{k}_0,r')$ and $\varphi_{\ell}(k_0,r) \propto \hat{f}_{\ell+}(k_0,r)$ (see Newton, eq. (12.150)), the bound state contributions can be incorporated in the following way

$$f_{\ell+}(k,r) = \hat{f}_{\ell+}(k,r) + \frac{1}{2\pi}\int_{-\infty}^{\infty} dk'\ f_{\ell+}(k',r)(-1)^{\ell}[S_{\ell}(k') - S_{\ell}^{B}(k') - (\hat{S}_{\ell}(k') - \hat{S}_{\ell}^{B}(k'))]$$

$$+\int_{r}^{\infty} \hat{f}_{\ell+}(k',r')\ \hat{f}_{\ell+}(k,r')\ dr' \qquad \text{(AII.16)}$$

where $S_{\ell}^{B}(k')$ and $\hat{S}_{\ell}^{B}(k')$ are suitably chosen functions (with poles at $k = k_0$ and $k = \hat{k}_0$ etc)

Equation (AII.16) is our final expression. It reduces to eq (4.17) upon taking $\ell=0$, $\hat{\mathcal{V}}(r) = 0$ and $S^B(k') = 1$. It can be taken as starting point for the conventional analysis of the inverse problem (bound state ambiguity etc).

So far, the momentum space formulation of section 5 has not been generalized to the case $\ell \neq 0$. On the other hand, if $\hat{\mathcal{V}}(r)$ is chosen in such a way that $\hat{S}_\ell(k')$ is close to the experimental $S_\ell(k')$ (e.g. corresponds to a suitably chosen Bargmann potential), one might hope that the iterative series solution to (AII.16) is rapidly converging. In such a case a practical momentum space formulation could be obtained by transforming each term in this series into momentum space in such a way that the r-integration (which only involves analytically known functions) is carried out before the k'-integration.

APPENDIX III

In this appendix we are going to investigate how the solution to the basic integral equation (4.17) can be simplified for the case of Yukawa type interactions, i.e. when $\mathcal{V}(r) = \int_m^\infty C(\mu)\, e^{-\mu r}\, d\mu$. In this case it can be shown that $f(k,r)$ is regular and analytic in the whole complex k-plane except for the half-axis $k \in [-i\infty, -i\frac{m}{2}]$. In such a case, the contour of integration in (4.17) can be shifted to the positive imaginary axis. If $2i\, \nu(\xi)\, \theta(\xi - m/2)$, $k = i\xi$, is the discontinuity of $S(k)$, (4.17) takes the form (we also take $k = i\xi$)

$$f_+(i\xi,r) = e^{-\xi r} - \frac{1}{\pi}\int_{\frac{m}{2}}^{\infty} \frac{\nu(\xi')}{\xi+\xi'}\, e^{-(\xi+\xi')r}\, f_+(i\xi',r)\, d\xi' \qquad \text{(AIII.1)}$$

This is a real integral equation for $f(i\xi,r)$ in terms of the discontinuity of $S(k)$ along the positive imaginary axis (this equation is closely related to Martin's inverse methods, see De Alfaro-Regge book). Its usefulness in the inverse problem hinges on whether

1) one can get back to $f_+(k,r)$ from $f_+(i\xi,r)$. But this is simple since one can chose $\xi = -ik$ in (AIII.1)

More troublesome is the problem to

2) construct $\nu(\xi)$ from $S(k)$, since this step involves the continuation of an analytic function to a cut. Even with stabilizers this is very difficult to carry out in a stable way.

Equation (AIII.2) can be further elaborated upon by using the representation

$$f_+(k,r)\, e^{-ikr} = 1 + \int_m^\infty \rho(k,\mu) e^{-\mu r}\, d\mu \qquad \text{(AIII.2)}$$

In terms of $\rho(k,\mu)$, (AIII.1) has the interesting form

$$\rho(i\xi,\mu) = -\frac{1}{2\pi}\,\frac{\nu(\frac{1}{2}\mu)}{\xi+\frac{1}{2}\mu}\,\theta(\mu-m)$$

$$-\frac{1}{2\pi}\int_m^{\infty} d\mu'\,\frac{\nu(\frac{1}{2}\mu')}{\xi+\frac{1}{2}\mu'}\,\rho(i\tfrac{1}{2}\mu',\mu-\mu')\,\theta(\mu-\mu'-m) \qquad \text{(AIII.3)}$$

For any fixed value of μ, the iterative series of (AIII.3) terminates after a finite number of terms (the number of terms increases linearly with μ).

REFERENCES

The literature on the inverse problem of scattering is extensive and we do not intend to give references to the original literature here, with a few exceptions. General references are

L.D. FADDEEV, Usp. Mat. Nauk 14, 57 (1959) (Translation J. Math. Phys. 4, 72 (1963))

V. De ALFARO and T. REGGE, Potential Scattering (North-Holland, Amsterdam, 1965), Chap. 12.

R.G. NEWTON, Scattering Theory of Particles and Waves, (Mc. Graw-Hill, New-York, 1966)

The unitarity constraints discussed in section 3 were first investigated by

M. BARANGER, B. GIRAUD, S.K. MUKHOPADHYAY and P.U. SAUER, Nucl. Phys. A138, 1 (1969)

A survey of the inverse problem for separable potentials is given in

K. CHADAN, in Mathematics of profile inversion, (NASA-Ames Research Center, Moffett Field, Calif. July 12-16, 1971)

The equation (4.17) and the momentum space methods in section 5 appeared in

B.R. KARLSSON, Phys. Rev. D 10, 1985 (1974).

Note that Eqns (6.16) and (6.18) of this reference are incorrect, and should be replaced by Eqn (5.19) of this paper.

UTILISATION DES GROUPES DE TRANSFORMATION POUR LA RESOLUTION DES EQUATIONS AUX DERIVEES PARTIELLES

J.R. BURGAN, M.R. FEIX, E. FIJALKOW,*
*J. GUTIERREZ,** A. MUNIER*
CRPE/CNRS, UNIVERSITE D'ORLEANS

INTRODUCTION

C'est le mathématicien Norvégien Sophus Lie qui dans la seconde moitie du 19e siècle introduit les groupes de transformation dans l'optique d'une résolution des équations differentielles. Toutefois, l'outil mathématique deviens rapidement un sujet d'études en soi et le lien avec les équations différentielles s'affaiblit.

Par contre, prolongeant des idées dues initialement à Boltzmann les physiciens développent le concept de solutions auto semblables qui s'apparente par certains aspects à l'analyse dimensionnelle dont on trouvera dans Sedov |2| un bon exposé. Malheureusement la méthode ne marche que rarement et corresponds à des solutions très "pauvres".

Eisenhart |3| puis Ovsjannikov |4| reprennent alors les travaux de Lie en réintroduisant le but initial (résoudre les équations). Le succès reste toutefois toujours très limité.

Le but de cet article est de proposer une nouvelle philosophie de l'emploi des groupes permettant de sortir quelque peu de l'impasse où l'on se trouve généralement. C'est d'une part en réduisant nos ambitions initiales et d'autre part en introduisant de nouveaux points de vue (en particulier le point de vue numérique) que nous pensons pouvoir dégager de nouvelles voies.

* U.E.R. Sciences
** Université d'Alcala, Madrid

Nous ne tenterons en aucune manière une présentation systématique de l'utilisation des groupes et encore moins une synthèse des idées nouvelles que nous présentons et qui d'ailleurs reste à faire. Nous nous contentons, à l'aide d'exemples pris dans des domaines variés de la physique mathématique de montrer les possibilités de la méthode en espérant inciter les physiciens dans ces domaines et dans d'autres, d'approfondir ces nouveaux concepts.

UTILISATION DES GROUPES ET SOLUTION GENERALE

Considérons une fonction

$u(\vec{x}) = u\ (x_1, x_2 \ldots x_n)$ dépendant de n variables

et une famille de m parametres réels

$$\vec{a} = \{a_j\} = \{a_1 \ldots . a_m\}$$

On envisage alors la transformation ponctuelle

(1) $$\bar{x}_i = \phi_i\ (\vec{x}, \vec{a})$$

$$x_i = \bar{\phi}_i\ (\vec{\bar{x}}, \vec{a})$$

L'équation aux dérivées partielles à laquelle satisfait

(2) $$L(u, p, \vec{x}) = 0$$

ou p désigne la famille des dérivées allant jusqu'à l'ordre k. Si l'on introduit dans (2) la transformation (1) il vient

(3) $$\bar{L}(\bar{u}, \bar{p}, \bar{x}) = L(u, p, x)$$

La dépendance fonctionnelle $\bar{L}$ est en général différente de L. Si cependant ces deux dépendances sont identiques on dit que l'équation est invariante.

De plus si on sait trouver une fonction continue f(x) tel que pour tout paramètre a on ait

$$f(x) = f(\overline{x})$$

On dira que f est invariant. Le résultat non trivial que nous ne démontrerons pas, voir |4|, est que toute solution particulière invariante de (2) est une fonction particulière des invariants du groupe. Une stratégie à priori attractive semble donc consister à :

- trouver les groupes qui laissent (2) invariante
- en déduire un certain nombre d'invariants $I_1(\vec{x}) \ldots . I_l(\vec{x})$
- utiliser ces invariants pour obtenir un nouveau système différentiel comportant un nombre inférieur de variables indépendantes, donc plus aisé à résoudre.

En particulier si on peut trouver n-1 invariants et k fonctions arbitraires de ces n-1 invariants on aura résolu l'équation (2) dans sa généralité.

Malheureusement cette méthode se révèle trop formelle et trop générale et mets en jeu des équations qui se révèlent encore plus complexes à résoudre que le système (2) lui même.

Ce point est fondamental et va nous obliger pratiquement à nous contenter de trouver quelques invariants ce qui nous permettra de faire décroitre (ordinairement d'une ou deux unités) le nombre de variables indépendantes, de résoudre la nouvelle équation (éventuellement par des méthodes numériques).

Tout cela au prix d'une particularisation des conditions initiales qui ne pourront plus être arbitraires mais qui seront imposées par notre choix du groupe.

La différence entre un calcul "classique" et un calcul à l'aide des groupes est indiquée sur la Fig. 1 où l'on considère un système évoluant dans le temps, caractérisé par la donnée des conditions initiales. Le groupe sélectionnera les conditions initiales qu'il veut bien traiter.

Ces conditions peuvent être :

- Triviales (c'est à dire qu'on se ramène à des conditions initiales sans intérêt).

- Non physiques (par exemple amenant à considérer des grandeurs physiques divergentes sur les limites).

- Singulières (c'est à dire amenant un comportement limite et particulier du système. Ce peut être un cas intéressant (par exemple l'obtention des solitons)).

On voit par conséquent qu'il y a intérêt à choisir le groupe le plus large possible ,c'est à dire comportant le plus de paramètre. Malheureusement les calculs sont inextricables et on est amené à se contenter de groupe à un paramètre. Le groupe infinitesimal |4| qui fournit en principe la classe maximale pour les groupes à un paramètre donne encore lieu à des calculs d'une lourdeur considérable et à des systèmes d'équations ordinairement non résolvables.

LES GROUPES AUTO SEMBLABLES |5| |6|

Il est alors curieux de constater qu'il peut exister toutefois des groupes extrêmement simples où tous les calculs deviennent triviaux et qui bien entendu ne fournissent que des résultats très "pauvres " (mais souvent guère plus pauvres que les groupes infinitésimaux). Ce sont les groupes auto semblables obtenus par des méthodes d'analyse dimensionnelle. Nous allons illustrer leur emploi sur un exemple.

Soit à résoudre

(4) $$\frac{\partial u}{\partial t} + \left(\frac{\partial u}{\partial x} \right)^2 = 1$$

On effectue la transformation

(5) $$\bar{t} = a^{\alpha} t \qquad \bar{x} = a^{\beta} x \quad \text{et} \quad \bar{u} = a^{\gamma} u$$

élément d'un groupe à un paramètre caractérisé par a. L'équation (4) deviens bien sûr

(6) $$a^{\alpha-\gamma} \frac{\partial \bar{u}}{\partial \bar{t}} + a^{2\beta-2\gamma} \left(\frac{\partial \bar{u}}{\partial \bar{x}} \right)^2 = 1$$

Ici l'invariance de (6) implique $\alpha = \gamma$ et $\beta = \gamma$ quant aux invariants ils sont obtenus en remarquant que

$$\frac{\bar{t}^{\beta}}{\bar{x}^{\alpha}} = \frac{t^{\beta}}{x^{\alpha}} \quad \text{et} \quad \frac{\bar{u}^{\beta}}{\bar{x}^{\gamma}} = \frac{u^{\beta}}{x^{\gamma}}$$

On introduit alors la nouvelle variable $\xi = t/x$ puisqu'ici $\alpha = \beta$ et la nouvelle fonction $\Phi = u/x$ $(\gamma = \beta)$. On peut affirmer que ϕ est une fonction de la seule variable ξ ce que l'on vérifie en portant $u = x\ \Phi(\xi)$ et $t = x\xi$ dans (4) et en constatant l'élimination de x. Il vient alors pour l'équation donnant Φ

(7) $$\frac{d\Phi}{d\xi} + \left(\Phi - \xi \frac{d\Phi}{d\xi} \right)^2 = 1$$

Les solutions de (7) sont

$$\Phi_1 (1-\lambda^2)\ \xi + \lambda \quad \text{et} \quad \phi_2 = \xi + \frac{1}{4\xi}$$

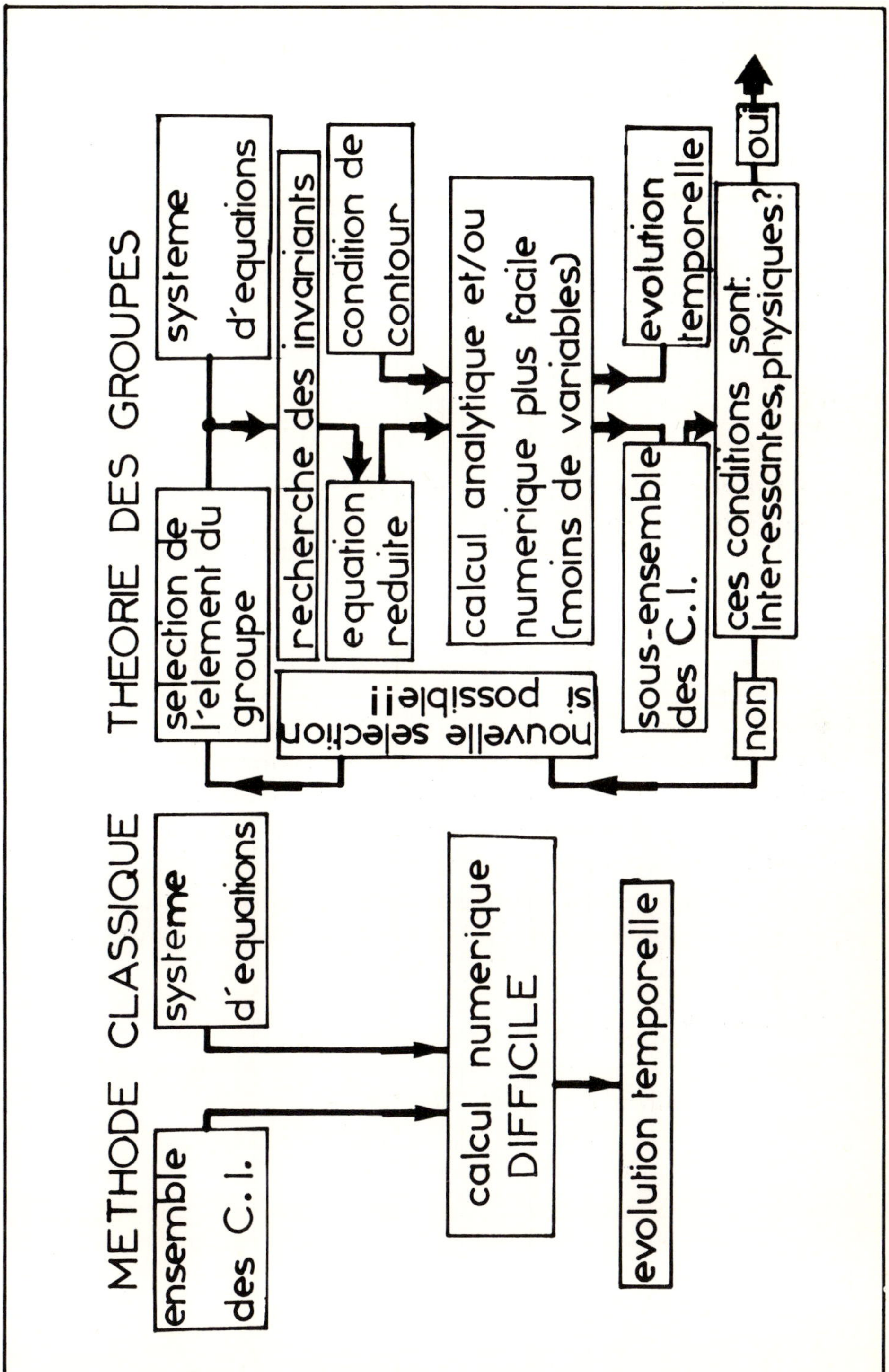
METHODE CLASSIQUE
ensemble des C.I.
systeme d'equations
calcul numerique DIFFICILE
evolution temporelle
THEORIE DES GROUPES
selection de l'element du groupe
systeme d'equations
recherche des invariants
equation reduite
condition de contour
nouvelle selection si possible!!
calcul analytique et/ou numerique plus facile (moins de variables)
sous-ensemble des C.I.
evolution temporelle
ces conditions sont interessantes, physiques?
oui
non

fig.1

Les solutions correspondantes $u(x,t)$ s'écrivent

$$(8) \qquad \begin{aligned} u_1(x,t) &= (1-\lambda^2)t + \lambda x \\ u_2(x,t) &= t + \frac{x^2}{4t} \end{aligned}$$

Si dans u_1 on fait $t = 0$ on obtiens une première série de conditions initiales possibles avec $u(x,0) = \lambda x$ ou λ est un paramètre arbitraire. Pour la solution u_2 on doit faire non pas $t = 0$ mais $t = T$ ou T est arbitraire et prendre $t = T$ comme nouvelle origine. On obtiens donc comme autres conditions initiales possibles $u(x,0) = T + (x^2/4t)$.

L'équation (4) n'ayant pas l'ambition de représenter un phénomène physique nous ne pourrons pas décider si les conditions initiales sont ou non intéressantes.

NOUVEAUX CONCEPTS, NOUVELLE PHILOSOPHIE

L'utilisation des groupes que nous venons de décrire très succinctement souffre en fait de deux défauts majeurs.

Sous sa forme la plus générale elle est trop formelle et remplace des équations difficiles par d'autres équations en général encore plus difficiles.

Sous la forme "auto semblable" elle réussit la simplification souhaitée des calculs mais au prix d'un appauvrissement considérable des solutions.

Nous pensons que l'idée de transformer les variables et par conséquent les équations reste une idée intéressante à condition de remarquer

- que la nouvelle équation *n'a pas besoin d'être absolument identique à l'ancienne* mais qu'elle doit simplement impliquer une physique et une mathématique à laquelle nous sommes habitués et que nous savons manipuler.

- que nous pouvons avoir comme but simplement de simplifier l'aspect numérique de l'équation, c'est un but suffisant en lui même.

- que l'élimination d'une variable est bien sur souhaitable mais qu'elle ne doit pas être forcée mais survenir au contraire d'une manière naturelle. Ceci peut arriver par exemple dans un traitement asymptotique ou n'apparait qu'une combinaison de deux variables.

- qu'il est intéressant, à partir d'une équation donnée et soluble, de générer une famille de nouvelles équations également solubles.

Bref cette nouvelle philosophie sera davantage basée sur l'aspect physique et numérique que sur les propriétés d'invariance stricte de l'équation. Nous l'avons baptisé quasi invariance. Il est également commode de définir les notions suivantes.

Le *reconditionnement* concerne un problème ou l'introduction du groupe de transformation permets de simplifier les conditions aux limites ou de supprimer un terme génant.

La *renormalisation* porte sur une variable, ordinairement le temps dont le domaine d'extension s'étend primitivement à l'infini et que l'on ramène sur un intervalle fini à l'aide de la transformation. Il faut bien entendu que l'infinité dont on s'est débarrassé ne réaparaisse pas sous une autre forme, en un autre endroit.

Enfin la *conformité* traduit le fait que le nouveau problème exprimé dans le nouvel espace est de même nature, et fait appel à une physique très analogue.

L'EXEMPLE DE L'OSCILLATEUR HARMONIQUE (EQUATION DIFFERENTIELLE)

Il s'agit du problème très classique de résolution de

$$(9) \qquad \frac{d^2x}{dt^2} + \omega^2 (t)x = 0$$

nous allons chercher une transformation du temps (c'est à dire une fonction $\theta(t)$ et une transformation de x de la forme

$$(10) \qquad x = \xi(\theta) \quad C(t)$$

(10) est une formule curieuse puisque nous utilisons deux fonctions C du temps et ξ de θ (c'est à dire du nouveau temps) pour décrire x. C(t) est une fonction arbitraire qui est à notre disposition . $\xi(\theta)$ est la fonction inconnue pour laquelle nous allons obtenir une nouvelle équation différentielle. Bien entendu nous espérons qu'un choix judicieux de C(t) nous simplifiera le calcul de $\xi(\theta)$. Pour le moment portons (10) dans (9) en exprimant toutes les dérivés de ξ par rapport à θ et en faisant bien sur intervenir $d\theta/dt, d^2\theta/dt^2$. Il vient

$$(11) \qquad \frac{d^2x}{dt^2} + \omega^2(t)x = \frac{d^2\xi}{d\theta^2} \; C \left(\frac{d\theta}{dt}\right)^2 + \left\{ \frac{d^2C}{dt^2} + \omega^2(t)C \right\} \xi + \frac{d\xi}{d\theta} \left[2 \frac{d\theta}{dt} \frac{dC}{dt} + C \frac{d^2\theta}{dt^2} \right] = 0$$

Pour garder à l'équation en θ la même structure (principe de conformité) nous annullons le coefficient du terme en $d\xi/d\theta$ ce qui nous donne

$$(12) \qquad \frac{d\theta}{dt} = \frac{1}{C^2}$$

et l'équation deviens

(13) $$\frac{d^2\xi}{d\theta^2} + \left[\omega^2 C^4 + C^3 \frac{d^2C}{dt^2}\right] \xi = 0$$

Le nouveau système peut être interprété comme un oscillateur soumis à la force $-\omega^2 C^4 \xi$ que nous appelerons nouveau champs physique auquel vient s'ajouter la force $-C^3 d^2C/dt^2$ que nous appelerons champs de transformation puisqu'il est lié du changement d'espace temps. Ceci dit la "règle du jeu" deviens la suivante.

Sélectionner C(t) avec deux objectifs :

- Premièrement et *prioritairement* ne faire apparaître aucune infinité ni dans de nouveaux champs physiques, ni dans le champ de transformation.

- Deuxièmement renormaliser le temps, ou si cet objectif se révèle impossible, simplifier le traitement asymptotique en θ par compression maximum du temps.

Les deux objectifs sont en fait antithétiques. La renormalisation du temps implique que C(t) croisse plus vite que $t^{1/2}$ pour $t \to \infty$ ce qui risque d'introduire des infinités dans les nouveaux champs. Tout dépends alors de la forme asymptotique de $\omega^2(t)$. Prenons l'exemple suivant

(14) $$\omega^2(t) = \frac{\omega_o^{\;2}}{(1+\Omega t)^{4\alpha}} \qquad \text{avec } \alpha > 0$$

Le choix de C(t) s'effectue selon le tableau suivant

Tableau 1

α	$C(t)$	$\omega^2 C^4$	transf. field	$\theta(t)$
$0<\alpha<\frac{1}{2}$	$(1+\Omega t)^{\alpha}$	ω_o^2	$\frac{\alpha(1-\alpha)\Omega^2}{[1+(1-2\alpha)\Omega\theta]^2}\vec{\xi}$	$\Omega\theta=\frac{(1+\Omega t)^{1-2\alpha}-1}{1-2\alpha}$
$\alpha=\frac{1}{2}$	$(1+\Omega t)^{\frac{1}{2}}$	ω_o^2	$(\Omega^2/4)\vec{\xi}$	$\Omega\theta=\log(1+\Omega t)$
$\frac{1}{2}<\alpha<1$	$(1+\Omega t)^{\frac{1}{2}}$	$\omega_o^2 \exp-(4\alpha-2)\Omega\theta$	$(\Omega^2/4)\vec{\xi}$	$\Omega\theta=\log(1+\Omega t)$
$\alpha=1$	$1+\Omega t$	ω_o^2	0	$\Omega\theta=\Omega t/(1+\Omega t)$
$\alpha>1$	$1+\Omega t$	$\omega_o^2(1-\Omega\theta)^{4(\alpha-1)}$	0	*if* $t\to\infty$ $\Omega\theta\to 1$

qui s'explique de la manière suivante.

Si $0<\alpha<1/2$ le terme $\omega^2 C^4$ ne devant pas croitre indéfiniment, nous le gardons constant par le choix $C(t) = (1+\Omega t)^{\alpha}$. Le champ de transformation, lui, décroit avec θ.

Si $1/2 < \alpha < 1$ le choix $C(t) = (1+\Omega t)^{\alpha}$ permettrait bien une renormalisation du temps, malheureusement le champ de transformation $-C^3\, d^2C/dt^2\, \xi$ qui vaudrait $-\alpha(\alpha-1)\,(1+\Omega t)^{4\alpha-1}$ divergerait On s'en tiens donc à $C(t) = (1+\Omega t)^{1/2}$. On remarque que le nouveau temps est à la limite de la renormalisation (l'intégrale $\int_0^t C^2\, dt$ ne diverge que logarithmiquement). Plus précisément encore

(15) $$\Omega\theta = \mathrm{Log}\,(1+\Omega t)$$

Ce qui implique une "compression énergique" du temps.

Si $1 < \alpha$ il y a possibilité de prendre $C(t) = 1+\Omega t$. Le champ physique ne diverge pas et le champ de transformation est identiquement nul en même temps que d^2C/dt^2. Alors le temps peut être renormalisé avec

$$\theta = \frac{t}{1+\Omega t} \tag{16}$$

Maintenant de ce tableau nous pouvons déduire le comportement asymptotique de la solution.

Si $\alpha < 1/2$ il apparaît un cycle limite puisque lorsque $\theta \to \infty$ la nouvelle fréquence est une constante et le champ de transformation tends vers zéro. Cela corresponds à une expansion en $(1+\Omega t)^\alpha$ de l'amplitude dans l'espace x c'est à dire comme $\omega^{-1/2}$. Il s'agit donc tout simplement de l'approximation WKB qui reste valable quelque soit le temps.

Bien que la fréquence de rappel tende vers zéro la particule n'échappe jamais au champ de force et oscille avec une fréquence bien sûr décroissante.

Si $\alpha > 1$ le temps θ est renormalisé et pour $\Omega\theta = 1$ la particule tends vers un point limite ξ_ℓ sur l'espace ξ. Le comportement asymptotique de x est donc donné par $x = (1+\Omega t)\xi$ et nous voyons que le comportement est finalement complètement pris en charge par la transformation. C'était le but poursuivi et nous voyons que pour obtenir le comportement asymptotique il nous suffit de résoudre

$$\frac{d^2\xi}{d\theta^2} + \omega_o^2 \, (1-\Omega\theta)^{4(\alpha-1)}\xi = 0 \tag{17}$$

Sur *l'intervalle finie* $0 < \theta < \Omega^{-1}$ (17) ne présente aucune difficulté spéciale.

Le cas $1/2 < \alpha < 1$ est un peu plus complexe. Nous n'avons pas pu renormaliser le temps mais nous avons trouvé une transformation telle que la nouvelle fréquence de rappel tende vers zéro alors que le champ de transformation est caractérisé par une force de répulsion $(\Omega^2/4)\xi$. Donc pour un θ suffisamment élevé ce dernier l'emporte et le champ physique peut être négligé ce qui signifie que le mouvement asymptotique de la particule est un mouvement libre dans l'espace caractérisé par $x = Kt$ ou K est une constante. On peut également retrouver ce résultat en remarquant simplement que $\xi \sim A \exp \frac{\Omega\theta}{2}$ pour θ suffisamment grand ce qui donne bien $x = (1+\Omega t)^{1/2}\xi = A \exp \Omega\theta = A(1+\Omega t)$.

Nous voyons donc que physiquement le cas $1/2 < \alpha < 1$ ne se distingue pas du cas $1 < \alpha$ et corresponds à un régime asymptotique ou l'oscillateur échappe au champ de force. Par contre le traitement mathématique est différent suivant que α est supérieur ou inférieur à 1.

Nous résumerons ce paragraphe en disant que les transformations (10) et (12) nous ont fourni *le cadre idéal* pour l'étude des *solutions asymptotiques* de (9). Un autre point que nous ne développons pas ici est l'intérêt numérique de la transformation. Par exemple pour $\alpha < 1/2$ le traitement direct conduisait à utiliser une délicate méthode à pas variable. En particulier, si α est voisin de 1/2 la période croit énormément à chaque révolution, puisqu'en fait l'oscillation est à la limite du décrochement. La transformation ramène une fréquence d'oscillation tendant vers une constante avec un haut degré de "compression" du temps. Cette utilisation des transformations en vue de simplifier le numérique est certainement un des aspects les plus prometteurs de ce travail.

EXEMPLES D'EQUATIONS AUX DERIVEES PARTIELLES. LA CHALEUR ET SCHROEDINGER

Il nous reste à montrer maintenant que la méthode est utilisable pour des équations à plusieurs variables indépendantes. Des considérations que nous n'exposerons pas ici mais dont on pourra trouver les détails dans |7| conduisent à envisager des applications systématiques à tout système dérivant d'un Hamiltonien de la transformation du groupe noté $G(C^+)$. 1 élément du groupe est caractérisé par une fonction $C(t)$ définie sur l'intervalle temporelle $0 \to \infty$

telle que

(18) $\quad C(t) > 0 \quad \forall t$

Parfois on particularisera les conditions initiales de la manière suivante :

(18bis) $\quad C(0) = 1 \qquad \frac{dC}{dt}(0) = 0$

A partir de la fonction $C(t)$ on définit un nouveau temps et un nouvel espace des phases par les relations.

$$\theta = \int_0^t \frac{d\sigma}{C^2(\sigma)}$$

(19)
$$\xi_i = q_i/C(t)$$

$$d\xi_i/d\theta = C\,\frac{dq_i}{dt} - \frac{dC}{dt}\,q_i$$

Il est alors remarquable de constater que si l'on introduit la fonction $D(\theta) = C^{-1}(t)$.

- $D(\theta)$ est également un élément de $G(C^+)$

- Les relations permettant de passer du nouvel espace à l'ancien s'expriment par les formules duales

$$t = \int_0^\theta \frac{d\sigma}{D^2(\sigma)}$$

(20)
$$q_i = \xi_i/D(\theta)$$

$$\frac{dq_i}{dt} = D\,\frac{d\xi_i}{d\theta} - \frac{dD}{d\theta}\,\xi i$$

On voit donc que les espaces anciens $(t, q, dq/dt)$ et nouveau $(\theta, \xi, d\xi/d\theta)$ sont équivalents pour le formalisme de la mécanique analytique.

Enfin le choix du groupe $G(C^+)$ s'explique par le fait que si C s'annulait pour un temps t_0 il serait impossible de définir la transformation au delà de ce temps qui dans l'espace dual correspondrait à $\theta = \infty$. Le choix des conditions initiales (18 bis) implique qu'à l'origine $t = \theta = 0$ les deux espaces des phases sont identiques.

Considérons l'équation de Schroedinger unidimensionnelle qui, nous le savons, dérive d'un formalisme Hamiltonien La nouvelle variable sera donc $\xi = x/C(t)$, le nouveau temps sera donné par la 1e des équations (19). Enfin quelques calculs montrent qu'il est nécessaire de transformer ψ avec

$$(21) \qquad \psi\,(x, t) = C^{-1/2} \left[\exp \frac{i}{2}\, \frac{1}{C}\, \frac{dC}{dt}\, x^2\right] \overline{\psi}(\xi, \theta)$$

Dans ces conditions l'équation de Schroedinger prise pour simplifier les notations à une dimension (avec $\hbar = m = 1$)

$$(22) \qquad i\,\frac{\partial\psi}{\partial t} = -\,\frac{1}{2}\,\frac{\partial^2\psi}{\partial x^2} + V(x,t)\,\psi$$

deviens

$$(23) \qquad i\,\frac{\partial\overline{\psi}}{\partial\theta} = -\,\frac{1}{2}\,\frac{\partial^2\overline{\psi}}{\partial\xi^2} + \overline{V}\,\overline{\psi}$$

avec

$$(24) \qquad \overline{V} = VC^2 + \frac{1}{2}\,C^3\,\frac{d^2C}{dt^2}\,\xi^2$$

Nous retrouvons l'utilisation du concept de conformité dans le nouvel espace : il s'agit de la même équation de Schroedinger mais dotée d'un nouveau potentiel. On notera en passant l'identité entre (13) et (24). Les deux formules indiquent que le nouveau potentiel est la somme d'un nouveau potentiel physique, produit de l'ancien potentiel par C^2 et d'un potentiel de transformation $(1/2)\,C^3\,\frac{d^2C}{dt^2}\,\xi^2$.

Donnons maintenant deux exemples d'application du concept de conformité.

Supposons que nous traitions le problème de l'oscillateur harmonique quantique à fréquence variable, c'est à dire que $V = 1/2\ \omega^2\ (t)\ x^2$. Il vient

$$\overline{V} = \frac{\xi^2}{2}\ \left[C^4\ \omega^2 + C^3\ \frac{d^2C}{dt^2}\right] \tag{25}$$

Nous pouvons choisir C de telle sorte que $\overline{V}$ corresponde à un problème soluble par exemple à un oscillateur quantique à fréquence fixe ω_o. Il suffit de résoudre

$$\frac{d^2C}{dt^2} + \omega^2\ (t)\ C = \frac{\omega_o^2}{C^3} \tag{26}$$

Nous voyons donc que le problème de l'oscillateur quantique à fréquence variable se ramène au problème de l'oscillateur à fréquences fixe dont on connait la solution analytique à l'aide des polynomes d'Hermite et à une transformation qui nécessite la résolution de l'équation différentielle (26). On pourrait d'ailleurs penser à annuler ω_o dans (26) mais on ne serait pas certain que la solution appartienne au groupe $G(C^+)$. La présence d'un ω_o fini nous assure cette propriété quelque soit $\omega(t)$ et donc la possibilité d'obtenir la solution pour tout temps. L'intérêt de (26) avait d'ailleurs été déjà signalé pour ce genre de problème par LEWIS et RIESENFIELD |8|.

Un autre exemple concerne le problème d'une particule quantique piégée entre deux barrières dont l'une fixe est située à l'origine et dont l'autre mobile avec une vitesse uniforme occupe la position $X(t) = L+\mu t$. On supposera en outre que le potentiel est nul à l'intérieur des 2 barrières et infini à l'extérieur. On admets que la barrière mise en mouvement à l'instant $t = 0$ est arrêtée à l'instant T et on désignera par Ω le rapport u/L. La mise en marche et l'arrêt de la barrière aux instants 0 et T permets de considérer pour $t < 0$ et $t > T$ des puits de potentiel immobiles ou l'on peut introduire des états stationnaires comme base de la première. En particulier, on s'intéressera au cas ou l'état pour des temps négatifs est un état stationnaire de nombre quantique k c'est à dire caractérisé par la fonction d'onde $\psi_k = \sqrt{2/L} \sin \pi kx/L$.

Il est clair que ce problème est très lié au concept de variation adiabatique. Si en effet la vitesse de la barrière tends vers zéro il est bien connu que la fonction d'onde ψ à un instant t quelconque deviens (à un facteur de phase près).

$$\psi(x,t) = \sqrt{2/X(t)}\,(\sin \pi\, kx/X(t)) \tag{27}$$

(27) exprime simplement que le nombre d'onde est conservé et que la fonction d'onde se dilate d'une manière affine avec le puit de potentiel, $X(t)$ pouvant être très différent de L ; la seule condition est que le mouvement soit infiniment lent.

L'utilisation des transformations (19) et (21) permets de résoudre élégamment le problème en le transformant en un problème à barrière fixe. On prendra $C(t) = 1+\Omega t$ pour $0 < t < T$ et $C(t) = 1+\Omega T$ pour $t > T$. Le nouveau potentiel n'est pas affecté par la multiplication par C^2 (il reste 0 ou $+\infty$). De même le potentiel de transformation est nul sauf au temps $t = 0$ et $t = T$. On a

$$\frac{d^2C}{dt^2} = \Omega \quad |\delta(t) - \delta(t-T)| \tag{28}$$

où δ est la distribution de Dirac. En utilisant la relation $t = \theta/(1-\Omega\theta)$ on obtient le potentiel de transformation dans l'espace ξ,θ

$$V(\xi,\theta) = \frac{\Omega}{2}\,\xi^2\,\delta(\theta) - \frac{\Omega}{2}\,\xi^2\,\frac{1}{1-\Omega\,\Theta}\quad \delta(\theta-\Theta) \tag{29}$$

formule dans laquelle $\Theta = T/1+\Omega T$

Il reste à remarquer que si un potentiel de Dirac $K(\xi)\ \delta(\theta)$ est appliqué à un système, la fonction d'onde ψ^+ tout de suite après $(\theta \to 0+)$ est reliée à la fonction d'onde ψ^- $(\theta \to 0-)$ par la relation

$$(30) \qquad \psi^+(\xi) = \psi^-(\xi) \exp -iK(\xi)$$

(29) et (30) permettent ainsi de calculer les variations de $\psi(\xi,\theta)$ aux instants $\theta = 0$ et $\theta = \Theta$. Il faut remarquer que les formules finales sont loin d'être triviales. Par exemple la probabilité de retrouver après l'arrêt la particule dans l'état m si l'on suppose que l'état initial est k est

$$(31) \quad P(k\to m) = \Bigg| \frac{4}{L^2} \int_0^L d\eta \sin \frac{\pi m\eta}{L} \exp \frac{i}{2} \frac{\Omega\eta^2}{1-\Omega\Theta} \sum_{l=1}^{\infty} \sin \frac{\pi l\eta}{L} \exp - \frac{i}{2} \frac{\pi^2 l^2 \Theta}{L^2} \int_0^L d\lambda \sin \frac{\pi l\lambda}{L} \exp - \frac{i}{2} \Omega\lambda^2 \sin \frac{\pi k\lambda}{L} \Bigg|^2$$

La figure (2) montre l'exemple de l'invariance de $P(k \to k)$ pour une particule initialement sur le niveau $k = 3$. On a pris $L = 1$ et $\Omega = .1$. La vitesse de la particule "classique" correspondante serait $\pi k/L = 3\pi$ et le rapport de la vitesse de la barrière à celle de la particule est par conséquent environ 10^{-2}. $P(k \to k)$ est donné en fonction du temps renormalisé $\Omega\Theta$ et on se rappelera que la longueur du puits est multiplié par $(1-\Omega\Theta)^{-1}$. Comme la vitesse de la particule dans l'approximation adiabatique est inversement proportionnelle à la longueur on voit que pour $\Omega\Theta = .97$ la vitesse "classique" serait 0.3 fois celle de la barrière et l'approximation adiabatique n'est plus valable. C'est bien ce que montre la courbe.

Disons pour terminer ce paragraphe un mot de l'équation de la chaleur

$$(32) \qquad \frac{\partial \psi}{\partial t} = \chi \frac{\partial^2 \psi}{\partial x^2}$$

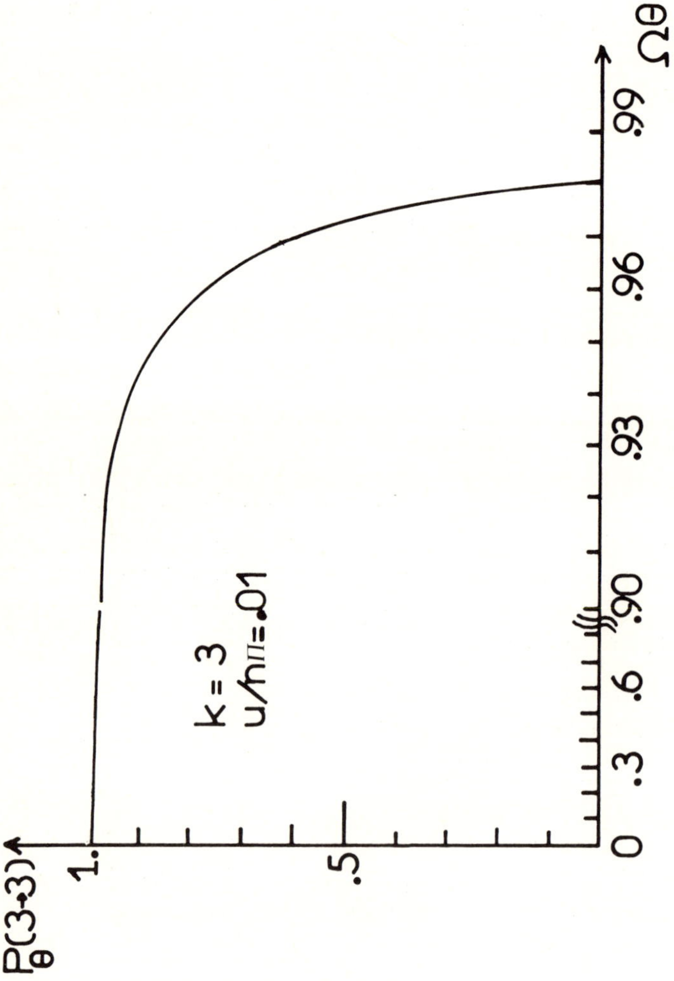

fig. 2

Bien que physiquement le phénomène de la diffusion de la chaleur soit très différent de ceux usuellement associé aux systèmes Hamiltoniens la similitude avec l'équation de Schroedinger invite à utiliser le même formalisme. On prendra de nouveau $C(t) = 1+\alpha t$ pour faire disparaître le terme de "champs de transformation" et l'équation de la chaleur restera donc invariante.

$$\frac{\partial \overline{\psi}}{\partial \theta} = \chi \frac{\partial^2 \overline{\psi}}{\partial \xi^2} \tag{33}$$

Le temps sera renormalisé avec $\alpha\theta = \alpha t/1+\alpha t$ et il suffira de résoudre (33) sur l'intervalle $0 - \alpha^{-1}$. L'analogue de la relation (21) qui relie ψ à $\overline{\psi}$ fait intervenir le facteur

$$\exp - \frac{\alpha x^2}{4\chi(1+\alpha t)} \tag{34}$$

Pour x et t tendant vers l'infini c'est le terme asymptotique bien connu $\exp - x^2/4xt$ que l'on obtient usuellement par l'intégration par la méthode du col des expressions

$$\begin{aligned} \psi(x,t) &= \int_{-\infty}^{\infty} \exp -ikx \exp -\chi k^2 t \ \ g(k)\, dk \\ g(k) &= (2\pi)^{-1} \int_{-\infty}^{\infty} \psi(x,o) \exp ikx\, dk \end{aligned} \tag{35}$$

On voit qu'ici comme pour l'exemple de l'oscillateur harmonique c'est le groupe de transformation qui prends à sa charge cet aspect plus ou moins délicat du traitement numérique. Ici nous avons une expression exacte sous réserve d'un traitement numérique sur un intervalle finie de (33).

EXEMPLE DANS L'ESPACE DES PHASES. LE PROBLEME GRAVITATIONNEL A N CORPS ET L'HYPOTHESE DE DIRAC

Reprenons les équations (19) d'une transformation à l'aide d'un élément de $G(C^+)$. Soit une particule i de fonction $\vec{x}_i$ soumise au champ d'accélération $\vec{E}_i$. L'équation du mouvement

$$\frac{d\vec{V}_i}{dt} = \vec{E}_i \tag{36}$$

Se transforme en

$$(37) \qquad \frac{d\vec{\eta}_i}{d\theta} = \vec{\varepsilon}_i$$

avec en recopiant (19)

$$(38) \qquad \theta = \int_o^t d\sigma/C^2(\sigma)$$

$$\vec{\xi}_i = \vec{x}_i/C(t) \qquad \vec{\eta}_i = d\vec{\xi}_i/d\eta = C(d\vec{x}_i/dt) - (dC/dt)\,\vec{x}_i$$

La relation liant $\vec{\varepsilon}_i$ et $\vec{E}_i$ s'écrit

$$(39) \qquad \vec{\varepsilon}_i = C^3\,\vec{E}_i - C^3\,\frac{d^2C}{dt^2}\,\vec{\xi}$$

On retrouve de nouveau la formule (24)

Si maintenant nous nous rappelons l'exemple de l'oscillateur harmonique nous savons que la renormalisation du temps n'était possible que si la force de rappel décroissait suffisamment vite avec le temps. Or il se trouve que dans le problème gravitationnel à N corps modèle très raisonable pour l'étude des structures galactiques, Dirac |9| pour diverses raisons a été amené à introduire l'hypothèse d'une "constante de gravitation" décroissant avec le temps sous la forme

$$(40) \qquad G(t) = \frac{G_o}{1+\Omega t}$$

Le temps étant compté à partir de l'instant présent et Ω étant une constante égale à l'inverse de l'âge de l'univers ($\Omega^{-1} \simeq 10^{10}$ ans). Nous serons amenés à généraliser cette loi et à considérer une variation

$$(41) \qquad G(t) = \frac{G_o}{(1+\Omega t)^{\alpha}} \qquad \alpha > 0$$

L'ensemble des particules i exerce sur la particule i supposée placée à l'origine le champ $\vec{E}_i$

(42) $$\vec{E}_i = \sum_{j \neq i}^{N} \mu_j \, G(t) \, \frac{\vec{x}_j}{|x_j|^3}$$

μ_j et x_j sont respectivement la masse et la position de la particule j. Le terme $C^3 \vec{E}_i$ deviens en utilisant (38)

(43) $$C^3 \vec{E}_i = G(t)\, C(t) \sum_{j \neq i}^{N} \mu_j \, \frac{\vec{\xi}_j}{|\xi_j|^3}$$

faisant apparaître dans le nouvel espace temps une "constante gravitationelle" $\overline{G}(\theta) = G(t)\,C(t)$.

Dressons donc pour la loi de variation (41) le tableau analogue au tableau (1) donné pour l'oscillateur harmonique en appliquant les mêmes règles. Prioritairement ni $G(\theta)$ ni le champ de transformation ne doivent tendre vers l'infini. Le temps doit être si possible renormalisé ou tout au moins compressé au maximum.

TABLEAU 2

α	C(t)	$\overline{G}(\theta)$	transf. field	$\theta(t)$
$0<\alpha<\frac{1}{2}$	$(1+\Omega t)^{\alpha}$	G_0	$\frac{\alpha(1-\alpha)\Omega^2}{[1+(1-2\alpha)\Omega\theta]^2}\vec{\xi}$	$\Omega\theta=\frac{(1+\Omega t)^{1-2\alpha}-1}{1-2\alpha}$
$\alpha=\frac{1}{2}$	$(1+\Omega t)^{\frac{1}{2}}$	G_0	$(\Omega^2/4)\vec{\xi}$	$\Omega\theta=\log(1+\Omega t)$
$\frac{1}{2}<\alpha<1$	$(1+\Omega t)^{\frac{1}{2}}$	$G_0 \exp-(\alpha-\frac{1}{2})\Omega\theta$	$(\Omega^2/4)\vec{\xi}$	$\Omega\theta=\log(1+\Omega t)$
$\alpha=1$	$1+\Omega t$	G_0	0	$\Omega\theta=\Omega t/(1+\Omega t)$ if $t\to\infty$ $\Omega\theta\to 1$
$\alpha>1$	$1+\Omega t$	$G_0(1-\Omega\theta)^{\alpha-1}$	0	

Nous voyons que les deux tableaux sont quasiment identiques et que pour les mêmes valeurs de α nous faisons le même choix de C(t) (c'est pour souligner cette identité que dans (14) nous avions pris le curieux coefficient 4α dans $\omega^2 = \omega_0{}^2/(1+\Omega t)^{4\alpha}$. Il est alors inutile que nous répétions les raisonnements sur les choix de C(t) Rappelons pour mémoire les résultats. Pour $\alpha > 1/2$ le système finalement "explose" c'est à dire que le champ gravitationnel ne maintiendra la cohésion du système et les particules finiront pas s'éloigner indéfiniment sur des trajectoires libres. Par contre pour $\alpha < 1/2$ dans le nouvel espace $G(\theta) = G_0$ et le champ de transformation tends à s'annuler pour θ grand. On obtiendra donc après la fuite des particules trop rapides et l'extinction du champ répulsif de transformation, une structure correspondant à un équilibre dynamique à laquelle dans l'espace réel corresponds une expansion en $(1+\Omega t)^{\alpha}$.

Enfin il faut souligner le caractère particulier de l'hypothèse de Dirac ($\alpha = 1$). On peut soit adopter l'image d'un univers en expansion avec une constante gravitationelle décroissant comme l'inverse du temps ou au contraire ne pas toucher au dogme de la constance de G mais admettre une limite sur θ. Enfin il ne faut pas oublier que l'hypothèse de Dirac n'est qu'une hypothèse et que la parole dans ce domaine passe aux expérimentateurs qui comparent le temps des éphémérides avec le temps d'une horloge atomique ou il n'y a pas lieu de penser que les forces gravitationelles, infinies par rapport aux forces électromagnétiques, perturbent les fréquences de base. C'est à la lettre comparer les temps θ et t.

CONCLUSION

Après avoir analysé les difficultés d'utilisation des groupes de transformations appliqué à la résolution des équations de la physique mathématique nous avons substitué à la notion trop rigide d'invariance celle plus souple d'invariance partielle (c'est à dire admettant l'apparition de nouveaux termes à condition que ceux ci ne rendent pas l'équation plus difficile à étudier et ne diminuent pas nos connaissances à priori sur les solutions). Le concept de conformité complémente celui d'invariance partielle en ce sens qu'il qualifie des transformations d'un problème en un autre problème sur lequel on possède soit plus d'informations, soit des techniques de résolution. Moyennant quoi nous avons pu remplacer dans les transformations les groupes à un paramètre par des groupes comprenant une fonction arbitraire (c'est à dire infiniment plus "riches").

Appliquant cette méthode à quelques problèmes de la physique mathématique nous en avons montré l'efficacité et dégagé trois concepts importants.

- Le premier est incontestablement le concept de renormalisation du temps, qui permet un traitement de l'équation "duale" sur un intervalle temporel borné, ou au moins "comprimé". Cette notion très simple de "compression" du temps qui facilite de plus les calculs numériques, permet une étude particulièrement simplifiée des solutions asymptotiques.

- Le deuxième concept important est celui de "champ complémentaire de transformation". Ce terme dû à l'invariance partielle des équations entre en compétition avec le nouveau champ physique, et le résultat de cette compétition déterminera la forme asymptotique de la solution.

Bien que nous n'ayons donné aucun exemple ici il faut mentionner un dernier concept celui de contamination |10||11|. Son intérêt essentiel est qu'il permets de faire le pont avec les méthodes des groupes auto semblables et de tourner parfois les difficultés associés aux conditions initiales et aux limites non physiques qui ne sont que trop fréquentes. On remarque en effet que c'est souvent à l'infini que les conditions initiales deviennent non physiques. On introduit alors une frontière au delà de laquelle on considerera que les conditions initiales données par les groupes auto semblables ne correspondent plus à la réalité. Le concept de contamination est alors basé sur l'idée physique d'une *vitesse de propagation* des intéractions (et est particulièrement utile dans les intéractions électrostatiques des plasmas qui se propagent à la vitesse des particules). Les "mauvaises" conditions initiales d'au delà de la frontière vont donc se propager vers l'intérieur mais avec une vitesse finie. Il sera parfois simple de suivre le mouvement de la "frontière de contamination". Cette frontière délimite la zone ou la modification des conditions initiales du système au delà de la frontière initiale ne s'est pas encore fait sentir. Cette contamination dépends bien sur beaucoup de la physique et pourra être totale ou partielle. Il existe en effet certains systèmes (faisceau d'électrons en particulier) ou la contamination n'affecte qu'une partie du système pour tous les temps, induisant la notion de sous invariance.

Nous avons montré, sur quelques exemples pris dans différents domaines, comment ces concepts nous permettaient de résoudre des problèmes nouveaux. Nous croyons fermement qu'elle doit permettre la résolution de *nombreux autres problèmes* de la physique mathématique. Il faut toutefois rappeler que la notion de conformité exige une parfaite connaissance du domaine de la physique concerné par la résolution d'un problème spécifique et que plutôt que de former une méthode générale, applicable mécaniquement, l'ensemble de ces notions qui conduisent l'utilisateur à s'interroger sur les méthodes et techniques qu'il peut utiliser, constitue plutôt une méthodologie.

BIBLIOGRAPHIE

|1| S. LIE, Théorie des transformations gruppen. I, II, III Chelsea Publication Company, New York, (1970).

|2| L.I. SEDOV, Similarity and dimensional methods 4th Ed. Academic Press, (1959).

|3| L.P. EISENHART, Continuous group of transformations DOVER, (1961).

|4| L.V. OVSJANNIKOV, Group properties of differential equations translated by G. Bluman, (1967).

|5| W.F. AMES, Non linear partial differential equations in engineering, Academic Press, Vol I et II, (1972).

|6| G.W. BLUMAN and J.D. COLE, Similarity models for differential equations, Springer Verlag, (1974).

|7| J.R. BURGAN, Thèse de Doctorat es Sciences, Février 1978, Université d'Orléans.

|8| H.R. LEWIS and W.B. RISENFELD, Journal of Mathematical Physics, 10, 1458, (1968).

|9| P.A.M. DIRAC, Proc. Roy. Society, A155, 199, (1938).

|10| J.R. BURGAN, J. GUTIERREZ, E. FIJALKOW, M. NAVET, M.R. FEIX, Journal de Physique Lettres, 38L, 161, (1977).

|11| J.R. BURGAN, J.GUTIERREZ, E. FIJALKOW, M. NAVET, M.R. FEIX, Journal of Plasma Physics, 19, 135, (1978).

SPECTRAL TRANSFORM AND NONLINEAR EVOLUTION EQUATIONS

F. CALOGERO and A. DEGASPERIS
Istituto di Fisica dell'Universita di Roma - 00185 Roma
Istituto Nazionale di Fisica Nucleare Sezione di Roma

Abstract

The spectral transform method for solving nonlinear evolution equations is tersely surveyed.

Contents

1. Introduction

In 1967 a novel technique to solve and investigate certain classes of nonlinear evolution equations appeared. Its invention by Gardner, Greene, Kruskal and Miura [1] occured in connection with the Korteweg - de Vries (KdV) equation, a nonlinear partial differential equation that has been introduced at the end of the last century in the context of hydrodynamics [2] (and has subsequently found many applications in other field). A few years later another important nonlinear partial differential equation, the so-called nonlinear Schroedinger (NLS) equation, was also shown to be solvable by the same technique (suitably modified by Zakharov and Shabat [9]). The way was thereby opened to the search and discovery of entire classes of nonlinear evolution equations solvable by these techniques (appropriately modified and extended). The relevant references are too many to be reported here ; we list only a few books that have appeared recently, or are due to appear in the nearest future [4 - 10].

The purpose and scope of this report is to describe one approach to identify nonlinear equations solvable by this technique -based on the spectral transform- and to analyze their properties. We focus on the class of nonlinear evolution equations solvable by the spectral transform associated to the Schoedinger spectral problem. Readers interested in other classes (in particular, that associated to the so-called generalised Zakharov-Shabat spectral problem), or in more detailed treatments of the topics outlined below, or in various extensions of this approach, are referred to the original literature [11 - 35].

The treatment given below is terse, and rather close to that of [23] ; but the attentive reader will find some novelties, including some (minor) results and some presentation angles, not previously published.

2. The spectral transform

Consider the (singular) Sturm-Liouville problem characterized, on the whole line $-\infty < x < \infty$, by the Schroedinger equation

$$(2.1) \qquad -\psi_{xx}(x,k) + q(x)\,\psi(x,k) = k^2\,\psi(x,k)$$

with the (real) "potential" $q(x)$ vanishing at infinity, say

$$(2.2) \qquad \lim_{x \to \pm\infty} [|x|^{1+\varepsilon}\, q(x)] = 0 \quad , \quad \varepsilon > 0 \ .$$

Here, and always in the following, subscripted variables indicate partial differentiations.

The continuous part of the spectrum of this problem, corresponding to k real so that $k^2 > 0$, can be characterised by the "reflection" and "transmission" coefficients $R(k)$ and $T(k)$ defined by the asymptotic behavior of an appropriate solution of (2.1) :

(2.3a) $$\psi(x,k) \to T(k)\exp(-ikx) \quad \text{as} \quad x \to -\infty ,$$

(2.3b) $$\psi(x,k) \to \exp(-ikx) + R(k)\exp(ikx) \quad \text{as} \quad x \to +\infty ,$$

The spectral problem (2. 1-2) may also admit a finite number of discrete negative eigenvalues ("bound states")

(2.4) $$k_n^2 = -p_n^2 \ , \quad k_n = i\,p_n \ , \quad p_n > 0 \ , \quad n = 1,2,\dots,N \ ;$$

the corresponding (real) wave functions $\psi(x,ip_n) \equiv \psi_n(x)$ define the quantities c_n , $\bar{c}_n$ through the formulas

(2.5) $$\int_{-\infty}^{+\infty} dx\, \psi_n^2(x) = 1$$

(2.6a) $$\psi_n(x) \to \bar{c}_n \exp(p_n x) \quad \text{as} \quad x \to -\infty ,$$

(2.6b) $$\psi_n(x) \to c_n \exp(-p_n x) \quad \text{as} \quad x \to +\infty .$$

The spectral transform S of the potential $q(x)$ is now defined as the collection of data

(2.7) $$S : \{R(k) \ , \ -\infty < k < \infty \ , \ p_n \ , \ c_n^2 \ , \ n = 1,2,\dots,N\} \ .$$

There is a biunivocal correspondance between a function $q(x)$ (vanishing at infinity) and its spectral transform S. The determination of S from q corresponds to the solution of the direct spectral problem, characterized by the equations written above ; it is clearly unique. The determination of q from S corresponds to the solution of the inverse spectral problem, characterized by the Gel'fand-Levitan-Marchenko equation

(2.8) $$K(x,x') + M(x+x') + \int_x^{\infty} dx''\, K(x,x'')\, M(x''+x) = 0 \ , \quad x \leqslant x'$$

This is a Fredholm integral equation for the dependence of $K(x,x')$ on the second variable (the dependence on the first variable is parametric) ; the Kernel M is defined, in terms of the spectral transform S, by the formula

(2.9) $$M(y) = \sum_{n=1}^{N} c_n^2 \exp(-p_n y) + (2\pi)^{-1} \int_{-\infty}^{+\infty} dk\, R(k) \exp(iky)$$

The unique solution $K(x,x')$ of (2.8) determines the function $q(x)$ through

(2.10) $$q(x) = -2\, d\, K(x,x)/dx$$

It is not the place here to specify in detail the conditions on the function $q(x)$ and on the spectral transform S, that are necessary and sufficient for the validity of the statements given above. Suffice to note that the spectral data can be given with large arbitrariness, provided of course $R(k)$ is Fourier-transformable (see (2.9)) and it satisfies the reflection property

(2.11) $$R(-k) = R^*(k) \quad ;$$

corresponding to the requirement that $q(x)$ be real. There are however special choices of the spectral data that produce potentials with particularly smooth behavior and fast asymptotic vanishing. These choices are functions $R(k)$ that admit analytic continuation off the real axis and have poles, in the upper half of the complex k-plane, only on the imaginary axis, the location and residues of these poles being related to the parameters characterizing the discrete part of the spectrum by the formula

(2.12) $$\lim_{k \to ip_n} [(k - ip_n) R(k)] = i c_n^2 \ .$$

But even spectral data that violate this rule can produce smooth potentials that vanish exponentially at infinity. An important instance is the spectral transform

(2.13) $$R(k) = 0 \ ; \ N = 1 \ , \ p_1 = p \ , \ c_1 = c \ ,$$

to which there corresponds the potential

(2.14) $$q(x) = - 2p^2 / \cosh^2 [p(x-x_0)]$$

with

(2.15) $$c^2 = 2p \exp(2p \, x_0) \ ,$$

as it is implied by the GLM integral equation (2.8) (that is in this case immediately solvable, the kernel $M(x''+x')$ being separable). A more general case (also obtained from the explicit solution of the GLM equation (2.8), having now a separable kernel of rank N) has a spectral transform with $R(k) = 0$ and N discrete eigenvalues ; the corresponding function $q(x)$ is given by the formula

(2.16) $$q(x) = -2 \, d^2 \{\log \det [I + C(x)]\} / dx^2 \ ,$$

where I is the unit matrix (of rank N) and $C(x)$ is the symmetrical matrix of rank N defined by

(2.17) $$C_{mn} = c_m c_n (p_m+p_n)^{-1} \exp [- (p_m+p_n)x]$$

The existence of a biunivocal correspondance between a function $q(x)$ and its spectral transform S implies of course that if q depends on another variable t ("time"), $q \equiv q(x,t)$, the spectral transform S also depends on time , $S \equiv S(t)$. The main idea of the spectral transform method for solving nonlinear evolution equations rests on the identification of classes of nonlinear evolution equations, characterizing the time evolution of $q(x,t)$, such that the corresponding time evolution of the spectral transforms is simple (generally described by a first-order linear ordinary differential equation). Then the determination of the time evolution of $q(x,t_0)$ can be obtained by first going, at the initial time t_0 , from a given $q(x,t_0)$ to $S(t_0)$, then letting S evolve (simply !) from t_0 to t , and finally recovering $q(x,t)$ from $S(t)$. This procedure is of course analogous to the technique to solve the (linear) partial differential equation

(2.18) $$u_t(x,t) = - i\, \omega(-i \frac{\partial}{\partial x}, t)\, u(x,t)$$

(where $\omega(z,t)$ is an entire function -say a polynomial- in the first argument z), by Fourier transforming of the x-dependence, namely setting

(2.19a) $$u(x,t) = (2\pi)^{-1} \int_{-\infty}^{+\infty} dk \exp(ikx)\, \hat{u}(k,t) \quad ,$$

(2.19b) $$\hat{u}(k,t) = \int_{-\infty}^{+\infty} dx \exp(-ikx)\, u(x,t)$$

and noting that (2.18) implies

(2.20) $$\hat{u}_t(k,t) = -i\, \omega(k,t)\, \hat{u}(k,t) \quad .$$

Then, given $u(x,t_o)$, one obtains $\hat{u}(k,t_o)$ from (2.19b), then from (2.20)

(2.21) $$\hat{u}(k,t) = \exp \left[-i \int_{t_o}^{t} dt' \ \omega(k,t')\right] \hat{u}(k,t_o)$$

and from this explicit expression $u(x,t)$ is computed using (2.19a).

To implement the program outlined above a technique is needed to relate changes in the potential to changes in the spectral data. A convenient tool to do this is provided by the following formulas, obtained from a generalized version of the wronskian theorem [11-13, 20] :

(2.22) $$2ik\, f(-4k^2)\, [R_2(k)-R_1(k)] = \int_{-\infty}^{+\infty} dx\, \psi_2(x,k)\, \psi_1(x,k)\, f(\underline{\Lambda}).[q_2(x)-q_1(x)] \quad ,$$

(2.23) $$(2ik)^2\, g(-4k^2)\, [R_2(k) + R_1(k)] = \int_{-\infty}^{+\infty} dx\, \psi_2(x,t)\, \psi_1(x,t)\, g(\underline{\Lambda})\, \underline{\Gamma}.1 \quad ,$$

(2.24) $$\underline{\Lambda}.F(x) = F_{xx}(x) - 2\, [q_2(x) + q_1(x)]\, F(x) + \underline{\Gamma}.\int_{x}^{\infty} dx'\, F(x') \quad ,$$

(2.25) $$\Gamma.F(x) = [q_{2x}(x) + q_{1x}(x)]\, F(x) + [q_2(x) - q_1(x)] \int_{x}^{\infty} dx'\, [q_2(x') - q_1(x')]\, F(x')$$

In these equations q_2 and q_1 are two different "potentials", R_2 and R_1 the corresponding "reflection coefficients", ψ_2 and ψ_1 corresponding "wave functions" (characterized by (2.1) and (2.3)), f and g arbitrary entire functions, and the integro-differential operators $\underline{\Lambda}$ and $\underline{\Gamma}$ are defined explicitly indicating, in (2.24-25), how they act on a generic function $F(x)$. Note that these operators depend on q_1 and q_2.

The significance of the operator $\underline{\Lambda}$ is displayed by the formula

(2.26) $$\int_{-\infty}^{+\infty} dx\, \psi_2(k,x)\, \psi_1(k,x)\, \Lambda\, F(x) = -4k^2 \int_{-\infty}^{+\infty} dx\, \psi_2(k,x)\, \psi_1(k,x)\, F(x) \quad ,$$

where $F(x)$ is an arbitrary function (such that the integrals do not diverge). Thus in some sense $\underline{\Lambda}$ may be viewed as the transpose of an operator having the product $\psi_2(k,x)\,\psi_1(k,x)$ as eigenfunction, with eigenvalue $-4k^2$. Note taht (2.26) also implies

(2.27) $$\int_{-\infty}^{+\infty} dx\, \psi_2(k,x)\,\psi_1(k,x)\, f(\underline{\Lambda})\, F(x) = f(-4k^2) \int_{-\infty}^{+\infty} dx\, \psi_2(k,x)\,\psi_1(k,x)\, F(x) ,$$

this last formula being instrumental in the derivation of (2.22-23).

The formulae (2.22-23) are clearly a convenient tool to correlate certain changes in the potential (or equivalently certain relations between two different potentials) to the corresponding changes of the reflection coefficients. In particular they imply that if the two potentials q_1 and q_2 are related by the formula

(2.28) $$f(\underline{\Lambda})\,[q_2(x) - q_1(x)] + g(\underline{\Lambda})\,\underline{\Gamma}.1 = 0 ,$$

the corresponding reflection coefficients R_1 and R_2 are related by

(2.29) $$f(-4k^2)\,[R_2(k) - R_1(k)] + 2ik\, g(-4k^2)\,[R_2(k) + R_1(k)] = 0 ,$$

or equivalently

(2.30) $$R_2(k) = R_1(k)\,[f(-4k^2) - 2ik\, g(-4k^2)] \,/\, [f(-4k^2) + 2iky\,(-4k^2)] .$$

We have in this manner identified a class of nonlinear integro-differential relations linking q_1 and q_2 -namely the equation (2.28), containing the arbitrary entire functions f and g such that the corresponding relation between the reflection coefficients takes the very simple form (2.30).

Let us end this section mentionning two amusing side remarks that are implied by these formulae.

The first concerns the possibility to derive explicitly pairs of potentials that yield the same reflection coefficient (following the treatment given, in the context of the radial Schroedinger equation, in [11]). Indeed setting $g = 0$, $f(\underline{\Lambda}) = \underline{\Lambda}$ in (2.28) and integrating, one gets with simple computations the explicit formula

(2.31) $$s(x) = \frac{1}{8} D^2(x) + D_{xx}(x)/D - \frac{1}{2} (D_x(x)/D(x))^2 ,$$

having set

(2.32) $$s(x) = q_2(x) + q_1(x) ,$$

(2.33) $$d(x) = q_2(x) - q_1(x) ,$$

(2.34a) $$D(x) = \int_x^{\infty} dx'\, [q_2(x') - q_1(x')]$$

(2.34b) $$d(x) = - D_x(x) .$$

Thus any choice of a function $D(x)$ that does not change sign, that vanishes as $x \to +\infty$, and that yields, through (2.34b) and (2.31), functions $d(x)$ and $s(x)$ that vanish as $x \to \pm\infty$, yields a pair of potentials that produce the same reflection coefficient -as implied by (2.30) with $g = 0$. Of course at least one of these two potentials must possess a bound state not to violate the property of biunivocal correspondance between potential and spectral transform. A simple instance that the interested reader may verify is the pair

(2.35) $$q_{1,2}(x) = (2x^2 \pm 4bx + b^2 - a^2)/(a^2 + x^2)^2$$

obtained from the choice

(2.36) $$D(x) = 4b/(a^2 + x^2) \quad .$$

The second remark consists of the derivation of a nonlinear operator identity, that generalizes the well-known (linear) formula

(2.37) $$F(x+a) = \exp\left(a \frac{d}{dx}\right) F(x) \quad .$$

It obtains setting

(2.38) $$q_1(x) = F(x) \quad , \quad q_2(x) = F(x+a)$$

Then clearly (2.1) and (2.3) imply

(2.39) $$R_2(k) = \exp(2ika)\, R_1(k) \quad .$$

But this equation coincides with (2.30) provided one sets

(2.40a) $$f(z) = \cos\left[\frac{1}{2}(-z)^{1/2} a\right]$$

(2.40b) $$g(z) = (-z)^{-1/2} \sin\left[\frac{1}{2}(-z)^{1/2} a\right] \quad .$$

Thus inserting these expressions in (2.28) and using (2.38) one gets the nonlinear operator identity [14]

(2.41) $$F(x+a) = F(x) + (-\underline{\Lambda})^{-1/2} \operatorname{tg}\left[\frac{1}{2}(-\underline{\Lambda})^{1/2} a\right] \underline{\Gamma}.1 \quad ,$$

where the integro differential operators $\underline{\Lambda}$ and $\underline{\Gamma}$ are now defined by the following formulae that detail their action on a general function $\phi(x)$:

(2.42) $$\underline{\Lambda}\, \phi(x) = \phi_{xx}(x) - 2\,[F(x+a) + F(x)]\, \phi(x) + \underline{\Gamma} \int_x^\infty dx'\, \phi(x')$$

(2.43) $$\underline{\Gamma}\, \phi(x) = [F_x(x+a) + F_x(x)]\, \phi(x) + \\ + [F(x+a) - F(x)] \int_x^\infty dx'\, [F(x'+a) - F(x')]\, \phi(x') \; .$$

Since $F(x)$ is arbitrary, one can replace it by $\lambda F(x)$ and expand in powers of λ the r.h.s. of (2.41), recovering the identity (2.37) (terms proportional to λ) and getting in addition an infinity of intriguing operator identities.

3. Nonlinear evolution equations solvable via the spectral transform

Assume now that the function q depend on another variable t ("time") ; this implies of course that R also depends on it. Now set $q_1 = q(x,t)$, $q_2 = q(x,t+\Delta t)$, and corresponding $R_1 = R(k,t)$, $R_2 = R(k,t+\Delta t)$, and consider the formulae that we obtain, in the limit $\Delta t \to 0$, inserting these positions in (2.22-25) :

$$(3.1) \qquad 2ik\, f(-4k^2,t)\, R_t(k,t) = \int_{-\infty}^{+\infty} dx\, \psi^2(k,x,t)\, f(\underline{L},t)\, q_t(x,t) \quad ,$$

$$(3.2) \qquad (2ik)^2\, g(-4k^2,t)\, R(k,t) = \int_{-\infty}^{+\infty} dx\, \psi^2(k,x,t)\, g(\underline{L},t)\, q_x(x,t) \quad ,$$

$$(3.3) \qquad \underline{L}\, F(x) = F_{xx}(x) - 4q(x,t)\, F(x) + 2q_x(x,t) \int_x^{\infty} dx'\, F(x') \quad .$$

In writing these formulae, we have explicitly indicated that the arbitrary functions f and g might be time-dependent, as it is implied by their derivation. These equations imply that, if the function $q(x,t)$ satisfies the nonlinear evolution equation

$$(3.4) \qquad q_t(x,t) = \alpha(\underline{L},t)\, q_x(x,t) \quad ,$$

the corresponding reflection coefficient $R(k,t)$ satisfies the linear ordinary differential equation

$$(3.5) \qquad R_t(k,t) = 2ik\, \alpha(-4k^2,t)\, R(k,t) \quad ,$$

that can be immediatelly integrated to yield

$$(3.6) \qquad R(k,t) = R(k,t_o)\, \exp\left[2ik \int_{t_o}^{t} dt'\, \alpha(-4k^2,t')\right] .$$

Because $\underline{L}$ is an integro-differential operator, it might appear that (3.4) is an integro-differential equation. It can however be shown that any power of $\underline{L}$, when applied to $q_x(x,t)$, yields only a (nonlinear) combination of q and its x-derivatives (all the integrations can be performed exactly). For instance

$$(3.7) \qquad \underline{L}\, q_x(x,t) = q_{xxx}(x,t) - 6q_x(x,t)\, q(x,t) \quad ,$$

$$(3.8) \qquad \underline{L}^2 q_x = q_{xxxxx} - 10q_{xxx}q - 20q_{xx}q_x - 30q_xq^2 \quad .$$

Thus for any choice of $\alpha(z,t)$ that is polynomial in the first variable z, the nonlinear evolution equation (3.4) is a pure partial differential equation. In particular, for $\alpha(z,t) = z$, it is the celebrated KdV equation [2] :

$$(3.9) \qquad q_t(x,t) = q_{xxx}(x,t) - 6q_x(x,t)\, q(x,t) \quad .$$

Rational choices of $\alpha(z)$ also yield equations that can be reduced to pure differential form, but only through a change of dependent variable. For instance insertion of $\alpha(z) = a/(b+z)$ in (3.4) yields the differential equation

$$(3.10) \qquad Q_{xxxt} + 4\, Q_{xt}\, Q_x + 2\, Q_{xx}\, Q_t + b\, Q_{xt} + a\, Q_{xx} = 0$$

for the dependent variable

$$(3.11) \qquad Q(x,t) = \int_x^{\infty} dx'\, q(x',t) \quad .$$

It is now clear how the class of nonlinear evolution equations (3.4) can be solved using the spectral transform method. Given $q(x,t_o)$, the corresponding reflection coefficient $R(k,t_o)$ is evaluated (direct spectral problem : eqs. (2.1) and (2.3)); then $R(k,t)$ is given by (3.6) ; finally $q(x,t)$ is recovered from $R(k,t)$ (inverse spectral problem : eqs. (2.8-10)). If bound states are present to perform the last step it is also necessary to know at time t the parameters of the discrete spectrum. The simpler way to obtain the time evolution of these parameters takes advantage of the relation (2.12), implying

$$p_n(t) = p_n(t_o) \equiv p_n \quad , \tag{3.12}$$

$$c_n(t) = c_n(t_o) \exp \left[-p_n \int_{t_o}^{t} dt' \, \alpha(4p_n^2,t')\right] \tag{3.13}$$

We also report the (extremely simple) equation describing the time evolution of the transmission coefficient, that can be obtain by techniques analogous to those indicated above in the case of the reflection coefficient :

$$T(k,t) = T(k,t_o) \equiv T(k) \quad . \tag{3.14}$$

Although this formula is not needed for the solution of the initial value problem of the nonlinear evolution equation (3.4), it clearly displays an important property of the flow (3.4) (see below). Attention should also be drawn to the time independence of the discrete eigenvalues (this isospectral property of the flow (3.4), first emphasized by Lax [36], has played a key rôle in the historical development of this subject).

We end this section noting that a straightforward extension of the approach that we have just described yields equations in more than one space variable [12, 16, 20] ; then the time evolution of the spectral parameters is some what more complicated. These classes of equations have not yet been fully investigated. In this report we shall always restrict consideration to functions of only two variables (one space variable x and one time variable t).

4. Behavior of the solutions of the nonlinear evolution equations solvable by the spectral transform : solitons

In the preceeding section we have identified the class of nonlinear evolution equations (3.4) and we have described how the initial value problem for these equations can be solved via the spectral transform. The quintessential message one gets from this analysis is that the time evolution, while it may be quite complicated in configuration space, is simple in the spectral space ; it is therefore the latter which provides the most convenient point of view , and suggests the most appropriate language, for understanding the b havior of the solutions of this class of equations.

This behavior may be analyzed in term of two components, one associated to the discrete spectrum of the corresponding Schroedinger spectral problem, the other to the continuum ; this separation being particularly evident in the definition (2.9) of the kernel M of the GLM integral equation (that provides, as explained above, an essential step in the solution of the initial value problem). In configuration space, these components get of course nonlinearly mixed (although they often separate asymptotically, as $t \to \infty$; see below) ; to understand the peculiarities of their behavior it is therefore convenient to focus attention firstly on special solutions with only one of the components present (note that the structure of the time evolution implies that, if one component is initially missing, it will remain missing throughout the time evolution).

Particularly interesting is the case of the special solutions whose spectral transform contains only the discrete component. The simplest case has a spectral transform with only one discrete eigenvalue ; the corresponding solution of the class of equations (3.4) is then

$$q(x,t) = -2p^2 / \cosh^2 \{p[x-x_o-v(t-t_o)]\} \tag{4.1}$$

with

$$v = -\alpha(4p^2) \tag{4.2}$$

as implied by (2.14-15) and (3.13). Note that the only dependence on the (generally polynomial) function $\alpha(z)$ that specifies the particular equation of the class (3.4) being considered, is in the definition (4.2) of the speed v ; clearly here, and always below, we are for simplicity restricting attention to nonlinear evolution equations with time independent coefficients, namely we assume $\alpha(z,t) = \alpha(z)$. The solution (4.1-2) represents the celebrated soliton : a smooth bump of constant shape moving with constant speed. It is characterized by the single parameter p, that determines its height $2p^2$, its width $1/p$ and its speed v ,-eq. (4.2)-, in addition to the parameters x_o and t_o, whose presence reflects the invariance of the equation under space and time translations.

The so-called multisoliton solutions correspond instead to a spectral transform that, again having no continuum component $(R = 0)$, contains N discrete eigenvalues. An explicit display of these solutions is provided by the formulas (2.16-17), with the parameters p_n independent of time, and the parameters c_n having the explicit time-dependence (3.13), that in the case with time independent $\alpha(z,t)$ takes the simple form

$$c_n(t) = c_n(t_o) \exp [p_n v_n(t-t_o)] \tag{4.3}$$

with

$$v_n = -\alpha(4p_n^2) . \tag{4.4}$$

It is easily seen that this solution describes asymptotically N separated solitons (provided all the v_n's are different, as it is generally the case), each characterized by its parameter p_n :

(4.5a) $$q(x,t) \underset{t \to -\infty}{} - 2 \sum_{n=1}^{N} p_n^2 / \cosh^2 \{p_n [x-x_n-v_n(t-t_o)]\}$$

(4.5b) $$q(x,t) \underset{t \to +\infty}{} - 2 \sum_{n=1}^{N} p_n^2 / \cosh^2 \{p_n [x-x_n-\Delta_n-v_n(t-t_o)]\} .$$

Thus the only permanent effect of the collision between solitons is to shift by the amount Δ_n the coordinate of each solitons ; it can moreover be shown that each shift Δ_n is simply the sum of the shifts resulting from the two-body collisions that each soliton experiences going though the others (factorization property; note that this happens even if the initial conditions are such to generate multiple collisions). The corresponding formula reads

(4.6) $$\Delta_n = \sum_{m=1}^{n-1} \Delta(p_n,p_m) - \sum_{m=n+1}^{N} \Delta(p_n,p_m) \quad , \quad n = 1,2,\ldots,N$$

where we have ordered the solitons so that

(4.7) $$v_n > v_{n+1} \quad , \quad n = 1,2,\ldots,N-1 \quad ,$$

and the quantity $\Delta(p_n,p_m)$ is the shift resulting from the collision of two solitons characterized by the parameters p_n and p_m. This quantity can be easily computed from the two-soliton solution (namely from the special case of the formula given above, corresponding to $N = 1$, $p_1 = p_m$, $p_2 = p_m$, $c_1 = c_n(t)$, $c_2 = c_m(t)$) :

(4.9) $$(p_n,p_m) = (p_n)^{-1} \ln |(p_n-p_m) / (p_n+p_m)|$$

This concludes our discussion of pure soliton solution. If instead the continuum component of the spectral transform is not missing $(R \neq 0)$, the behavior of the solutions of (3.4) in configuration space is more complicated. Suffice here to say qualitatively that one can then analyze the solution in terms of a soliton part and a background. The background behaves roughly as the solution of the linearized equation

(4.10) $$q_t(x,t) = \alpha\left(\frac{\partial^2}{\partial x^2}\right) q_x(x,t)$$

a dispersive equation whose associated group velocity is

(4.11) $$v_g = - d\,[k\,\alpha(-k^2)] / dk \quad ;$$

while the soliton part retains the stability property described above, namely each soliton maintains its separated identity throughout the motion (this is evident from the spectral space analysis of the time evolution ; of course it does not imply that these solitons are distinguishable throughout the motion in configuration space ; indeed this need not be the case even is for pure soliton solutions). If all the solitons have velocities v_n of the same sign, while the group velocity (4.11) has, for all values of k, the appropriate sign (this is for instance the case for the KdV equation (3.9), with $v_n = -4p_n^2$ and $v_g = 3k^2$), then asymptotically (both in the past and in the future) there occurs a complete separation of the soliton and background parts, the former giving rise asymptotically to N separated solitons, each characterized by its parameter p_n, the latter dispersing away essentially in the fashion appropriate to the solution of the linear evolution equation (4.10), that can be conveniently rewritten as

$$(4.12) \qquad q_t(x,t) = -i\,\omega(-i\frac{\partial}{\partial t})\, q(x,t) \quad ,$$

with

$$(4.13) \qquad \omega(k) = -k\,\alpha(-k^2)$$

(note the consistency between the fact that the background component disperses asymptotically and that it satisfies the linearized equation (4.10) or (4.12) ; indeed when q becomes small all nonlinear components in (3.4) become negligeable).

5. Bäcklund transformations, nonlinear superposition principle, resolvent formula

Let $q(x,t)$ be a solution of (3.4) ; then the corresponding reflection coefficient $R(k,t)$ satisfies (3.5). Set now, in (2.28), $q_1 = q(x,t)$, $q_2 = q'(x,t)$, setting

$$(5.1) \qquad f(\underline{\Lambda})\,[q'(x,t) - q(x,t)] + g(\underline{\Lambda})\,\underline{\Gamma}\,.\,1 = 0$$

where of course now

$$(5.2) \qquad \underline{\Lambda}\,F(x) = F_{xx}(x) - 2\,[q'(x,t) + q(x,t)]\,F(x) + \underline{\Gamma}\int_x^\infty dx'\,F(x') \quad ,$$

$$(5.3) \qquad \underline{\Gamma}\,F(x) = [q'_x(x,t) + q_x(x,t)]\,F(x) + \\ + [q'(x,t) - q(x,t)]\int_x^\infty dx'\,[q'(x',t) - q(x',t)]\,F(x') \quad .$$

Then the reflection coefficient $R'(k,t)$ corresponding to $q'(x,t)$ is related to $R(k,t)$ by (2.30), reading now

$$(5.4) \qquad R'(k,t) = R(k,t)\,[f(-4k^2) - 2ik\,g(-4k^2)]\,/\,[f(-4k^2) + 2ikg(-4k^2)]$$

This formula clearly implies that, if $R(k,t)$ satisfies (3.5), so does $R'(k,t)$. But then, just as the fact that $q(x,t)$ satisfies (3.4) implies that $R(k,t)$ satisfies (3.5), the fact that $R'(k,t)$ satisfies (3.5) implies that $q'(x,t)$ satisfies (3.4) (to be sure, this conclusion would also require an analysis of the time dependence of the parameters of the discrete spectrum, that is omitted here for simplicity).

Thus the (nonlinear integro differential) equation (5.1) relating $q'(x,t)$ to $q(x,t)$ has the property that, if $q(x,t)$ satisfies (3.4), so does $q'(x,t)$. Such relations, for historical reasons that need not be elaborated here, are called Bäcklund transformations. Note that the dependence of $q'(x,t)$ from t implied by (5.1) obtains parametrically through the dependence (if any) of $q(x,t)$ on t, and moreover from the "constants of integration", that obtain solving (5.1) for $q'(x,t)$ (note that the integro differential character of (5.1) refers only to the x-dependence) ; the time dependence of these quantities can be ascertained inserting $q'(x,t)$ in (3.4) and integrating the ordinary first-order differential equations thus obtained.

It is perhaps worth emphasizing that the Bäcklund transformation (5.1), while nonlinear (and not too simple) in the configuration space in which it has been written, has the extremely simple counterpart (5.4) in spectral space. And clearly it is this latter equation that displays most transparently the effect of the Bäcklund transformation.

Note that the Bächlund transformation (5.1-4) contains the arbitrary functions f and g. The simpler case obtains when this functions are constant, in which case (5.1) and (5.4) yield respectively $(f/g = 2p)$

(5.5) $$Q'_x(x,t) + Q_x(x,t) = -\frac{1}{2}\left[Q'(x,t) - Q(x,t)\right]\left[h_p + Q'(x,t) - Q(x,t)\right]$$

(5.6) $$R'(k,t) = - R(k,t)\ (k + ip) \ / \ (k-ip) \quad .$$

In writing the first equation, we have used the more convenient dependent variable defined by (3.11) ; note that it has the form of a Riccati equation.

As it is seen by (5.6) , $R'(k,t)$ has generally one more pole that $R(k,t)$, at $k = ip$. This suggests that the $q'(x,t)$ related to $q(x,t)$ by (5.5) and (3.11) has one more soliton ; although to pin this point down a more detailed discussion should be given than it is possible here of the effect of the Bäcklund transformation (5.5) on the discrete spectrum parameters. To illustrate this point, and also to display one of the uses of Bäcklund transformations, let us mention the simplest case corresponding to the choice $q(x,t) = 0$, that is clearly a (trivial) solution of (3.4). Then (5.5) is easily integrable and it yields

(5.7) $$q'(x,t) = - 2p^2 \ / \ \cosh^2 \{p[x - \xi(t)]\}$$

and insertion of this expression in (3.4) yields

(5.8) $$\xi_t(t) = \alpha(4\ p^2)$$

so that

(5.9) $$\xi(t) = x_o - \alpha(4\ p^2)(t-t_o) \quad .$$

The single-soliton solution (4.1-2) has thus been reproduced.

Assume next to perform sequentially two Bäcklund transformations characterized by parameters p_1 and p_2. Then, with obvious notations

(5.10) $$Q_{1x} + Q_x = -\frac{1}{2}(Q_1 - Q)(4p_1 + Q_1 - Q)$$

(5.11) $$Q'_x + Q_{1x} = -\frac{1}{2}(Q' - Q_1)(4p_2 + Q' - Q_1)$$

(5.12) $R'(k,t) = R(k,t) \{(k+ip_1)(k+ip_2) / [(k-ip_1)(k-ip_2]\}$

(5.13) $Q_{2x} + Q_x = - \frac{1}{2} (Q_2 - Q)(4p_2 + Q_2 - Q)$

(5.14) $Q''_x + Q_{2x} = - \frac{1}{2} (Q'' - Q_2)(4p_1 + Q'' - Q_2)$

(5.15) $R''(k,t) = R(k,t) \{(k+ip_2)(k+ip_1) / [(k-ip_2)(k-ip_1)]\}$,

the first(last) 3 formulas referring to the case when the transformation with parameter p_1 is performed first (last). But clearly (5.12) and (5.15) imply $R''(k,t) = R'(k,t)$, and this in turn implies

(5.16) $Q''(x,t) = Q'(x,t)$

(to be sure this conclusion requires a verification, that also the parameters of the discrete spectrum can be identified). Inserting (5.16) in (5.14) and then subtracting (5.11) and (5.13) from the sum of (5.10) and (5.14) yields, after a little trivial algebra

(5.17) $$Q'(x,t) = Q(x,t) - (p_1+p_2) [Q_1(x,t)-Q_2(x,t)] / \{p_1-p_2+ \frac{1}{2} [Q_1(x,t) - Q_2(x,t)]\}$$

This formula is referred to as a "nonlinear superposition principle" : it yields explicitly (also using (3.11)) a new solution $q'(x,t)$ of the nonlinear evolution equation (3.4), in terms of the 3 solutions $q(x,t)$, $q_1(x,t)$ and $q_2(x,t)$, the second and third of these 3 solutions being related to the first by (5.5) and (3.11), with $p = p_1$ and $p = p_2$. Note that generally $q_1(x,t)$ resp. $q_2(x,t)$ has one more soliton (with parameter p_1 resp. p_2) than $q(x,t)$, and the $q'(x,t)$ given by (5.17) has two more solitons.

Starting from the trivial solution $q = 0$ and from the single soliton solution $q_1(x,t)$ resp. $q_2(x,t)$, given by (4.1-2) with $p = p_1$ resp. $p = p_2$, this formula yields the pure two solitons solution (an appropriate choice of the constants x_{01} and x_{02} must be made, to avoid the vanishing of the denominator in the r.h.s. of (5.17) ; one of them must have an imaginary part) ; and repeating the process one can construct, by a sequence of purely algebraic steps, also the multisoliton solution, that shall of course eventually coincide with the multisoliton formula (2.16-17) and (4.3-4), obtained by solving the Gel'fand-Levitan equation (2.8) with the kernel (2.9) containing only the contribution of the discrete spectrum.

Let us finally return to the formulae (2.28-30), but considering a more general case, with f and g depending explicitly on time and in particular

(5.18a) $f(z,t) = \cos [\frac{1}{2} (-z)^{1/2} \alpha(z) (t-t_o)]$,

(5.18b) $g(z,t) = (-z)^{-1/2} \sin [\frac{1}{2} (-z)^{1/2} \alpha(z) (t-t_o)]$.

A comparison of (2.30) with (3.6) (with $\alpha(z,t) = \alpha(z)$) then implies that, if in (2.30) $R_1 = R(k,t_o)$, then $R_2 = R(k,t)$. This implies that the same identification,

(5.19) $$q_1 = q(x,t_o) \quad , \quad q_2 = q(x,t)$$

should be made in (2.28) ; that provides then a direct, if highly complicated, relationship , between $q(x,t)$ and $q(x,t_o)$. It is natural to term this formula generalized resolvent ; it constitutes the generalization, for the nonlinear evolution equation (3.4), of the formal resolvent formula

(5.20) $$q(x,t) = \exp\left[-i(t-t_o)\ \omega(-i \frac{\partial}{\partial x})\right] q(x,t_o)$$

for the linear evolution equation (4.12).

The result that we have just displayed provides an explicit illustration of the fact that the time evolution implied by (3.4) (with $\alpha(z,t) = \alpha(z)$) may be viewed as a Bäcklund transformation. Indeed $q(x,t)$ and $q(x,t+T)$ both satisfy (3.4), and therefore any nonlinear relation linking them can be considered, by definition, to be a Bäcklund transformation. Moreover it is easily seen that, in the limit in which t_o and t differ only infinitesimally, the resolvent formula that we have just introduced reproduces, as indeed it should, the evolution equation (3.4).

Let us finally note that an analogous, but more general, treatment leads to the identification of "generalized Bäcklund transformations" relating solutions of different nonlinear equations of the class (3.4) (corresponding to different functions $\alpha(z,t)$) [17].

6. Conservation laws

An important property of the class of equations (3.4) is the existence of an infinite number of conserved quantities. Indeed we have already seen that the flow (3.4) is isospectral, namely that the discrete eigenvalues corresponding, in the Schroedinger spectral problem (2.1), to the time dependent potential $q(x,t)$, are time independent. Moreover the transmission coefficient $T(k)$ corresponding to $q(x,t)$ is time independent, when $q(x,t)$ evolves according to (3.4). Thus the discrete eigenvalues p_n ,and the transmission coefficient $T(k)$ (for all values of k) provide a constant of the motion. They are, however, complicated functionals of the potential $q(x,t)$.

A standard technique to obtain conserved quantities that have a simpler, and explicit, expression in terms of $q(x,t)$, is through an asymptotic expansion in k of $T(k)$ (or some appropriate function of $T(k)$) , whose coefficients are clearly time independent (since $T(k)$ itself is time independent) and have simple expressions in terms of $q(x,t)$.

A series of conserved quantities obtained in this manner can be written in the compact form

(6.1) $$c_n = \int_{-\infty}^{+\infty} dx\, q(x,t)\, M^{2n}.1 \quad , \quad n=0,1,2,...$$

the operator M being given by the formula

(6.2) $$M\, F(x) = F_x(x) - \int_{-\infty}^{x} dx'\, q(x',t)\, F(x') \quad ,$$

that specifies its action on the generic function F. The formula (6.1) is a direct consequence of the results detailed in the last part of section 5 of [23]. An equivalent formula providing essentially the same series of conserved quantities is given by the compact formula [34]

(6.3) $$C_n = \int_{-\infty}^{+\infty} dx\, L^n\, [x\, q_x(x,t) + 2q(x,t)] \quad , \quad n=0,1,2,...$$

the integro differential operator L being the same one, defined by (3.3), that enters in the definition of the class (3.4) of solvable evolution equations. The first 3 conserved quantities that are identified in this manner can be most simply written as follows :

$$c_o = \int_{-\infty}^{+\infty} dx\, q(x,t)$$

$$c_1 = -3 \int_{-\infty}^{+\infty} dx\, q^2(x,t)$$

$$c_2 = 5 \int_{-\infty}^{+\infty} dx\, [2q^3(x,t) + q_x^2(x,t)]$$

Another constant of the motion for the KdV equation (3.9) is given by the formula

(6.4) $$c = \int_{-\infty}^{+\infty} dx\, [x\, q(x,t) - 3t\, q^2(x,t)]$$

as can be easily verified by direct computation. This formula, together with the already mentionned constancy of the space integrals of q and q^2, implies that for the generic solution of the KdV equation the "center of mass"

(6.5) $$X(t) = \int_{-\infty}^{+\infty} dx\, x\, q(x,t) \Big/ \int_{-\infty}^{+\infty} dx\, q(x,t)$$

moves with constant speed,

(6.6) $$X(t) = X_o + V\, t$$

(6.7) $$V = 3 \int_{-\infty}^{+\infty} dx\, q^2(x,t) \Big/ \int_{-\infty}^{+\infty} dx\, q(x,t) \quad .$$

7. Nonlinear evolution equations solvable by the spectral transform associated to the matrix Schroedinger spectral problem : boomerons, trappons, zoomerons

Up to this point, we have discussed the class of nonlinear evolution equations solvable by the spectral transform associated to the Schroedinger spectral problem

(2.1-7). A larger class has also been identified and discussed, in which the spectral transform associated to the matrix Schroedinger spectral problem plays the key role. In this class the dependent variable is also a matrix, and essentially all the properties and results described above can be duplicated. In special cases the matrix can be reduced and relatively simple equations can be written directly for one of its matrix elements, so that again a nonlinear partial differential equation for a single field can be obtained, whose initial value problem can be completely solved by techniques analogous to those described above.

The interested reader may find a complete account of these results in the litterature [18-21, 24-25]. Here we merely mention the main novel feature that characterizes this larger class of solvable nonlinear equations, namely the fact that the solitons move generally with variable speed.

To illustrate this point we display the simpler equations contained in this class and their single soliton solutions.

Consider first the nonlinear coupled equations (termed "boomeron" equations ; see below)

(7.1a) $$U_t = \vec{b} \cdot \vec{V}_x$$

(7.1b) $$\vec{V}_{xt} = U_{xx}\vec{b} + \vec{a} \wedge \vec{V}_x - 2\,\vec{V}_x \wedge (\vec{V} \wedge \vec{b}) \quad .$$

Here $\vec{a}$ and $\vec{b}$ are two given three vectors (the equations (7.1) are actually solvable even if $\vec{a}$ and $\vec{b}$ are time dependent, although for simplicity we assume hereafter that they are constant) ; $U \equiv U(x,t)$ is a scalar field and $\vec{V} \equiv \vec{V}(x,t)$ is a three-vector field ; they satisfy the boundary conditions

(7.2) $$U(\pm\infty,t) = U_x(\pm\infty,t) = 0 \quad , \quad \vec{V}(\pm\infty,t) = \vec{V}_x(\pm\infty,t) = 0$$

It is in fact the x-derivative of these quantities, rather than the quantities themselves, that are the analogs of the "potentiel" q of the preceeding section ; as suggested by (7.2). Accordingly we write the single-soliton solution for the x-derivatives, to evidence the analogies to (and the differences from) the case of the preceeding sections :

(7.3a) $$U_x(x,t) = p^2 / \cosh^2 \{p[x - \xi(t)]\} \quad ,$$

(7.4.a) $$\vec{V}(x,t) = \hat{n}(t)\, U(x,t) \quad .$$

In the last equation, $\hat{n}(t)$ is a unit vector ; its time evolution, as well as that of the soliton coordinate $\xi(t)$, are known in completely explicit form [18]. Here we report only the differential equations they satisfy :

(7.4) $$\hat{n}_t(t) = \vec{a} \wedge \hat{n}(t) + 2p\, \hat{n}(t) \wedge [\hat{n}(t) \wedge \vec{b}] \quad ;$$

(7.5) $$\xi_t(t) = -\vec{b} \cdot \hat{n}(t) \quad .$$

The behavior of these quantities can be easily discussed on the basis of their explicit representation, or on the basis of the differential equations we have just written. In the (most interesting) special case with the two vectors $\vec{a}$ and $\vec{b}$ orthogonal, to which the analysis is hereafter limited, the motion of the soliton is described by the following

Theorem : the soliton position $\xi(t)$ evolves in time exactly as the coordinate of a unit mass nonrelativistic particle of total energy

(7.6) $$E = V(0) + (1/2)(\vec{b}\cdot\hat{n}_0)^2 = (1/2)v^2 \text{ sign } (2pb-a)$$

acted upon by the external force associated to the potential energy $V(x-\xi_0)$, where

(7.7) $$V(x) = A \exp(-4px) - B \exp(-2px)$$

with

(7.8) $$A = (1/2)\, a^{-2} \{[(b^2(\vec{a}.\hat{n}_0)^2 + (\vec{a}\wedge\vec{b}.\hat{n}_0)^2 + (a^2/2p)(\vec{a}\wedge\vec{b}.\hat{n}_0))^2 + \\ + (a^2b/2p)^2(\vec{a}.\hat{n}_0)^2] / (b^2(\vec{a}.\hat{n}_0)^2 + (\vec{a}\wedge\vec{b}.\hat{n}_0)^2)\}$$

(7.9) $$B = (2p)^{-1}\, [(a^2/2p) + \vec{a}\wedge\vec{b}.\hat{n}_0)]$$

and of course $\hat{n}_0$ and ξ_0 indicate the initial values of $\hat{n}$ and ξ .

The potential energy (7.7) diverges to positive infinity as $x \to -\infty$, it vanishes as $x \to +\infty$, and it has a single negative minimum at $x = \bar{x} \equiv (2p)^{-1} \ln (2A/B)$ provided $B > 0$; while the total energy (7.6) is positive or negative depending whether the soliton parameter p is larger or smaller than the characteristic quantity $\bar{p} = (a/2b)$. Thus in the former case $(p > \bar{p})$ the soliton comes in from the right in the remote part and boomerangs back in the remote future ; in the latter it is trapped around $\bar{\xi} = \bar{x} + \xi_0$ (note that in this case the potential energy (7.7) always has a negative minimum at $\bar{x}$; and it can be shown that the motion of the trapped soliton is periodic in time, with period $T = 2\pi(a^2 - 4p^2b^2)^{-1/2}$). These behaviors have suggested the names "boomeron" and " trappon" for these kinds of solitons ; it should be emphasized that both can be simultaneously present in a multisoliton solution whose analytic form is easily obtained (the two solitons solution is displayed explicitly, in analytic form, in [19], and its behavior in a variety of cases can be seen in a computer-produced film [37]).

Another remarkable nonlinear evolution equations belonging to this family (in fact, closely related to the boomeron equation discussed above) is the "zoomeron" equation

(7.10) $$[\partial^2/\partial t^2 - \partial^2/\partial x^2][z_{xt}(x,t) / z(x,t)] + 2\,[z^2(x,t)]_{xt} = 0 \quad .$$

The solitons of this equation have a behavior closely analogous to that described above as far as their motion is conserved, but then amplitude and width also changes with time. Also in this case the multisoliton solutions can be obtained in closed form, and a variety of two-soliton cases is displayed in a computer-produced film [38].

8. Recent developments

The approach described in the preceeding sections has been recently extended. The main tool to perform this extension is the equation

$$(8.1) \qquad (2ik)\, h(-4k^2,t)k\, R_k(k,t) = -\int_{-\infty}^{+\infty} dx\, \psi^2(k,x,t)\, h(\underline{L},t)\, [x\, q_x(x,t) + 2\, q(x,t)]$$

supplementing the formulae (3.1-3). It has thus been possible to identify and discuss the larger class of equations

$$(8.2) \qquad q_t(x,t) = \alpha(\underline{L},t)\, q_x(x,t) + \beta(\underline{L},t)\, [x\, q_x(x,t) + 2q(x,t)] \quad .$$

For details, the reader is referred to the original papers [29-35].
Another extension proceeds through the consideration of a different spectral problem characterized by the Schroedinger equation

$$(8.3) \qquad \psi_{xx}(x,\lambda) = [q(x) + x - \lambda]\, \psi(x,t) \quad ,$$

(always with $q(x)$ vanishing asymptotically. The corresponding (direct and inverse) spectral problem is now well understood. The interest of this problem is, that via the spectral transform associated to it one can solve a nonlinear evolution equation of applicative relevance, namely the so-called "cylindrical KdV equation"

$$(8.4) \qquad q_t(x,t) + q(x,t)/(2t) = q_{xxx}(x,t) - 6q_x(x,t)\, q(x,t) \quad .$$

The interested reader is referred to the papers where these results are reported [39].

References

[1] C.S. Gardner, J.M. Greene, M.D. Kruskal and R.M. Miura : "Method for solving the Korteweg-de Vries equation", Phys. Rev. Lett. 19, 1095-1097 (1967).

[2] D.J. Korteweg and G. de Vries : "On the change of form of long waves advancing in a rectangular canal, and on a new type of long stationary waves", Phil. Mag. 39, 422-443 (1895).

[3] V.E. Zakharov and A.B. Shabat : "Exact theory of two-dimensional self-focusing and one-dimensional self-modulation of waves in nonlinear media", Sov. Phys. JETP 34, 62-69 (1972) [Russian original : Zh Eksp. Teor. Fiz. 61, 118-134 (1971)].

[4] A.C. Scott, F.Y.F. Chu and D.W. McLanghlin : "The soliton : a new concept in applied science", Proc. IEEE 61, 1443-1483 (1973).

[5] Nonlinear Wave Motion (A.C. Newiell, et.), Lect. Appl. Math. 15, American Mathematical Society, Providence, K.I. 1974.

[6] Dynamical Systems, Theory and Applications. (J. Moser, ed.), Lect. Notes in Physics 38, Springer, 1975.

[7] Bäcklund Transformations. (R.M. Miura, ed.), Lect. Notes in Mathematics 515, Springer, 1976.

[8] Nonlinear Evolution Equations Solvable by the Spectral Transform. (F. Calogero, ed.), Research Notes in Mathematics, Pitman, 1978.

[9] Solitons. (R.K. Bullough, ed.), Lect. Notes in Physics, Springer, 1978.

[10] Yu.I. Manin : "Algebraic aspects of nonlinear differential equations", in Contemporary Problems of Mathematics 11, VINITI, Moscow 1978 (in Russian).

[11] F. Calogero : "Generalized wronskian relations : a novel approach to Bargmann-equivalent and phase-equivalent potentials", in Studies in Mathematical Physics (Essays in honor of Valentine Bargmann), edited by E.H. Lieb, B. Simon and A.S. Wightman, Princeton University Press, 1976.

[12] F. Calogero : "A method to generate solvable nonlinear evolution equations", Lett. Nuovo Cimento 14, 443-448 (1975).

[13] F. Calogero : "Generalized wronskian relations, one-dimensional Schroedinger equation and nonlinear partial differential equations solvable by the inverse-scattering method", Nuovo Cimento 31B, 229-249 (1976).

[14] F. Calogero : "Bäcklund transformations and functional relation for solutions of nonlinear partial differential equations solvable via the inverse-scattering method", Lett. Nuovo Cimento 14, 537-543 (1975).

[15] F. Calogero and A. Degasperis : "Nonlinear evolution equations solvable by the inverse spectral transform associated with the multichannel Schroedinger problem, and properties of their solutions", Lett. Nuovo Cimento 15, 65-69 (1976).

[16] F. Calogero and A. Degasperis : "Nonlinear evolution equations solvable by the inverse spectral transform. I", Nuovo Cimento 32B, 201-242 (1976).

[17] F. Calogero and Degasperis : "Transformations between solutions of different nonlinear evolution equations solvable via the same inverse spectral transform, generalized resolvent formulas and nonlinear operator identites", Lett. Nuovo Cimento 16, 181-186 (1976).

[18] F. Calogero and A. Degasperis : "Coupled nonlinear evolution equations solvable via the inverse spectral transform and solitons that come back : the boomeron", Lett. Nuovo Cimento 16, 425-433 (1976).

[19] F. Calogero and A. Degasperis : "Bäcklund transformations, nonlinear superposition principle, multisoliton solutions and conserved quantities for the "boomeron" nonlinear evolution equation", Lett. Nuovo Cimento 16, 434-438 (1976).

[20] F. Calogero and A. Degasperis : "Nonlinear evolution equations solvable by the inverse spectral transform. II", Nuovo Cimento 39B, 1-54 (1977).

[21] F. Calogero and A. Degasperis : "Nonlinear evolution equations solvable by the inverse spectral transform associated to the matrix Schroedinger equation", in [9].

[22] F. Calogero and A. Degasperis : "Special solution of coupled nonlinear evolution equations with bumps that behave as interacting particles", Lett. Nuovo Cimento 19, 525-533 (1977).

[23] F. Calogero : "Nonlinear evolution equations solvable by the inverse spectral transform", to appear in the Proceedings of the Rome Conference on Mathematical Physics, June 6-15, 1977, edited by G.F. Dell'Antonio, S. Doplicher and G. Jona-Lasimio, Lect. Notes in Physics, Springer, 1978.

[24] A. Degasperis : "Solitons, boomerons and trappons", in [8].

[25] A. Degasperis : "Spectral transform and solvability of nonlinear evolution equations ; in Proceedings of the Advanced Study Institute on Nonlinear equations in physics and mathematics, August 1977, edited by A.O. Barut, Reidel, 1978.

[26] M. Bruschi, D. Levi and O. Raguisco : "Evolution equations associated to the triangular matrix Schroedinger problem solvable by the inverse spectral transform", Nuovo Cimento (in press).

[27] K.M. Case and S.C. Chu : "Some remarks on the wronskian technique and the inverse scattering transform", J. Math. Phys. 18, 2044-2052 (1977).

[28] S.C. Chu and J.F. Ladik : "Generating exactly soluble nonlinear discrete evolution equations by a generalized wronskian technique", J. Math. Phys. 18, 690-700 (1977).

[29] F. Calogero and A. Degasperis : "Extension of the spectral transform method for solving nonlinear evolution equations", Lett. Nuovo Cimento.

[30] F. Calogero and A. Degasperis : "Exact solution via the spectral transform of a nonlinear evolution equation with linear x-dependent coefficients", Lett. Nuovo Cimento

[31] F. Calogero and A. Degasperis : "Extension of the spectral transform method for solving nonlinear evolution equations. II", Lett. Nuovo Cimento

[32] F. Calogero and A. Degasperis : "Exact solution via the spectral transform of a generalization with linearly x-dependent coefficients of the modified Korteweg-de Vries equations", Lett. Nuovo Cimento

[33] F. Calogero and A. Degasperis : "Exact solution via the spectral transform of a generalisation with linearly x-dependent coefficients of the nonlinear Schroedinger equation", Lett. Nuovo Cimento

[34] F. Calogero and A. Degasperis : "Conservation laws for classes of nonlinear evolution equations solvable by the spectral transform", Commun. Math. Phys. (submitted to).

[35] D. Levi and O. Reguisco : "Extension of the spectral transform method for solving nonlinear differential difference equations", Lett. Nuovo Cimento (in press).

[36] P.D. Lax : "Integrals of nonlinear equations of evolution and solitary waves", Comm. Pure Appl. Math. 21, 467-490 (1968).

[37] J.C. Eilbeck : "Boomeron", a computer-produced film (Mathematics Dept., Heriot-Watt University, Edinburgh).

[38] J.C. Eilbeck : "Zoomerons", a computer-produced film (Mathematics Dept., Heriot-Watt University, Edinburgh).

[39] F. Calogero and A. Degasperis (to be published).

WHAT YOU ALWAYS WANTED TO KNOW ABOUT THE APPLICATION OF INVERSE PROBLEMS TO NONLINEAR EQUATIONS
(or what you would like to do with the I.S.T.)

I. MIODEK
Département de Physique Mathématique
Université des Sciences à Techniques du Languedoc
34060 Montpellier Cédex, FRANCE

Abstract

The inverse scattering method for solving nonlinear evolution equations is discussed in the framework of lax's approach.

Contents

1. WHAT THE I.S.T. CAN'T DO.

Everybody who has gone into the physical background of almost any mathematical model knows that most linear models are approximations to nonlinear ones which probably describe the physical situation more accurately but have the nasty defect of being insoluble. Not only are they insoluble but quite often i's not certain that the linear "approximation" gives results which are close to those of the nonlinear model. We know and awful lot about linear equations. We know awfully little about general solutions to nonlinear equations. I will tell you the most embarassing thing about the recent advances in our ability to solve certain nonlinear equations by those inverse scattering methods called I.S.T. (Inverse Scattering Transform). Wereas previously we know that we coultdn't solve almost any nonlinear equation presented to us, today it is very difficult to decide if we can or can't solve a given nonlinear equation unless, with enormous luck, that equation just happens to be written down in a form we have seen before. Some progress has been made to overcome this difficulty : as I said it is very difficult but it is not impossible to decide.

2. WHAT THE I.S.T. DOES DO.

Now that I have stressed the major shortcomming of the I.S.T. let me get on with describing what the I.S.T. is. How can inverse scattering methods be used to solve those nonlinear equations for which they work ? The answer is quite simple. If $Q(x)$ is the potential of some linear scattering problem and $\mathcal{S}$ is the inverse scattering data for that same scattering problem then we let $\mathcal{S}$ evolve from some initial value, $\mathcal{S}(t = 0)$ (determined by $Q(t = 0,x)$), to some later value $\mathcal{S}(t)$ from which $Q(t,x)$ is then constructed by the inverse method. "Brilliant!" I hear you say doubtfully. "But isn't there a vital flaw in this reasoning ? Even supposing that you know the appropriate scattering problem, how do you know what the correct evolution of $\mathcal{S}$ should be ?" A good and profound question. The answer is that the evolution equation in question was actually generated by causing $\mathcal{S}$ to evolve in a known way, so the question was answered before the problem was posed. The remarkable thing from this point of view is not that we know how $\mathcal{S}$ evolves with t but that when $\mathcal{S}(t)$ evolves according to anyone of a certain restricted class of well specified ways (yet to be specified) the associated potential $Q(t,x)$ satisfies one of a family of nonlinear evolution equations, say $Q_t = w(t,Q,Q_x,..)$. Here w is a well determined polynomial in Q and a finite number of its x-partial derivatives. We stress : not only does $Q(t,x)$ satisfy such a "local" polynomial evolution equation for one given t-dependence of $\mathcal{S}(t)$, which is remarkable enough, but there exists a whole family of such evolution equations.

Well at this stage I hope you are convinced that the I.S.T. could work, and probably you are suitably amazed that polynomial nonlinear evolution equations can be obtained by such a wiley trick. (If you're not you should be.) But could such nonlinear equations possibly be of any interest ? Yes! It turns out that such well known and physically interesting equations as the Korteweg De Vries, the sine-Gordon, the cubic Schrödinger, the Toda Lattice and many of their generalizations are generated and solved by the I.S.T. . Not only that but their solutions contain essentially nonlinear properties referred to as solitons which are intimitly associated with the point spectrum of the linear scattering operator and which do not exist at all in the linearized approximation to nonlinear equation.

3. WHAT IS AN INVERSE SCATTERING PROBLEM.

You might say : "Well, O.K. You've convinced me that the I.S.T. could work and that it is interesting and useful. Could you now tell me what is meant by an inverse scattering problem in this context and why is the method called the Inverse Scattering (or Spectral) Transform ?" (This is a most convenient question because it just happens to fit into the logical development I had in mind.)

The inverse scattering problem is the inverse of a direct scattering problem which is of course called direct because it was studied first. It is in fact easier to first define a direct problem. Given some linear operator $H(Q)$, which depends on one or several functions of say x and denoted by $Q(x)$, the direct problem consists of determining how the spectrum and (generalized) eigenfunctions of H depend on Q. The scattering data of the direct problem is the set of x-independent "constants" which characterise the "asymptotic" behaviour of these eigenfunctions ; ie in those regions of x-space where all the admissible potentials take known values, usually zero. Each bit of scattering data is by definition independent of x, but depends on the eigenvalue parameter, say k. In fact the analytic nature of the eigenfunctions and scattering data's dependence on k (for x fixed) is the basic ingredient which makes it possible to recover $Q(x)$ from the scattering data, and that is the goal of the inverse problem.

More precisely, the inverse problem seeks bo reconstruct the unique $Q(x)$ in some given class of permissible potentials, given the operator structure of $H(Q)$ and given the "inverse scattering data", $\mathcal{S}$, which is a "suitable" subset of the scattering data - by "suitable" we mean that $\mathcal{S}$ can in some sense be freely specified within some open set called the permissible scattering data. Thus $\mathcal{S}$ and $Q(x)$ are in one-to-one correspondence and can be considered as "Scattering Transforms" of one another.

A natural variational question is now posed :
if $\mathcal{S}$ is varied in some known way, say $\mathcal{S}(0) \rightarrow \mathcal{S}(t)$, how will its "Inverse Scattering Transform", $Q(t,x)$, vary ? Part of the answer is the family of nonlinear evolution equations already mentioned. But the full answer is richer(eg. Prof. Calogero's discrete transformations) and not completely known because, so far, the answer is only known for a subset ot the permissible variations of $\mathcal{S}$.

4. ANALOGY WITH FOURIER TRANSFORMS.

"Can you tell me in some concise and fairly simple way what such a family of evolution equations would look like ? Do such families have analogues for linear evolution equations ?" As usual, just the right question, posed in just the right way.

First we obtain a family of linear p.d.e. 's by asking : what happens to $f(t,x)$ when its Fourier x-transform, $\hat{f}(t,k)$, is subjected to the known evolution given by $w(t,k)$ as follows ?

$$(\ln \hat{f}(t,k))_t = w(t,k) \qquad \text{where}$$

$$x \equiv (x_1, x_2, \ldots, x_n) \in \mathbb{R}^n, \; k \in \mathbb{R}^n, \; t \in \mathbb{R},$$

$f_t \equiv \partial f/\partial t$ and $w(t,k)$ is polynomial in the $\{k_j\}$. The well known result is that

$$f_t(t,x) = w(t, -i\nabla)f(t,x), \text{ where } \nabla \equiv (\frac{\partial}{\partial x_1}, \ldots, \frac{\partial}{\partial x_n}),$$

$$= \int (d^n k)\, e^{ik\cdot x}\, \hat{f}_t(t,k).$$

Thus under the Fourier transform such evolutions $\hat{f}(t,k)$ map into the family of linear evolution equations with $w(t, -i\nabla)$ being a linear partial differential operator .

Replacing $\{-i \frac{\partial}{\partial x_j}\}$ with some other family of self-adjoint commuting t-independent operators say $\{-iD_j\}$, works just as well if we develop $f(t,x)$ in the common eigenfunctions of this new family rahter than $e^{ik.x}$. A point spectrum may now be present, but the evolution equations remain linear because the D_j do not depend on f.

It is also possible to generalize to ratios $w(t,k) = w_1(t,k)/w_2(t,k)$ of polynomials providing that the poles do not invalidate the spectral integral.

5. THE A.K.N.S. RESULT - AN EXAMPLE OF A NONLINEAR FAMILY.

This linear transformation method fails as soon as we consider the nonlinear equation,

$$f_t(t,x) = w(t,L)f(t,x)$$

where linear operator $L = L(f)$ itself depends on f. But, and this essentially is the remarkable and elegant discovery of Ablowitz, Kaup, Newell & Segur (A.K.N.S.)(1974), if the inverse scattering data $\mathcal{S} \equiv \{R(k) | k \in \Sigma \subset \mathbb{C}\}$, of the Zakharov-Shabat-A.K.N.S. eigenvalue problem,

$$F_x = (Q - ik\sigma)F \equiv ZF \tag{5.1}$$

(where $Q \equiv \begin{pmatrix} 0 & q_2 \\ q_1 & 0 \end{pmatrix}$, $\sigma \equiv \begin{pmatrix} 1 & 0 \\ 0 & -1 \end{pmatrix}$, $F \equiv (F^{(1)} \vdots F^{(2)})$ is a square matrix),

evolves according to

$$(\ln R(t,k))_t \equiv -w(t,k)\sigma, \tag{5.2}$$

then there exists a linear integro-differential operator

$$L^* \equiv L^*(Q) \equiv \frac{1}{2i} \begin{pmatrix} \partial_x - 2q_1 \int_{-\infty}^{x} q_2 & + 2q_1 \int_{-\infty}^{x} q_1 \\ - 2q_2 \int^{x} q_2 & - \partial_x + q_2 \int_{-\infty}^{x} q_1 \end{pmatrix} \tag{5.3}$$

such that

$$\begin{pmatrix} q_1 \\ q_2 \end{pmatrix}_t \equiv \sigma\, w(t,L^*) \begin{pmatrix} q_1 \\ q_2 \end{pmatrix}. \tag{5.4}$$

This is the family of evolution equations obtained by A.K.N.S. .

Let us deter the precise definition of $R(k)$ for the moment and make some remarks about this family evolution equations. We will do so in the context of the important special case $w(t,k) = \alpha/k$. We first state the more general A.K.N.S. result

$$w(t,k) = w_1(t,k)/w_2(t,k), \tag{5.5}$$

where the $w_j(t,k)$ are polynomials in k, then the family enlarges to

$$[w_2(t,L^*)\,\sigma\,\partial_t - w_1(t,L^*)] \begin{pmatrix} q_1 \\ q_2 \end{pmatrix} = 0, \tag{5.6}$$

providing that the poles of $w(t,k)$ do not coincide with the point spectrum of (5.1). The sine-Gordon equation, $U_{xt} = \sin U$, $U(-\infty) = 0$, is obtained when

$$- q_1(x) = q_2(x) \equiv \frac{1}{2} U_x(x) \quad \text{and} \quad w(k) = \frac{1}{2ik}. \tag{5.7}$$

It does not however just drop out, but requires the succesive steps :

$$U_{xxt} + U_x \int_{-\infty}^{x} U_x U_{xt} = U_x \tag{5.8a}$$

$$\Leftrightarrow (\partial_x + U_x \int_{-\infty}^{x} U_x)(U_{xt} - \sin U) = U_x(1-\cos U(-\infty)) = 0 \qquad (5.8b)$$

$$\Leftrightarrow U_{xt} = \sin U,\ U(-\infty) = 0 \qquad (5.8c)$$

6. SOME REMARKS ABOUT GOING BACKWARDS.

Equations (5.6) through(5.8) should illustrate well enough that is extremely difficult to know if a given nonlinear p.d.e. belongs to a known family : not only must you proceed from (5.8c) to (5.8a) but from (5.8c) to the particular form of (5.6) (given by (5.6-7) together) ; then there remains the problem of seeing if that particular form of (5.6) is a member of a family of evolution equations.

Whenever an evolution is solvable by some I.S.T. it must be a member of a family of evolution "equations" each member of which is obtained by choosing one of the many permissible paths of evolution for the inverse scattering data. "Equations" has been written in inverted comma's because not all these equations need be of the same kind as the equation started with-not all such equations need be p.d.e.'s for instance.

The search for such a family is essentially the study of the symmetry group of (or commuting flows for) the original p.d.e. . This is because each point in the set of permissible initial values for $\mathcal{S}(t)$ generates a new solution of the n.l.p.d.e. in question. If for each value of s (another t-like parmeter) :

(i) $\mathcal{S}(s,0)$ is a permissible value for $\mathcal{S}(t=0)$;

(ii) $\mathcal{S}(s,0) \xrightarrow{w} \mathcal{S}(s,t)$

(iii) $\mathcal{S}(0,t) \xrightarrow{\nu} \mathcal{S}(s,t)$

(ν plays the same role for s with t fixed as w plays for t when s is fixed);

(iv) $\mathcal{S}_{st}(s,t) = \mathcal{S}_{ts}(s,t)$ (compatibility of ∂_s and ∂_t) ;

then we may expect that the induced evolution equations,

$$Q_t = w(Q) \quad \text{and} \quad Q_s = \nu(Q),$$

are compatible in the sense $Q_{ts} = Q_{st}$. Given $w(Q)$, the search for compatible $\nu(Q)$'s which are non-trivial is the kind of problem we are dealing with. Several kinds of approach have been made including : Lax (1968), Magri (1977), Kumei (1977) fairly directly ; by Eastabrook & Wahlquist (1975), Morris (1977) via differential forms ; Corones (1977) uses a bit of both.

If I am now permitted to make two short remarks (subremarks ?) within this remark notice that first the compatability of Q_x and Q_t is already implicitly assumed when the evolution equation is written down. Second once several such compatible nonlinear operators have been found we can generate higher order nonlinear p.d.e.'s in several t-like variables - eg.

$$Q_{st} = (\nu(Q))_t \equiv \nu_Q \, Q_t = \nu_Q \, w(Q)$$

or

$$Q_{ss} + Q_{tt} = \nu_Q \, \nu(Q) + w_Q \, w(Q) \ .$$

Here the linear operator ν_Q (and w_Q) is a kind of functional derivative of $\nu(Q)$ (resp. $w(Q)$) of which we will give an explicit example shortly, when we examine conditions on $\nu(Q)$ and $w(Q)$ such that $Q_{ts} = Q_{st}$. The compatability (or integrability) condition $Q_{tx} = Q_{xt}$ will also be seen to play an explicit role when we derive the A.K.N.S. family of evolution equations.

Let us first fill the hole we have letf-let us define the inverse scattering data in the A.K.N.S. example.

7. THE A.K.N.S. INVERSE SCATTERING DATA.

The A.K.N.S. inverse scattering data is most conveniently defined in terms of a pair of matrix fundamental solutions, $J_\pm$, of this eigenvalue problem which are defined by the complementary Jost-like boundary conditions,

$$\lim_{x \to \pm\infty} e^{ikx\sigma} J_\pm(x,k) = I \quad \text{for} \quad k \in \mathbb{R}. \tag{7.1}$$

If we now define $S(k)$ by :

$$J_+(x,k) \equiv J_-(x,k) S(k) \quad \text{for} \quad k \in \mathbb{R}, \tag{7.2}$$

we know that $S(k)$ is independent of x and that

$$\det S(k) = 1 \ [\text{since } (\ln \det F)_x = \text{trace } (Q - ik\sigma) = 0 \Rightarrow \det J_\pm = 1].$$

Thus we can put

$$S(k) \equiv \begin{pmatrix} a^-(k) & a^+(k) r^+(k) \\ a^-(k) r^-(k) & a^-(k) \end{pmatrix}, \quad a^+ a^- (1 - r^+ r^-) = 1. \tag{7.3}$$

Note that if $Q(x)$ is short range enough then

$$e^{-ikx\sigma} \underset{x \to \infty}{\leftarrow} J_+(x,k) \underset{x \to -\infty}{\sim} e^{-ikx\sigma} \, S(k) \tag{7.4}$$

so $(a^\pm(k))^{-1}$ play the roles of "transmission coefficients" and $r^\pm(k)$ the roles of reflection coefficients to the left. Similarly S^{-1} gives us the scattering data to the right. The inverse scattering data is given by

$$R(k) \equiv \begin{pmatrix} r^-(k) & 0 \\ 0 & r^+(k) \end{pmatrix} \text{for } k \in \mathbb{R} \tag{7.5}$$

plus the residues of its "extension" to the bound state energies at

$$k \in \{k_n\} \equiv \{k_m^\pm\} \equiv \{k_m^+, k_m^- | a^\pm(k_m^\pm) = 0, \pm \operatorname{Im} k_m^\pm \geq 0 \text{ resp.}\}, \tag{7.6}$$

ie. the respective zeros of $a^\pm(k)$ when analytically continued respectively to $\pm \operatorname{Im} k \geq 0$. We will justify this analytic continuation later if time permits. At the bound states, say at $k = k_n^+$ (where $a^+(k_n^+) = 0$, $\operatorname{Im} k_n^+ \geq 0$), the second column of J_+, $J_+^{(2)}$, is a well defined column solution of the matrix d.e. (5.1) and satisfies

$$e^{-ikx}\binom{c_n^+}{0} \xleftarrow[x \to -\infty]{} J_+^{(2)} \xrightarrow[x \to +\infty]{} e^{ikx}\binom{0}{1}, \quad k = k_n^+, \; c_n^+ \neq 0. \tag{7.7}$$

Similarly at $k = k_n^-$ (where $a^-(k_n^-) = 0$, $\operatorname{Im} k_n^- \leq 0$),

$$e^{ikx}\begin{pmatrix} 0 \\ c_n^- \end{pmatrix} \xleftarrow[x \to -\infty]{} J_-^{(1)} \xrightarrow[x \to +\infty]{} e^{-ikx}\binom{1}{0}, \quad k = k_n^-, \; c_n^- \neq 0. \tag{7.8}$$

It is notationally convenient to combine these putting

$$S_n(k_n) \equiv \begin{pmatrix} 0 & C^+(k_n) \\ C^-(k_n) & 0 \end{pmatrix} \text{with } \{k_n\} \equiv \{k_n^+, k_n^-\}, \tag{7.9}$$

and where

$$C^\pm(k_n^\pm) = c_n^\pm \neq 0 = C^\pm(k_n^\mp) .$$

Now supposing simple bound state zeros, $a_n^\pm \equiv \lim_{\epsilon \to 0} a^\pm(k_n^\pm + \epsilon) \neq 0$. It can be shown that $\lim_{\epsilon \to 0} a^\pm r^\pm(k_n^\pm + \epsilon) \equiv a_n^\pm r_n^\pm = c_n^\pm$ <u>if the limit exists</u>. The bound state scattering data then is $\{k_n^\pm, a_n^\pm, c_n^\pm\}$, while the bound state part of the inverse scattering data is $\{c_n^+/a_n^+, c_n^-/a_n^-\}$ or equivalently

$$\{R_n(k_n) \equiv \begin{pmatrix} c_n^-(k_n)/a_n^- & 0 \\ 0 & c_n^+(k_n)/a_n^+ \end{pmatrix}\} . \tag{7.10}$$

We summarize this by saying that the inverse scattering data is given by

$$\mathcal{S} \equiv \{R(k),\ k \in \Sigma\},\quad \Sigma = \mathbb{R} \cup \{k_n^+, k_n^-\}, \tag{7.11}$$

it being understood that at $k = k_n^{\pm}$ it is the residue of $R(k)$ (or of its "extension") which enters $\mathcal{S}$. (We use the word "extension" with reserve because $R(k)$ need not be defined off the real k-axis when $Q(x)$ doesn't decrease fast enough as $x \to \pm\infty$.)

Let us now go on to derive the above family of evolution equations for the Zak.-Shab.-A.K.N.S. eigenvalue problem. (I wonder if the right question can be asked again ?)

8. OBTAINING THE A.K.N.S. FAMILY OF EVOLUTION EQUATIONS.

"I can see that it may indeed be very difficult to obtain either the scattering equation or the whole family of evolution equations given just a single member of such a family. But maybe some light can be thrown on this difficulty by examining the essentials in the passage from the scattering equation to the evolution equations. Is the derivation of the evolution equations simple and straightforward enough to be examined here ?"

Yes and no. The evolution equations can be obtained with remarkable simplicity. (But I know that simplicity depends much on the eye of the beholder and even more on what he is familiar with so I leave you to judge the credentials of this derivation with respect to simplicity.) Why yes and no then ? Because the simple derivation I will now give assumes some k-analiticity properties which look natural enough but they are very hard to derive. I will not derive these k-analyticity properties.

Let us start by rewriting the eigenvalue equation (5.1) as follows :

$$F_x F^{-1} = Q - ik\sigma \quad \text{where } \det F \neq 0 \tag{8.1}$$

It is convenient to define

$$\text{(a) } Z \equiv F_x F^{-1}\ ,\ \text{(b) } B \equiv F_t F^{-1} \tag{8.2}$$

in terms of which we write out the (F_x, F_t)-compatibility condition

$$\text{a) } F_{xt} = F_{tx} \Leftrightarrow \text{(b) } Z_t = B_x + [B,Z]\ . \tag{8.3}$$

Substituting equation (1) into (3b) and using $\partial k\sigma/\partial t = 0$ now gives us an ordinary d.e. to be solved for $B(t,x,k)$:

$$\text{a) } Q_t = B_x + [B, Q-ik\sigma],\ \text{(b) } (k\sigma)_t = 0. \tag{8.4}$$

That, in essence, is all there is to the derivation. The explicit forms of the evolution equations are now found by exploiting the k-independence and "off-diagonality",

$Q \equiv \begin{pmatrix} 0 & q_2 \\ q_1 & 0 \end{pmatrix}$, in order to solve (4a) for $B(t,x,k)$. The A.K.N.S. family is readily obtained if we assume that for some entire scalar function of k, say $w_2(t,k)$, the product $w_2(t,k)B(t,x,k)$ entire in k, given the technical proviso that $B(t,x,k)$ is regular in k at the bound state energies of (1). What we will do now to obtain the solution B of (4a) is the natural generalization of the recurrence arguments on powers of k which may be used when $w_2(t,k)B(t,x,k)$ is polynomial in K. What is perhaps of particular interest is the way the operator L* makes a natural appearance. The derivation of the k-analyticity assumptions (for some solution of (4a) is however another matter which will be touched on again but not derived.

Note first that we may assume, w.l.o.g., that trace $B = 0$. This is equivalent to assuming that $(\det F)_t = 0$ [since $(\ln \det F)_t \equiv \text{trace}\, F_t F^{-1} \equiv \text{trace}\, B$]. Thus we may put

$$B \equiv \sigma\, B_{11} + B^{off} \; , \; B^{off} \equiv \begin{pmatrix} 0 & B_{12} \\ B_{21} & 0 \end{pmatrix}. \tag{8.4c}$$

Note that $[\sigma, B^{off}] = 2\sigma B^{off} = -2B^{off}\sigma$.

Substituting into (8.4a) gives us the coupled equations :

$$\sigma B_{11x} = [Q, B^{off}] \text{ (diagonal part)} \tag{8.5a}$$

$$Q_t = B^{off}_x + 2ik\sigma B^{off} - 2Q\,\sigma B_{11} \text{ (off-diag.part)}. \tag{8.5b}$$

Defining $B_{11}(t, x \to -\infty, k) \equiv \frac{1}{2} w(t,k)$ we then can eliminate B_{11} to obtain a purely off-diagonal equation :

$$Q_t = (\partial_x + 2ik\sigma - 2Q \int_{-\infty}^{x} [Q,.])\, B^{off} + w\sigma Q \tag{8.6a}$$

$$\equiv -2i\sigma(L' - k)B^{off} + w(t,k)\sigma Q \tag{8.6b}$$

where

$$L' \equiv \frac{1}{2}\, i\partial_x - i\sigma Q \int_{-\infty}^{x} [Q,.] \tag{8.6c}$$

is a linear integro differential operator in the space of 2×2 off-diagonal matrices. Now for $w_1(t,k) \equiv w_2(t,k)w(t;k)$, rearrange (8.6b) as follows :

$$w_2(t,L')\sigma Q_t - w_1(t,L')Q \tag{8.7}$$

$$= -\ 2i(L'-k)w_2(t,k)B^{off}$$

$$-\ (w_1(t,L') - w_1(t,k))Q$$

$$+\ (w_2(t,L') - w_2(t,k))\sigma Q_t$$

$$\equiv (L'-k)\ \{-\ 2iw_2(t,k)B^{off} + (\Delta w_1)(t,k,L')Q \tag{8.8a}$$

$$-\ (\Delta w_2)(t,k,L')Q_t\}.$$

where

$$(\Delta w_j)(t,k,k') \equiv (k'-k)^{-1}(w_j(t,k') - w_j(t,k)),\ j = 1 \text{ or } 2, \tag{8.8b}$$

are both polynomials (or entire functions) in both k and k' providing that $w_j(t,k)$ are polynomials (entire) in k. For the polynomial cases recurrence on the powers of k is now enough to show that both sides must be zero because all terms inside $\{...\}$ are polynomials in k. For the entire function cases $\{...\}$ is entire in k if we may ignore the convergence problems for the infinite series in L'. If $(L'-k)^{-1}$ is well defined for $|k|$ large enough (ie. it L' 's point spectrum lies in a finite region of the k-plane) it then follows that both sides (8.7) are zero because

$$(L'-k)^{-1}\ (\text{L.H.S. of (8.7)}) \to 0 \text{ as } |k| \to \infty$$

and is equal to the entire function of k denoted by $\{...\}$.

Thus we have the family of evolution equations,

$$w_2(t,L')\sigma Q_t = w_1(t,L')Q, \tag{8.9}$$

defined on off-diagonal matrices Q. The A.K.N.S. vector form, (5.3), can be obtained by applying (8.9) to $\binom{1}{1}$.

The question of how the scattering data evolves is answered by "finding" the "correct" right hand normalization for F. Define :

$$F_+ \equiv J_-(x,k)\begin{pmatrix} \alpha^+ & a^+r^+/\alpha^+ \\ 0 & a^+/\alpha^+ \end{pmatrix}$$

$$F_- \equiv J_-(x,k)\begin{pmatrix} a^-/\alpha^- & 0 \\ a^-r^-/\alpha^- & \alpha^- \end{pmatrix}$$

The conditions $\lim_{x\to-\infty} (F_\pm)_t(F_\pm)^{-1} = \frac{1}{2} w(t,k)\sigma$ are now satisfied iff $(a^\pm)_t = 0$, $((r^\pm)^2/\alpha^\pm)_t = 0$ and finally

$$(\ln r^\pm)_t = \pm w(t,k), \text{ ie. } (\ln R)_t = -w(t,k)\sigma \ . \qquad (8.10)$$

These conditions also permit us to use either of $F_\pm$ for F since :

$$B = (F_\pm)_t(F_\pm)^{-1} \text{ for } k \in \mathbb{R}. \qquad (8.11)$$

This is an ingredient which makes it possible to extend B to both the upper and lower halves of the k-plane. But we will not go into this question further here-as we said it is too complicated. Suffice it to say that one should now derive the assumed analyticity properties of $B(t,x,k)$ because, given $w(t,k)$, $r^\pm(0,k)$ and (8.10), everything is determined. It is conceivable that, on the one hand, no solution B exists with these k-analyticity properties and on the other that egn. (8.9) will not have any solutions. [The question of existence of solutions of (8.9), has been studied elsewhere (Miodek 1977), but this is not central to what we are doing here.]

I would like to conclude this section by drawing attention back to the derivation of (8.4) because apart from technical details equation (8.4) already contains the whole family of evolution equations. [Notice also that (8.4) follows from (8.1) even for $n\times n$ matrices.] The derivation of (8.4) from (8.1) is simple by any standards. It is also easy to go from (8.4) back to (8.1). That is what I was trying to show.

"Very good. But can one systematically start with a given evolution equation and find out if it can be put into the form of equation (8.4) ?"

I think the answer is a qualified yes. One promising method is due to Franco Magri (1977-and I don't understand it completely as yet. However in searching for compatible operators Magri's method seems to give criteria for the existence of a pair of operators which play roles similar to those of $\partial_x - [Q,.]$ and $i[\sigma,.]$ in equation (8.4). Let me indicate very briefly, in an example, the starting point of compatibility-type arguments.

The Korteweg de Vries (KdV) equation

$$u_t = -u_{xxx} + buu_x \equiv K(u) \qquad (8.12)$$

is part of the family $u_s = \omega(\mathcal{L}^*)u_x \equiv P(u)$ where $\mathcal{L}^* \equiv -\frac{1}{4}\partial_x^2 + u + \frac{1}{2}u_x\int_{-\infty}^x$ and $\omega(k)$ is polynomial in k. This family may be obtained by solving the following ordinaty d.e. for β :

$$\tfrac{1}{2} u_t = (-\tfrac{1}{4} \beta_{xxx} + u\beta_x + \tfrac{1}{2}u_x\beta - k\beta_x) \equiv (\Lambda - k\partial_x)\beta \; . \qquad (8.13)$$

Now, assuming that $u = u(s,t,x)$ depends on (s,t,x) in a compatible way we have :

$$u_{ts} \equiv (K(u))_s = (-\partial_x^3 + 6\, u\partial_x + 6\, u_x)u_s \equiv K_u\, u_s \qquad (8.14)$$

where K_u is the indicated linear differentiel operator ; K_u is a kind of functional derivative of $K(u)$. Similarly

$$u_{st} \equiv (P(u))_t = P_u\, u_t \qquad (8.15)$$

Now

$$u_{st} = u_{ts} \Leftrightarrow [P,K](u) \equiv P_u\, K(u) - K_u\, P(u) = 0. \qquad (8.16)$$

Given $K(u)$ it is this last "commutator" equation for $P(u)$ that needs to be solved to obtain the above family, or equivalently by to obtain the antihermitian linear operator Λ and ∂_x of equation (8.13). Magri (1977) gives a rather complicated algorithm for doing this.

[Note : eqn (8.13) is readily obtained from the Schrödinger equation, $\psi_{xx}\psi^{-1} = (u-k)$, given $\psi_t = \alpha\psi + \beta\psi_x$, $2\alpha_x = \beta_{xx}$; cf. (8.1) through (8.4).]

CONCLUSION

At this point I hope I have explained what I set out to explain: that the I.S.T. is a method for generating solvable nonlinear equations but does not yet give satisfactory criteria for recognizing the classes of nonlinear equations to which it may be applied.

Section 9 has been added for complefemess as an example of how inverse scattering problems may be solved and may be thought of as an appendix to section 7.

9. HOW THE k-ANALYTICITY PROPERTIES SOLVE THE INVERSE PROBLEM.

The inverse problem for the Zakharov-Shabat-A.K.N.S. eigenvalue problem seeks to reconstruct the k-independent potential

$$Q(x) = e^{-ikx\sigma}(e^{ikx\sigma}J_\pm(x,k))_x\, J_\pm(x,k)^{-1}, \; Q_{jj}(x) = 0, \qquad (9.1)$$

given the asymptotic properties of the two fundamental Jost like solutions $J_\pm$:

$\lim_{x\to\pm\infty} e^{ikx\sigma}J_\pm(x,k) = I$ for $k \in \mathbb{R}$. These asymptotic properties are contained in

$S(k) \equiv J_+(x,k)J_-(x,k)^{-1}$. Since $(\ln \det J_\pm)_x = \text{trace}\,(Q-ik\sigma) = 0$ we see that $\det J_\pm(x,k) = 1$ and thut that $\det S(k) = 1$. Consequently we may put

$$S(k) \equiv \begin{pmatrix} a^- & a^+r^+ \\ a^-r^- & a^+ \end{pmatrix}, \quad a^+a^-(1-r^+r^-) = 1.$$

The reason for this choice of notation is that

$$e^{-ikx\sigma}S(k) \underset{x \to -\infty}{\sim} J_-(x,k)S(k) = J_+(x,k) \xrightarrow[x \to \infty]{} e^{-ikx\sigma} \equiv \begin{pmatrix} e^{-ikx} & 0 \\ 0 & e^{ikx} \end{pmatrix}$$ so $r^\pm(k)$ are reflection coefficients to the left. Obviously the scattering data to the right is given by $S(k)^{-1}$ and is not independent.

The reconstruction of $Q(x)$ amounts to reconstructing some family of fundamental solutions, say $\{J_-(x,k) | k \in \Sigma \subset \mathbb{C}\}$, at each fixed x such that the R.H.S. of 1) is independent of k and decreases rapidly enough as $|x| \to \infty$ so that $S(k)$ is well defined for $k \in \Sigma \supset \mathbb{R}$, Σ will be specified later. The inverse scattering data $\mathcal{S}$ will be equivalent to some subset of $\{S(k))k \in \Sigma\}$.

The properties we have at our disposal are those analyticity properties of $J_\pm(x,k)$ in the k-plane (for x fixed) which are consequences of equation (9.1). These analyticity properties can be readily obtained from the integral form of equation (9.1) :

$$e^{ikx\sigma}J_\pm(x,k) - I = \mp\int_{\pm\infty}^{x} dy\; e^{iky\sigma}Q(y)J_\pm(y,k) \quad \Leftrightarrow \qquad (9.2a)$$

$$J_\pm(x;k)e^{ikx\sigma} - I \equiv A_\pm(x,k) = \mp\int_{\pm\infty}^{x} dy\; e^{ik(y-x)\sigma}Q(y)(A_\pm(y,k)+I)e^{ik(x-y)\sigma}$$

(9.2b)

The equations for the columns $A_\pm^{(j)}$ are uncoupled and so :

$$A_\pm^{(j)}(x,k) = \mp\int_{\pm\infty}^{x} dy\; e^{ik(y-x)(\sigma-\sigma_{jj}I)}Q(y)(A_\pm^{(j)}(y,k)+I^{(j)}).$$

(9.2c)

Here σ_{jj} are the diagonal entries of σ. Since $(y-x)$ doesn't change sign during this integration it follows that if the itterative series converges for $k \in \mathbb{R}$ then it must also converge in the half-plane $(\sigma-\sigma_{jj}I)(x-y)\mathrm{Im}k \geq 0$. Thus, given that $(\sigma_{jj}-\sigma_{11}) \leq 0$, $(\sigma_{jj}-\sigma_{22}) \geq 0$ it follows that :

$A_-^{(1)}(x,k)$ and $A_+^{(2)}(x,k)$ are analytic [cts] in k for $\mathrm{Im}k > 0$ [≥ 0] ;

$A_-^{(2)}(x,k)$ and $A_+^{(1)}(x,k)$ are analytic [cts] in k for imk < 0 [≤ 0] .

Furthermore since $Q_{jj}(x) \equiv 0$, the inhomogeneous term in (9.2c) tends to zero as $|k| \to \infty$ in the appropriate half-planes. Thus $A_\pm^{(j)}(x,k) \to 0$ as $|k| \to \infty$ in the appropriate half-plane. The analyticity properties are immediately translated into k-analyticity properties for the columns of $J_\pm(x,k)$ by using the definition in (9.2b). It is thus convenient to introduce the notation :

$$J_+ \equiv (\psi_r^- \quad \psi_r^+), \; J_- \equiv (\psi_l^+ \quad \psi_l^-) \tag{9.3}$$

where subscripts l and r stand for left and right and the columns ψ_r^+, ψ_l^+ (resp. ψ_r, ψ_l) are analytic for Imk > 0 (resp. Imk < 0).

The x-independance of the determinants of matrix solutions of (9.1) can now be used, as Wronskians are used in the Schrödinger equation, to obtain the analytic properties of S(k).

Thus :

$$\left\{ \begin{aligned} & a^\pm(k) = \pm\det(\psi_l^\pm \quad \psi_r^\pm) \text{ respectively analytic [cts] for } \pm \text{ Imk} > 0 \; [\geq 0] \; ; \\ & a^\pm(k)r^\pm(k) = \pm\det(\psi_r^\pm \quad \psi_l^\mp) \text{ for } k \in \mathbb{R} \end{aligned} \right. \tag{9.4}$$

only unless $Q(x)$ decreases exponentially as $|x| \to \infty$. We will assume, without specifying explicit conditions on $Q(x)$, that $r^\pm(k)$ can be somehow extended to at least partial neighborhoods of the respective zeros $\{ k_n^\pm | \pm \text{Im } k_n^\pm \geq 0\}$ of $a^\pm(k)$. The equation $J_+(x,k) = J_-(x,k)S(k)$ can then also be extended to such nbhds as a definition of those columns which have not yet been defined there. We will also assume that the zeros of $a^\pm(k)$ are simple so that

$$a_n^\pm \equiv \lim_{\epsilon \to 0} a^\pm(k_n^\pm + \epsilon)/\epsilon \neq 0 \; .$$

At these zeros (the bound state energies) we have linear dependence,

$$\psi_r^\pm(x,k_n^\pm) = \psi_l^\pm(x,k_n)c_n^\pm \neq 0 \quad ,$$

of the indicated columns. Thus for $\text{Imk}_n^\pm \neq 0$ the indicated columns decrease exponentially as $|x| \to \infty$ and these bound state energies off the real k-axis are the point spectrum of 1). To the above assumption about $r^\pm(k)$ we now add

$$\lim_{\epsilon \to 0} (a^\pm \; r^\pm)(k_n^\pm + \epsilon) = a_n^\pm \; r_n^\pm$$

and we deduce that $r_n^{\pm} = c_n^{\pm}/a_n^{\pm}$ by using appropriate columns of $J_+(x,k) = J_-(x,k)S(k)$ as $k \to k_n^{\pm}$.

Now we want to obtain equations (of the Marchenko type) which can be used to construct $\{J_-(x,k)|k \in \Sigma \equiv \{k_n^+, k_n^+\} \cup \mathbb{R}]$ given $\mathcal{Y} \equiv \{r^+(k), r^-(k)|k \in \mathbb{R}\} \cup \{r_n^+(k_n^+), r_n^-(k_n^-)\}$ where it is to be understood that $r^{\pm}(k_n^{\pm}) = c_n^{\pm}/a_n^{\pm}$. Such equations can be obtained by explicitly introducing the deduced k-analytic properties into the what is essentially the Fourier k-transform of $J_+(x,k) = J_-(x,k)S(k)$. To this end we introduce the Fourier k-transform of $A_-(x,k)$.
Define :

$$2\pi \hat{A}_-(x,x+y) \equiv \int_{-\infty}^{+\infty} dk\, A_-(x,k)e^{iky\sigma} \quad . \tag{9.5}$$

This particular way of writing the Fourier transform has been chosen because use of Cauchy's theorem shows that the k-analytic properties of the columns of $A_-(x,y)$ are now exactly equivalent to the <u>triangularity property</u> $\hat{A}_-(x,x+y) = 0$ for $y > 0$. Thus by inverting this Fourier transform we obtain the "triangular" integral transformation $e^{-ikx\sigma}$ (the solution when $Q \equiv 0$) $\to J_-(x,k)$,

$$J_-(x,k) - e^{ikx\sigma} = \int_{-\infty}^{x} dy\, \hat{A}_-(x,y)e^{-iky\sigma} \quad . \tag{9.6}$$

Triangular integral transformations of this general type are the starting points of many inverse methods and are equivalent to appropriate k-analyticity properties.

We now take the same Fourier k-transform of $J_+ = J\text{-}S$ after rearranging it as follows to take advantage of our bound state properties as well :

$$J_+(x,k)N(k)e^{ikx\sigma} - I = J_-(x,k)(I + \sigma_1 R(k))e^{ikx\sigma} - I \tag{9.7}$$

where $N(k)^{-1} \equiv \begin{pmatrix} a^-(k) & 0 \\ 0 & a^+(k) \end{pmatrix}$, $\sigma_1 \equiv \begin{pmatrix} 0 & 1 \\ 1 & 0 \end{pmatrix}$,

$$R(k) \equiv \begin{pmatrix} r^-(k) & 0 \\ 0 & r^+(k) \end{pmatrix} .$$

Note that the first and second columns of $J_+(x,k)N(k)$ are respectively meromorphic for $\text{Im}k < 0$ and $\text{Im}k > 0$, the residues at the supposedly simple poles being given by

$$J_-(x,k_n)\sigma_1 \begin{pmatrix} c_n^-(k_n)/a_n^- & 0 \\ 0 & c_n^+(k_n)/a_n^+ \end{pmatrix} \equiv J_-(x,k_n)\sigma_1 R_n(k_n) \tag{9.8}$$

where $c_n^{\pm}(k_n^{\pm}) \equiv c_n^{\pm}$ but $c_n^{\pm}(k_n^{\pm}) \equiv 0$.

Now for $y \leq 0$ the Fourier k-transform of (9.7) gives us :

$$2\pi i \sum_{k_n} J_-(x,k_n)\sigma_1 R_n'(k_n)\, e^{ik_n(x+y)\sigma}$$

$$= 2\pi \hat{A}_-(x,x+y) + \int_{-\infty}^{+\infty} dk\; J_-(x,k)\sigma_1 R(k) e^{ik(x+y)\sigma} \qquad (9.9)$$

where $R_n'(k_n^{\pm}) \equiv \pm R_n(k_n^{\pm})$ if $\text{Im}k_n^{\pm} \neq 0$,

$R_n'(k_n^{\pm}) \equiv \mp \frac{1}{2} R_n(k_n^{\pm})$ if $\text{Im}\; k_n^{\pm} = 0$.

Finally use (9.6) to eliminate $J_-(x,k)$ to obtain the (Marchenko-like) Fredholm equation :

$$\hat{A}_-(x,y) + \hat{R}(x+y) + \int_{-\infty}^{x} dy'\; \hat{A}_-(x,y')\hat{R}(y'+y) = 0 \;, \qquad (9.10)$$

for $y \leq x$, when

$$2\pi\sigma_1 \hat{R}(y) \equiv \int_{-\infty}^{+\infty} dk R(k) e^{iky\sigma} - 2\pi i \sum_{k_n} R_n'(k_n) e^{ik_n y\sigma}$$

Equation (9.10) is to be solved for $\hat{A}_-(x,y)$ for $y \leq x$ with x fixed - it is not a Volterra equation. Equations (9.6) and (9.1) must then respectively yield $\{J_-(x,k) | k \in R\}$ and $Q(x)$, providing that (9.10) has a unique solution.
[It may be possible to work directly with $\{A_-(x,k)\}$ rather than $\{\hat{A}(x,y) | y \leq x\}$ if the Cauchy principle value integral, $\int_{-\infty}^{+\infty} dk \frac{P.V}{k-k'}$ .... for $k' \in \mathbb{R}$, is used for (9.7) rather then its Fourier transform.].

Restrictions on $\hat{R}(x)$ (i.e. onthe inverse scattering data) must now be added to (9.10) such that a unique solution $\hat{A}_-(x,y)$ exists. The A.K.N.S. paper (1974) gives some sufficient restrictions.

REFERENCES

MAGRI (Franco) (1977) Preprint of Instituto di Matematica del Politecnico, Milana, Italy.

KUMEI (Sukeyuki) J.M.P. 18 (Feb. 1977) p. 256.

F.B. ESTABROOK & H.D. WAHLQUIST J.M.P. 16 (Jan. 1975) p. 1 ; 17(1976) p. 1293.

CORONES (James) J.M.P. 18 (January 1977) p. 163.

H.C. MORRIS J.M.P. 18 (Feb. 1977) p. 285.

ABLOWITZ, KAUP, NEWELL & SEGUR (A.K.N.S.) Studies in App. Math. L III n°4 (1974) p. 249.

P.D. LAX, Comm. on Pure and App. Math. XXI (1968) p.467, XXVIII (1975) p. 141.

I. MIODEK, J.M.P. 19 (January 1978) p 19.

PART II

FIVE LECTURES ON SPECIAL APPLICATIONS

and

ONE THEORETICAL LECTURE

ON SOLUTIONS OF INVERSE PROBLEMS

ON THE INVERSE PROBLEM OF LOCAL SEISMIC FOCI

Georges Jobert, Armando Cisternas
Laboratoire d'étude géophysique de structures profondes, associé au C.N.R.S., I.P.G.
Université P. et M. Curie, 75230 Paris.

Abstract

A fast method to solve the inverse problem of local earthquakes foci is presented. The introduction of a quadratic hyperbolic norm in space-time permits the treatment of the non-linearity of the problem by means of a simple scalar equation. The duration of the process is comparable to that of one iteration of Geiger's method. A complete analytical study is given.

Introduction.

The problem of determination of the hypocenters of earthquakes from the arrival times of seismic waves at a network of seismic stations is a simple kind of inverse problem that may be either overdetermined or over- and underdetermined at the same time.

The method of Geiger (e.g. Flinn,1960) is the standard least squares technique to obtain the hypocenters end origin times of earthquakes. Since the problem is non-linear the solution is found by iterations, starting from a guess trial. This process is rather slow since a typical determination requires about 5 or 6 iterations and at each step the matrix coefficients must be calculated again due to non-linearity.

If the amount of data to be processed becomes large, an algorithm faster than Geiger's becomes desirable. Such is the case for example when a seismo-volcanic crisis is being observed, in the recording of aftershocks after a big earthquake or in studies of microearthquakes in a tectonic region. In all these situations a small local network is required and therefore the effect of lateral heterogeneities may be considered as a small perturbation about a homogeneous model.

In this paper we present a detailed analysis of a method that, we claim, is at least about 5 times faster than Geiger's. Since we need the hypothesis of homogeneity we shall suppose that the focus is very close to the network ; if necessary, stations corrections should be added to the arrival times. Even when no such corrections are made, preliminary determinations prior to a more refined treatment are so obtained. Such a method should also be useful for an automatic real time hypocenter determination.

The same norm has already been employed by Miyamura (1960) for the location of earthquakes in Japan. We develop here the theory for the regular and singular cases and give an approximate solution for the case when P and S waves are observed.

1. OBSERVATION OF ONE PHASE.

1.1. Choice of a norm.

We suppose an earthquake with focus F and origin time t_o. Seismic waves propagate with velocity V. The arrival times t_i are recorded by a network of n stations S_i. The commonly used norm is defined by:

$$R_o = \Sigma \left[\| \overrightarrow{FS_i} \| - V\,(t_i - t_o) \right]^2 \qquad (1)$$

where Σ notes a summation from 1 to n. The square roots present in this expression forbid any algebraic treatment and oblige to solve the problem - i.e. to find (F,t_o) such that R_o be a minimum - by linearization near a given starting point and successive iterations, each of which implies a 4x4 matrix inversion. Moreover uniqueness and conveniency of the found solution are difficult to discuss.

We have introduced (Cisternas and Jobert, 1977) the norm defined by :

$$R = \Sigma\, p_i \left(\| \overrightarrow{FS_i} \|^2 - V^2 (t_i - t_o)^2 \right)^2 \qquad (2)$$

with positive or null weights p_i . As we may write :

$$R = \Sigma\, p_i \left(\| \overrightarrow{FS_i} \| + V(t_i - t_o) \right)^2 \left(\| \overrightarrow{FS_i} \| - V(t_i - t_o) \right)^2$$

and as, for the solution, we have :

$$\| \overrightarrow{FS_i} \| + V(t_i - t_o) \approx 2 \| \overrightarrow{FS_i} \|$$

we see that comparable results are found by using (1) or(2) if we take as weight the quantity :

$$p_i = 2 \| \overrightarrow{F_1 S_i} \|^{-2}$$

where F_1 is a fixed point in the focal region, the position of which may be modified if necessary, as the proposed method is very fast. Other weights may be used, for example:smaller for larger distances as the hypothesis of homogeneity is probably less valid when distance to the focus increases.

1.2. Evaluation of R and of its derivatives.

Using the weights p_i we define the barycenters G of S_i and τ of t_i and take them as origin to define the vectors:

$$X = \begin{bmatrix} \overrightarrow{GF} \\ V(t_o - \tau) \end{bmatrix} \qquad Y_i = \begin{bmatrix} \overrightarrow{GS_i} \\ V(t_i - \tau) \end{bmatrix} \qquad (3)$$

which are considered as one-column matrices. By definition :

$$\Sigma\, p_i\, Y_i = 0 \quad . \qquad (4)$$

We then may write :

$$R = \Sigma\, p_i \left[(X - Y_i)^T\, \Gamma\, (X - Y_i) \right]^2 \qquad (5)$$

where Γ = diag (1 1 1 -1) and ^{T} notes a transposition. Developing this expression and taking (4) into account we obtain :

$$R = p\,(X^T \Gamma X)^2 + 4\, \Sigma\, p_i (X^T \Gamma\, Y_i)^2 + 2m\, X^T \Gamma X - 4\, .\, p_i (Y_i^T\, \Gamma Y_i) Y_i^T \Gamma\, X + r$$

where : $r = \Sigma\, p_i (Y_i^T \Gamma\, Y_i)^2 \qquad m = \Sigma\, p_i (Y_i^T \Gamma\, Y_i) \qquad p = \Sigma\, p_i \qquad (6).$

From this expression we deduce the gradient of R, ∇R , defined by :

$$R(X + H) - R(X) = \nabla R^T . H + O(\|H\|^2) \quad . \quad (7)$$

We find :

$$\nabla R = 4\,\Gamma[(M + \lambda I)X - B] \quad (8)$$

$$M = 2\,\Sigma\, p_i Y_i\, Y_i^T\, \Gamma = 2\,J\,\Gamma \quad (9)$$

$$\lambda = p\,X^T\,\Gamma X + m \quad (10)$$

$$B = \Sigma\, p_i\, Y_i (Y_i^T\, \Gamma Y_i) \quad (11)$$

$$\mathrm{tr}\, M = 2\,m\,.$$

I is the identity matrix, J the matrix of inertia of the vectors Y_i .

The Hessian Ω of R will be similarly defined by :

$$\nabla R(X + H) - \nabla R(X) = \Omega\, H + O(\|H\|^2) \quad (12)$$

From (8) we deduce :

$$\Omega = 4\,\Gamma\,(M + \lambda I + 2pXX^T\Gamma)$$

$$= 4\,\Gamma\,(2J + 2\,p\,XX^T + \lambda\Gamma)\,\Gamma \quad . \quad (13)$$

This expression shows that Ω is a positive definite matrix if $\lambda = 0$ and $(J + pXX^T)$ is regular. In such conditions an extremum of R will be a true minimum.

1.3. Causality.

Causality induces some conditions for the solution. Using the triangular inegality, from :

$$p\,\overrightarrow{FG} = \Sigma\, p_i\, \overrightarrow{FS_i}$$

we deduce :

$$p\|\overrightarrow{FG}\| \leqslant \Sigma\, p_i \|\overrightarrow{FS_i}\| \quad .$$

Equality obtains only if all stations are aligned with the focus, a case we shall exclude. In absence of errors we have :

$$\|\overrightarrow{FS_i}\| = V(t_i - t_o)$$

and therefore:

$$\Sigma\, p_i\|\overrightarrow{FS_i}\| = p\,V(\tau - t_o)$$

so that ;

$$\|\overrightarrow{FG}\| \leqslant V(\tau - t_o)$$

or

$$X^T\Gamma X \leqslant 0 \quad . \quad (14)$$

On the other hand, always in absence of errors, we may write :

$$\|\overrightarrow{FG} + \overrightarrow{GS_i}\|^2 = V^2(t_i - \tau + \tau - t_o)^2$$

from which we deduce, taking (4) into account :

$$p\|\overrightarrow{FG}\|^2 + \Sigma\, p_i \|\overrightarrow{GS_i}\|^2 = p\,V^2(\tau - t_o)^2 + \Sigma\, p_i(t_i - \tau)^2$$

$$\lambda = p\,X^T\Gamma X + m = 0 \quad . \quad (15)$$

Comparing (14) and (15) we see that m is positive.

1.4. Extrema of R.

The extrema of R are given by the zeroes of ∇R and correspond to solutions of :

$$(M + \lambda I)X = B \quad . \quad (16)$$

This is a vectorial third degree equation. Its peculiar form allows one to reduce it to a scalar ninth degree equation.

If λ is not opposite to one of the eigenvalues m_i of M, supposed regular the solution of (16) is given by :

$$X = (M + \lambda I)^{-1} B \quad . \quad (17)$$

Taking this expression into (10) we obtain an algebraic equation for λ :

$$\lambda = f(\lambda) \quad (18)$$

with : $$f(\lambda) = p\, B^T (M + \lambda I)^{-1} \Gamma (M + \lambda I)^{-1} B + m \quad . \qquad (19)$$

Let us now introduce the unitary matrix P diagonalizing the symmetric inertia matrix J :

$$2J = P\, \Sigma P^T \quad .$$

Σ is a positive diagonal matrix. Let us introduce also the matrix N :

$$N = \Sigma^{-1/2} P^T M P \Sigma^{1/2} = \Sigma^{-1/2} P^T P\, \Sigma\, P^T \Gamma P \Sigma^{1/2} = \Sigma^{1/2} P^T \Gamma P \Sigma^{1/2} \quad .$$

N is thus symmetric and can be diagonalized by the unitary matrix R :

$$N = R^T \Lambda R \quad .$$

Inversely we have :

$$M = P \Sigma^{1/2} N \Sigma^{-1/2} P^T = Q \Lambda Q^{-1} \quad , \qquad (20)$$

with

$$Q = P \Sigma^{1/2} R^T \quad .$$

Let us remark that : $$Q^T \Gamma Q = R\Sigma^{1/2} P^T \Gamma P \Sigma^{1/2} R^T = R N R^T = RR^T \Lambda RR^T = \Lambda \quad .$$

As $$(Q^T \Gamma Q X , X) = (\Gamma Q X , Q X) = (\Lambda X, X)$$

we see, using the inertia theorem for quadratic forms, that Λ has the same signature as Γ , i.e. 3 positive eigenvalues and one negative. $\Lambda = \mathrm{diag}\,(m_1, m_2, m_3, m_4)$.

Using in (19) the expression found for M we obtain :

$$f(\lambda) = p\, B^T Q^{-1T} (\Lambda + \lambda I)^{-1} Q^T \Gamma Q (\Lambda + \lambda I)^{-1} Q^{-1} B + m$$

$$= p\, \bar{B}^T (\Lambda + \lambda I)^{-2} \Lambda \bar{B} + m \quad , \qquad (22)$$

where : $$\bar{B} = Q^{-1} B \quad .$$

Equation (18) becomes :

$$\lambda = p \sum_1^4 \frac{\bar{b}_i^2\, m_i}{(m_i + \lambda)^2} + m \quad . \quad (23)$$

In absence of errors we have shown (equ. 15) that $\lambda = 0$. In this ideal case the graph of $f(\lambda)$ pass through the origin. If there exist errors equation (18) may be solved by classical numerical methods. The (small) solution will always be on the segment $[-m_3, -m_4]$ where m_3 is the smallest positive eigenvalue and m_4 the negative one.

1.5 Approximate solution.

If errors are present λ may differ from zero. Let us assume however that it remains small compared to $\| M \|$. For a regular matrix M equation (17) may be linearized :

$$X = M^{-1} B - \lambda M^{-2} B \quad . \qquad (24)$$

Similarly to the second order in $\lambda / \| M \|$ we have :

$$f(\lambda) = f_0 + \lambda f_1$$

with :

$$f_0 = p\, B^T M^{T-1} \Gamma M^{-1} B + m = p\, B^T \Gamma M^{-2} B + m$$

$$f_1 = -2p\, B^T M^{T-1} \Gamma M^{-2} B = -2p\, B^T \Gamma M^{-3} B \quad .$$

The solution of (18) is then given by :

$$\lambda = f_o / (1 - f_1) \qquad . (25)$$

Once verified that this quantity is small compared to $\| M \|$, X is obtained from (24) This method needs only one inversion of a 4x4 matrix. In the case when the stations and the weights are kept constant , a 3x3 matrix inversion is sufficient. We may indeed write :

$$M = 2 \begin{bmatrix} J_S & V A \\ V A^T & pV^2\theta^2 \end{bmatrix} \Gamma$$

where J_S is the matrix of inertia of the stations, θ the r.m.s. of times of arrival, A some vector. Its inverse is given by :

$$M^{-1} = \frac{1}{2} \Gamma \begin{bmatrix} K & V^{-1}C \\ V^{-1}C^T & kV^{-2} \end{bmatrix}$$

where :

$$k^{-1} = p\,\theta^2 - A^T J_S^{-1} A \qquad\qquad C = - k J_S^{-1} A$$

$$K = J_S^{-1} + k J_S^{-1} A A^T J_S^{-1} \quad .$$

This formula could also be used to discuss the effect of an error in the velocity used for the determination of the focus, but we shall not develop this point here.

2. OBSERVATION OF TWO PHASES.

Let us suppose now that two kinds of phases with velocities V and V' are observed respectively at n stations S_i and n' stations S_i' at times t_i and t_i' . From now on we shall note with a dash parameters related to the second set and write only equations related to the first with a (&) when they have to be repeated for dashed variables (two dashes will be equivalent to no dash). Introducing the barycenters G,G', τ, τ' , we note :

$$X = \begin{bmatrix} \overrightarrow{GF} \\ V(t_o - \tau) \end{bmatrix} \ (\&) \qquad Y_i = \begin{bmatrix} \overrightarrow{GS_i} \\ V(t_i - \tau) \end{bmatrix} \ (\&) \quad .$$

Between the vectors X,X' there exists a relation :

$$L X - L' X' = \begin{bmatrix} \overrightarrow{GG'} \\ \sqrt{VV'}\,(\tau' - \tau) \end{bmatrix} = 2 A = - 2 A' \quad , (26)$$

Where : $L = \mathrm{diag}\,(1 \ 1 \ 1 \ \rho)$ (&) $\qquad \rho = 1/\rho' = \sqrt{V'/V}$.

Evidently $LL' = I$.

The minimum of $R(X) + R'(X')$ is now to be searched for with the condition (26) . Using a vectorial Lagrange multiplier $Z = - Z'$ we obtain the equations :

$$\nabla R + L Z = 0 \qquad (\&)$$

Eliminating the multipliers Z ,Z' between them we deduce the equation :

$$L'\nabla R + L \nabla R' = 0 \qquad . (27)$$

We thus obtain the new equation in X,X' :

$$L' (M + \lambda I) X + L (M' + \lambda' I) X' = L' B + L B' \quad ; (28)$$

with the same definitions (9)(10)(11)(&). We shall introduce a new unknown vector W defined by :

$$2 W = L X + L' X' \qquad (29)$$

so that : $X = L'(W + A)$ (&) (30).

W is thus solution of :

$$(N + \mu \Sigma) W = C \qquad (31)$$

where :

$$N = L' M L' + L M' L$$

$$\mu = \lambda + \lambda'$$

$$\Sigma = \mathrm{diag}(1 \;\; 1 \;\; 1 \;\; \sigma)$$

$$\sigma = (\lambda V^2 + \lambda' V'^2) / VV'(\lambda + \lambda') \quad \text{if } \lambda^2 + \lambda'^2 \neq 0,$$

$$= 1 \quad \text{if} \quad \lambda = \lambda' = 0 .$$

$$C = L' B + L B' - L'(M + \lambda I) L'A - L(M' + \lambda' I) L A' .$$

The method presented in § 4 cannot be used simply here as Σ is not the unit matrix. We shall therefore adapt only the linear approximation of §5 . If $|\mu| \ll \|N\|$ we may write the solution of (31) :

$$W = (N^{-1} - \mu N^{-1} \Sigma N^{-1}) C$$

and

$$X = L'(N^{-1} C - \mu N^{-1} \Sigma N^{-1} C_0 + A) \quad (\&) \; (32)$$

where :

$$C_0 = L'B + L B' - L'M L'A - L M' LA' .$$

Taking (32)(&) into (10)(&) we obtain two first degree equations in λ, λ' which allow one to solve the problem. As the formulae are rather complex the solution will not be given here.

3. CASES OF SINGULARITY.(ONE PHASE OBSERVED)

Till now we have supposed the matrix M regular. If $\mathcal{Y}$ is the subspace spanned by the vectors Y_i , if M is singular $\dim \mathcal{Y} < 4$.

3.1. $\dim \mathcal{Y} = 3$.

When the dimension of $\mathcal{Y}$ is reduced to 3, there exists a vector Z such that :

$$\forall i, \quad Z^T Y_i = 0 . \qquad (33)$$

This happens when all the seismic stations have the same altitude, then $Z = e_3$ the vertical unit vector, when the times of arrival at all stations are the same, then $Z = e_4$ the time unit vector, or when the focus is at very large distance from the barycenter, a case we have discarded ; but other cases are theoretically possible. We shall discuss only the first one and develop in Appendix a method valid when the differences of altitudes are small compared to differences in horizontal distances.

From (33) (9) (11) we deduce :

$$Z^T B = Z^T M = M \Gamma Z = 0 \qquad (34)$$

Multiplying both sides of equation (16) at left by Z^T we obtain :

$$\lambda Z^T X = 0 .$$

Two cases are possible :

- $Z^T X \neq 0$, then $\lambda = 0$. Equation (16) reduces to :

$$M X = B$$

the general solution of which is given by :

$$X = M^{(-1)} B + z \Gamma Z = M^{(-1)} B + z e_3$$

$M^{(-1)}$ is the generalized inverse of M, easily deduced by inversion of a reduced 3x3 matrix and such that : $\mathrm{Im}\, M^{(-1)} = \mathcal{Y}$ $\quad \mathrm{Ker}\, M^{(-1)} = \{y\, e_3\}$.

z is the depth of the focus. It is deduced from (10) :

$$0 = \lambda = p\,(B^{T} M^{T(-1)} + z\, e_3^{T})\,\Gamma\,(M^{(-1)} B + z\, e_3) + m \quad , \text{ or}$$

$$z^2 = -(B^{T} \Gamma M^{(-2)} B + m/p) \quad . \qquad (35)$$

Evidently only one solution is convenient. According to the remark at the end of § 2 it corresponds to a true minimum of R ; the inertia matrix is singular but X is not in $\mathcal{Y}$ and therefore the matrix $J + p\,XX^{T}$ is regular.

- $\lambda \neq 0$, then $Z^{T} X = 0$, $X \in \mathcal{Y}$.

The solution would be in the hyperplane of the stations- times of arrival ; thus the focus would be in the plane of the stations. X may be found by the procedure described in § 4 after suppressing the null eigenvalue. But such solution , the nearest to the true minimum found above, can correspond only to a maximum or a saddle point of R .

3.2. dim $\mathcal{Y}$ = 2 .

As an example of such case we shall discuss only the case of stations aligned in a direction taken as support of e_1 . There exist then 2 orthogonal vectors $Z_1 = e_2$, $Z_2 = e_3$ orthogonal to each Y_i. The solution with $\lambda = 0$ is given by :

$$X = M^{(-1)} B + z_2\, e_2 + z_3\, e_3 \quad .$$

The coordinates z_2, z_3 are such that :

$$\lambda = 0 = p\,(B^{T} M^{T(-1)}\,\Gamma\, M^{(-1)} B + z_2^2 + z_3^2) + m .$$

The solutions are therefore on a circle with e_1 as axis, as was evident beforehand.

References

Cisternas A. , Jobert G., Problème inverse des foyers sismiques pour un réseau local tridimensionnel, Comptes- R. Ac. Sc., 1977,B,284,69-72.

Flinn E.A., Local earthquake location with an electronic computer. Bull. Seism. Soc. Am., 1960,50,467-470.

Miyamura S. , Local earthquakes in Kii peninsula, Central Japan,Part 4. Bull. Earthqu. Res. Inst., 1960,38,71-112.

APPENDIX

NEARLY COPLANAR STATIONS

We shall limit ourselves to the case where only one phase is observed.The method could be extended to the case of 2 phases to the price of rather boring calculations. Let us suppose that the stations have altitudes only slightly different, these differences being taken as first-order infinitesimals. In the following the vectors will be written as the sum of their vertical component (on the e_3 unit vector) and of a "horizontal" space-time vector. So Y_i will be replaced by $H_i + z_i e_3$, and the solution X by $U + u e_3$. We may write the elements of equation (16) as sums or series of terms of different orders in ε , a quantity related to the differences of altitudes.

a) $$M = M_o + \varepsilon M_1 + \varepsilon^2 M_2$$

where $$M_o = 2 \Sigma p_i H_i H_i^T$$

$$\varepsilon M_1 = 2 \Sigma p_i z_i (e_3 H_i^T \Gamma + H_i e_3^T) = \varepsilon (e_3 W^T \Gamma + W e_3^T)$$

$$\varepsilon^2 M_2 = 2 \Sigma p_i z_i^2 e_3 e_3^T \quad .$$

b) $$m = m_o + \varepsilon^2 m_2$$

c) $$B = B_o + \varepsilon \beta_1 e_3 + \varepsilon^2 B_2 + \varepsilon^3 \beta_3 e_3$$

where $$B_o = \Sigma p_i H_i (H_i^T \Gamma H_i) \qquad \varepsilon \beta_1 = \Sigma p_i z_i (H_i^T \Gamma H_i)$$

$$\varepsilon^2 B_2 = \Sigma p_i H_i z_i^2 \qquad \varepsilon^3 \beta_3 = \Sigma p_i z_i^3$$

d) $$X = X_o + \varepsilon X_1 + \varepsilon^2 X_2 + \dots .$$

$$\lambda = \lambda_o + \varepsilon \lambda_1 + \varepsilon^2 \lambda_2 + \dots .$$

Taking these expressions into equation (16) we obtain two series of equations :

$$(M_o + \lambda_o I) X_o = B_o \qquad \lambda_o = p X_o^T \Gamma X_o + m_o$$

$$(M_o + \lambda_o I) X_1 + (M_1 + \lambda_1 I) X_1 = \beta_1 e_3 \qquad \lambda_1 = 2p X_o^T \Gamma X_1$$

$$(M_o + \lambda_o I) X_2 + (M_1 + \lambda_1 I) X_1 + (M_2 + \lambda_2 I) X_o = B_2$$

$$\lambda_2 = p X_1^T \Gamma X_1 + 2 X_o^T \Gamma X_2 + m_2 \quad .$$

By identifying respectively the horizontal and the vertical components we obtain from the first equations of each system :

$$(M_o + \lambda_o I) U_o = B_o \qquad \lambda_o u_o = 0 = p (U_o^T \Gamma U_o + u_o^2) + m_o \cdot$$

The situation is exactly the same as that discussed in § 3.1 . The solution is therefore :

$$U_o = M_o^{(-1)} B_o \qquad u_o^2 = - (U_o^T \Gamma U_o + m_o /p) \quad .$$

Taking these expressions into the second equations we obtain :

$$M_o U_1 + (e_3 W^T \Gamma + W e_3^T + \lambda_1 I)(U_o + u_o e_3) = \beta_1 e_3 \quad , \text{ or}$$

$$W^T \Gamma U_o + \lambda_1 u_o = \beta_1$$

which gives λ_1 , and

$$M_o U_1 + W u_o + \lambda_1 U_o = 0$$

which gives :

$$U_1 = - M_o^{(-1)} (u_o W + \lambda_1 U_o) .$$

The first order perturbation in the depth of focus, u_1 , is given by the second equation of the second system :

$$\lambda_1 = 2 p (U_o^T \Gamma U_1 + u_o u_1) .$$

We shall stop at this point but the following terms would be deduced by the same procedure.

AN INVERSE PROBLEM FOR ELECTROMAGNETIC PROSPECTION

V. BARTHES, G. VASSEUR
Centre Géologique et Géophysique
Université des Sciences et Techniques du Languedoc
Place E. Bataillon
34060 MONTPELLIER CEDEX (France)

ABSTRACT : From ground based electromagnetic field measurements, the integrated conductivity (conductance) of a thin, horizontal, inhomogeneous, conducting sheet, embedded in a stratified medium can be computed using an iterative method. A numerical example demonstrates a good convergence towards the solution for synthetic data.

INTRODUCTION

Electromagnetic prospecting methods are designed to study the electrical conductivity $\sigma(M)$ of the subsoil from the observation of an electromagnetic (e.m.) field generated by some external sources of natural or artificial origine. The inverse problem consists of determining the conductivity $\sigma(M)$ from ground measurements of one or several components of this field. Several studies have been proposed for solving this problem in the one dimensional case, i.e. when the conductivity only depends on the depth z (Bailey, 1970 ; Parker, 1971, Weidelt, 1972). When the electrical conductivity also varies with horizontal coordinates, the direct problem and moreover the inverse one are much more complex (Weidelt, 1975).

However large simplifications occur when the domain where σ depends upon horizontal coordinates x,y, can be handled in the thin sheet approximation (Price, 1949). This approximation may be applied to thin plane inhomogeneous structures whose thickness and conductivity are such that the tangential electric current flowing in the sheet can be approached by a surface current. The electrical property of such a non uniform structure can be described by its conductance τ (i.e. conductivity integrated over the thickness).

The conductivity model under consideration (Fig. 1) consists of a non uniform horizontal thin sheet at depth z_s, with conductance $\tau(x,y)$, embedded in a stratified conducting medium (M layers with constant conductivity $\sigma_1, \sigma_2, \ldots \sigma_M$) The direct

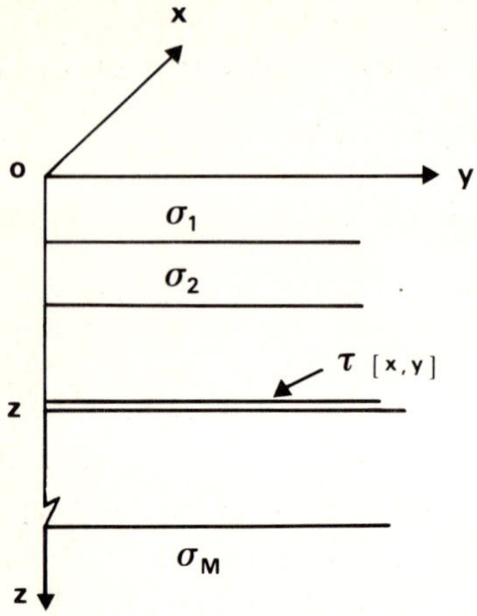

Figure 1 : Model of conductivity structure.

problem of computing the e.m. field may be studied using an integral equation, the integration domain being limited to parts of the plane $z = z_s$ where the conductance τ deviates from its normal value τ_n (Vasseur and Weidelt, 1977). This algorithm also permits to tackle the following inverse problem : assume the normal (z dependant) conductivity distribution is known (i.e. $\sigma_1, \sigma_2, \ldots \sigma_M$ with corresponding thicknesses and the normal conductance τ_n) ; then, using ground measurements of some components of the e.m. field generated by some source in $z < 0$, find the unknown conductance distribution $\tau(x,y)$ in the sheet at $z = z_s$.

INTEGRAL EQUATION FORMULATION

Neglecting the displacement current, for a time variation with pulsation ω, the complex amplitudes $\underline{E}$ and $\underline{H}$ of the electric and magnetic fields are related for $z \neq z_s$ by :

$$\underline{\nabla} \times \underline{H} = \sigma \, \underline{E} + j_e \qquad (j_e \text{ : current intensity of external sources}) \quad (1a)$$

$$\underline{\nabla} \times \underline{E} = - i\omega\mu_0 \, \underline{H} \quad (1b)$$

For $z = z_s$, (1) is replaced by a relation between the discontinuity of the magnetic field and the surface current in the sheet :

$$\underline{z} \times \underline{H} \Big|_-^+ = \tau \, \underline{E}_s \quad (2)$$

where $\underline{z}$ is vertical unit vector, $\Big|_-^+$ the jump of a quantity across $z = z_s$, and the subscript s the horizontal projection of a vector.

Then, using the Green's dyad $\underline{\underline{\mathcal{G}}}(M|M_0)$ relative to the normal medium (i.e. for $\tau = \tau_n$) and $\underline{E}_n$ the normal electric field (which should exist in such a normal medium), it may be shown (Vasseur and Weidelt, 1977) that, for M_0 included in the sheet :

$$\underline{E}_s(M_0) = \underline{E}_{ns}(M_0) - i\omega\mu_0 \int_s (\tau(M) - \tau_n) \, \underline{E}_s(M) \, \underline{\underline{\mathcal{G}}}_s(M|M_0) \, ds \quad (3)$$

which is a vectorial Fredholm integral equation of the second kind for the horizontal components of the electric field in the sheet, the domain of integration s being the

the plane $z = z_s$.

Similarly, for M_0 elsewhere, the e.m. field may be computed from :

$$\underline{E}(M_0) = \underline{E}_n(M_0) - i\omega\mu_0 \int_s (\tau(M) - \tau_n)\, \underline{E}_s(M)\, \underline{\underline{\mathcal{G}}}(M|M_0)\, ds, \qquad (4)$$

$$\underline{H}(M_0) = \underline{H}_n(M_0) + \int_s (\tau(M) - \tau_n)\, \underline{E}_s(M)\, \underline{\underline{\mathcal{H}}}(M|M_0)\, ds \quad ; \qquad (5)$$

(5) is obtained from (4) via (1b) and $\underline{\underline{\mathcal{H}}}$ is a dyad obtained through derivation of $\underline{\underline{\mathcal{G}}}$ with respect to M_0.

For a given $\tau(x,y)$, the direct problem can be solved through the numerical solution of (3) for $\underline{E}_s$ followed by the computation of $\underline{E}$ and $\underline{H}$ from (4) - (5). The inverse problem for determination of $\tau(x,y)$ from $\underline{E}$ or $\underline{H}$ measurements can be handled in the same way since the characteristics of normal medium - and therefore $(M|M_0)$, $\underline{E}_n$, etc... - are assumed to be known. Would $\underline{E}_s(M)$ for M in the sheet be known, then Eq.(4) (or (5)) is a Fredholm equation of the first kind for the unknown $\tau(x,y)$. In fact $\underline{E}_s(M)$ is not known and an iterative procedure has to be used : $\underline{E}_s(M)$ is first approximated by its normal value $\underline{E}_{ns}$ and a first approximation $\tau(x,y)$ is obtained from (4) (or (5)). This function $\tau(x,y)$ is used in (3) to derive a new approximation to $\underline{E}_s(M)$ and the previous procedure is iterated.

A NUMERICAL EXAMPLE

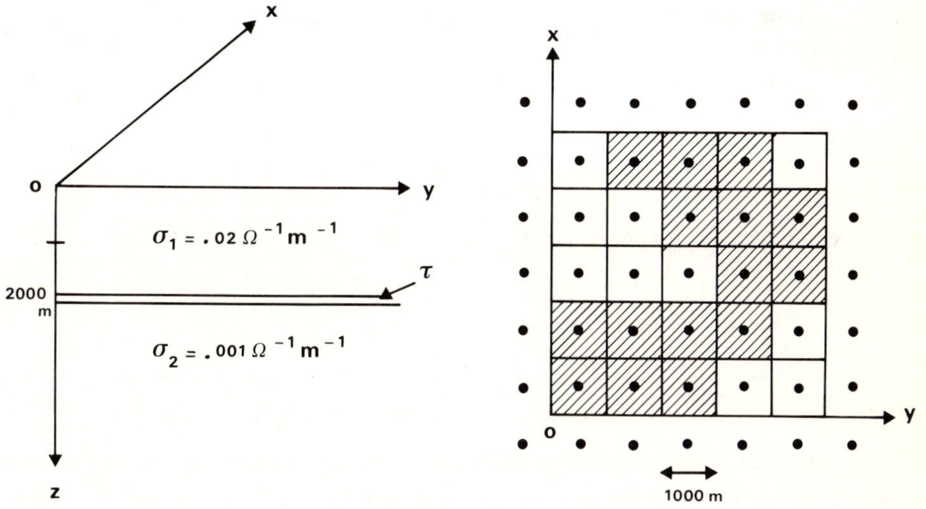

Figure 2 - left part : assumed model ; right part : sketch of the conductance in the plane $z = z_s$. Hatched squares correspond to $\tau \neq \tau_n$ ($0.1\ \Omega^{-1}$ for model A and $1000\ \Omega^{-1}$ for model B). Elsewhere, $\tau = \tau_n = 10\ \Omega^{-1}$. Dots indicate measurement stations in the plane $z = 0$.

The example shown on Fig. 2 consists of an inhomogeneous thin sheet at $z_s = 2000$ m surrounded by two conducting layers ($\sigma_1 = .02\ \Omega^{-1}m^{-1}$; $\sigma_2 = 10^{-3}\Omega^{-1}m^{-1}$).

The anomalous zone of the layer (Fig. 2b) is described by 5 x 5 square elements $\Delta_j (j = 1, \ldots, N = 25)$ with size 1 x 1 km^2. In 15 of these 25 elements, τ has a value which deviates from its normal one $\tau_n = 100\ \Omega^{-1}$ (model A is a resistive inhomogeneity and model B a conducting one). Synthetic data are given by 49 complex values of the horizontal H_x field in the plane z = 0, measured at points $P_i (i = 1, \ldots, N = 49)$. The field source is a uniform horizontal current sheet flowing along y direction in z < 0 with period 1 s.

In numerical computation, for $M \in \Delta_j$, $\underline{E}_s(M)$ is approached by its value at the center of Δ_j, $\underline{E}_{sj}$. Equations (3) and (5) become :

$$E_{x_j} = E_{n\,x_j} - i\omega\mu_0 \sum_{k=1}^{M} (\tau(k) - \tau_n)\left[E_{x_k} g_{xx}^{kj} + E_{y_k} g_{xy}^{kj}\right]$$

$$E_{y_j} = E_{n\,y_j} - i\omega\mu_0 \sum_{k=1}^{M} (\tau(k) - \tau_n)\left[E_{x_k} g_{yx}^{kj} + E_{y_k} g_{yy}^{kj}\right] \qquad (6)$$

for j = 1, ..., M = 25

$$H_{x_i} = H_{nx_i} + \sum_{k=1}^{M} (\tau(k) - \tau_n)\left[E_{x_k} h_{xx}^{ki} + E_{y_k} h_{xy}^{ki}\right] \qquad (7)$$

for i = 1, ..., N = 49

with :

$$g_{xx}^{kj} = \int_{\Delta_k} \mathcal{G}_{xx}(M|M_j)\, ds, \quad \text{etc...} \qquad (8)$$

For solving the inverse problem, $\underline{E}_{s_j}$ is first approached by $\underline{E}_{nsj}$; (7) is then an overdeterminated linear system of 2N (=98) real equations for M (=25) real unknowns $\tau(k)$. This system is solved in the mean square sense ensuring $\tau(k) \geq 0$ through quadratic programming (Boot, 1967). With the $\tau(k)$ value, (6) is then solved for $\underline{E}_{s_j}$ using Gauss-Seidel inversion and a new iteration is initiated.

The values of $\tau(k)$ at each step are shown on Fig. 3. The convergence is rapid for $0 < \tau \ll \tau_n$ but slow for $\tau \gg \tau_n$; nevertheless the geometry of the anomalous zone is clearly shown.

The proposed algorithm could be applied to various geological structures with a large extent compared to thickness. However it is necessary before using actual data to ensure the stability of computation with respect to errors on data or on the normal structure parameters.

A ↓ ITER = 1

10.76	0.	0.	0.	5.00
2.99	6.56	0.	0.	0.
5.41	4.74	0.87	0.	0.
0.	0.	0.	0.	6.89
0.	0.	0.	6.43	10.16

ITER = 2

10.07	0.	0.	0.05	9.77
9.93	9.48	0.	0.	0.
9.74	9.37	9.29	0.42	0.05
0.	0.	0.	0.	10.10
0.	0.	0.	9.90	10.02

ITER = 10

9.99	.10	.10	.10	9.98
9.99	10.0	.10	.10	.10
10.0	10.0	10.0	.10	.10
.10	.10	.10	.10	9.93
.10	.10	.10	10.03	9.92

B ↓

12.	50.	53.	59.	13.
0.	6.	48.	76.	62.
0.	4.	12.	90.	65.
13.	14.	56.	69.	12.
15.	14.	58.	11.	0.

13.	93.	95.	108.	7.
8.	8.	104.	149.	143.
7.	8.	3.	184.	124.
72.	85.	97.	142.	9.
120.	99.	98.	5.	6.

11.	543.	637.	862.	9.
10.	9.	590.	776.	645.
9.	10.	10.	916.	608.
587.	520.	576.	821.	9.
647.	757.	738.	9.	9.

Figure 3 : Obtained solution for several steps of iteration (Iter = 1, 2, 10)
Upper : model A ; lower : model B

BIBLIOGRAPHY

BAILEY R.C., 1970, Inversion of the geomagnetic induction problem. Proc. Roy. Soc. Lond., A315, 185-194.

PARKER R.L., 1971, The inverse problem of electric conductivity in the mantle. Geophys. J. Roy. Astr. Soc., 22, 121-138.

WEIDELT P., 1972, The inverse problem of geomagnetic induction. Zeit. für Geophys., 38, 257-289.

BOOT,J.C.G., 1967, Quadratic Programming. North Holland, Amsterdam.

PRICE A.T., 1949, The induction of electric currents in non-uniform thin sheets and sheets. Q.J. Mech. appl. Math., 2, 283-310.

VASSEUR G., WEIDELT P., 1977, Bimodal electromagnetic induction in non uniform thin sheets with an application to the northern Pyrenean induction anomaly. Geophys. J. Roy. Astr. Soc., 51, 669-690.

WEIDELT P., 1975, Inversion of two dimensional conductivity structures. Phys. Earth and Planet. Int., 10, 282-291.

Survey of the phenomenological approach to the Inverse Problem in Elementary Particles Scattering.

Virginio Pelosi (*)

Abstract

A review of different approaches to the phenomenological inverse problem in particle scattering is presented, with the aim of a possible application in meson nucleon scattering. Typical results and hints from the Optical Model are discussed to serve as a guide-line to the problem and as source of background expectations. Examples in nucleon-nucleon and absorptive pion nucleon cases are examined as in use in Nuclear Physics. Finally a naive pion nucleon potential extracted from data with a more general tool is forward as well some possible improvements.

(*) Istituto di Fisica dell'Università ,Milano.

Sezione di Milano dell'Istituto Nazionale di Fisica Nucleare.

Contents:

Introduction

It is well known that complex phase shifts allow a complete and subtile parametrization of elementary particle interactions. In particular, for two-body scattering, an enormous bulk of experimental information was aimed to give them numerical values, and, at least up to a not too high energy, different kinds of analysis disentangled phases from data in agreement with each others (Particle Data Group, 1976). Actually, if we neglect particular analytical ambiguities (for a review see for instance Berends 1976 and Berends 1977), they represent the most detailed and realistic description of the experimental facts ever reached in scattering. The interpretation is usually made in terms of static properties, namely the quotation of quantum numbers of resonances. If we try to give them a dynamical interpretation, we are faced with the problem of correlating, for each particular wave, the values with energy, and the different waves with the interaction itself. The simplest way to do so (but perhaps the most naive one) should be through the old fashoned concept of potential, and as this concept has been largely elaborated in nuclear physics, we should try to learning something there. As I am substancially an experimentalist, I think that my task here will be better accomplished if I put out questions, than If I tried to forward answers. So the paper discusses only those general ideas on the inverse problem which were realised in an extended analysis of data. Section 1 presents some topics from the very well cultivated field of the Optical Model. Section 2 makes a short comment on nucleon nucleon potential. Section 3 will be devoted to separable potentials, and Section 4 deals with more general approaches, which will be given a naive application. Some conclusions will be drawn tentatively in Section 5.

Section 1. Optical Model.

The Optical Model was started by Feshbach and Loman in the early fifties. Formal aspects were later developed by Brueckner.

The section is concerned with the very flowered nucleon-nucleus interactions. For technical details on pion-nucleus interactions see for instance Sternheim 1974. Moreover, I will leave out topics like J (Besson 1970) or nuclear spin (Feshback 1960) dependence which are of little relevance here, to concentrate on general concepts and results.

1.1. Local Optical Model

The physical interpretation of the Optical Model may be clarified (Greenless 1966) considering the effective potential V(r) as a folding of an elementary nucleon nucleon potential through a form-factor of the nucleus itself:

(1,1) $$V(\underline{r}) = \int \rho_m(\underline{r})\, v(|\underline{r} - \underline{r}'|)\, d\underline{r}'$$ +higher order terms

Assuming that nucleon nucleon potential v is of a very short range, the dominant term in the integral is the form-factor $\rho_m(\underline{r})$, which is often given the so called Saxon-Wood (or Fermi) shape for spherical nuclei:

(1,2) $$\rho_m(r) = \frac{\rho_m(0)}{1 + \exp\frac{r - R_m}{a_m}}$$

where R_m, a_m measure the mean radius and the diffuseness parameters respectively.

It will be of use in the following to recall also the Gaussian distribution:

(1,3) $$\rho_G(r) = \frac{\rho_G(0)}{1 + \exp\left(\frac{r - R_G}{a_G}\right)^2}$$

The different terms (1,1), (1,2), (1,3) are depicted in fig. 1

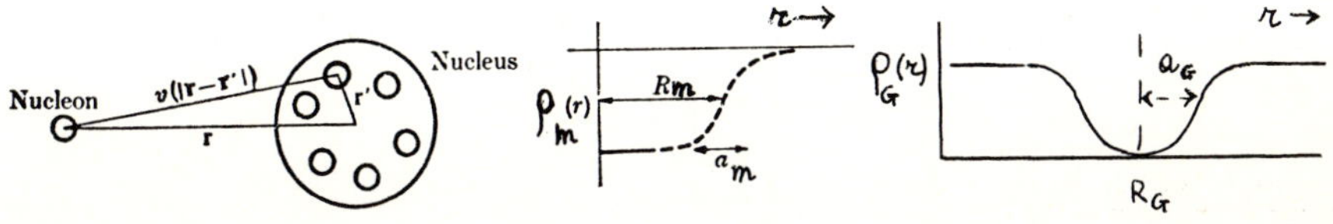

Fig. 1 – (left) Sketch of a collision in the Greenlees interpretation of the Optical Model form factor. (middle) Shape of a Saxon-Wood-type form-factor. (right) Shape of a Gaussian-type form factor.

If the nucleus has a spin O , the local form (1,1) becomes:

$$(1,4) \qquad V(\underline{r}) = V_c(\underline{r}) + V_I(\underline{r}) + V_S(\underline{r})$$

where the labels in the r.h.s. of (1,4) indicate a decomposition in terms which depend respectively only on the (real) central part, isospin and spin. $V_c(r)$ has the shape (1,2) with a coefficient related to the incident energy.

The form (1,4) is real if elastic scattering is the only actual process. Were absorption present, the contribution of reaction channels may be shown to be equivalent to the introduction of an imaginary component of $V_c(r)$, to which a form is given either of volume absorption (usually in a Saxon-Wood form) or of surface absorption (in a Gaussian or in a Saxon-Wood derivative form).

The absorption potential depth shifts from a purely surface $W_S(r)$, to a volume form $W_V(r)$ with increasing energy (fig. 2a,b)

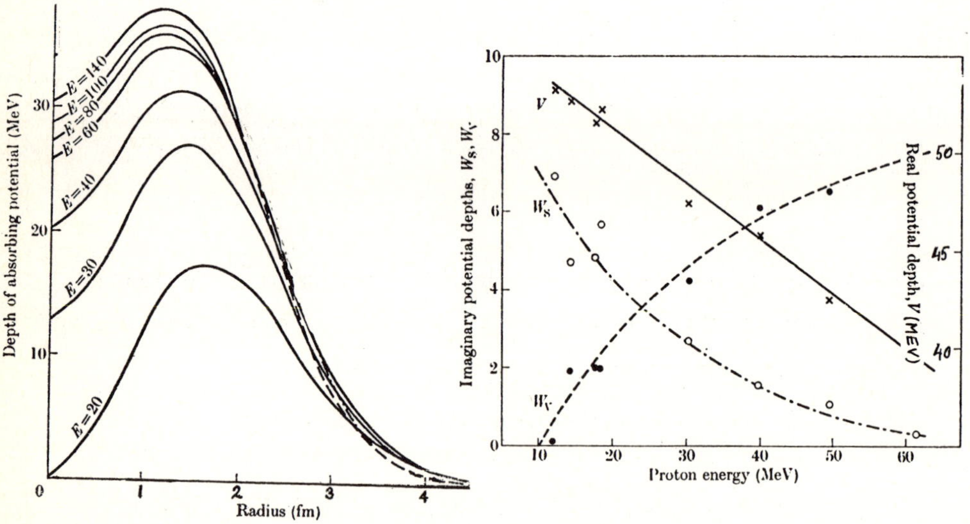

Fig. 2 (left) Radial variation of absorbing potential for ^{16}O for several incident energies.(right)Energy variation of the real and imaginary central parts of the Optical Potential for protons elastically scattered by ^{58}Ni.

The term V_S has to be introduced to describe the coupling of incident spin with the angular momentum of the interaction. Usualy it is assumed with a form suggested by the Thomas model of atoms, namely:

$$V_S(r) = \left(\frac{\hbar}{m_\pi c}\right)^2 U_S \frac{1}{r} \frac{d f(r)}{dr} \vec{L}\cdot\vec{\sigma} \tag{1,5}$$

where f(r) is the assumed form factor for the real part of the potential.

No compelling evidence there is for a corresponding imaginary part of $V_S(r)$, which however may be allowed. In contrast with $V_C(r)$, $V_S(r)$ (or its strenght U_S) turns out to be quite independent from incident energy (fig. 3)

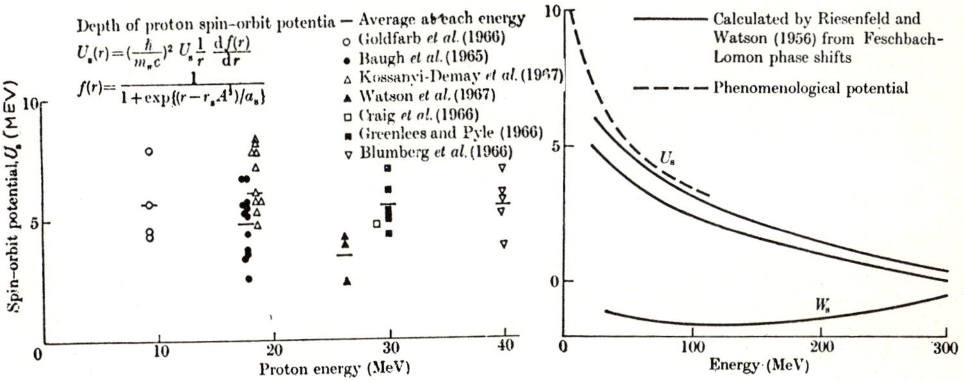

Fig. 3 (left)Depth of the spin-orbit potential. (right) Comparison of calculations of the energy variation of the spin-orbit of the Optical Potential with the variations found by the phenomenological analysis of experimental data.

As to isospin term $V_I(r)$, the dependence may be formulated so that

$$V_{proton} = V_0 + \frac{1}{4} U_I \frac{N-Z}{A} \; ; \; V_{neutron} = V_0 - \frac{1}{4}\frac{N-Z}{A} \tag{1,6}$$

and the difference $V_{proton} - V_{neutron}$ is proportional to the isospin contribution.

In conclusion, the Optical Potential may be given the expression (1,7):

(1,7) $$V(r) = V \cdot f(r) + i W \cdot g(r) + (U_S + i W_S)\left(\frac{\hbar}{m_\pi c}\right)^2 \frac{1}{r} \frac{d f_s(r)}{dr} \vec{L} \cdot \vec{\sigma}$$

Usually, as it has been said before, the form factors f(r), g(r), $f_s(r)$ are assumed in a fixed "a priori" analytical form, and the parameter V, W, U_S, W_S (and eventually some of the radii and diffuseness parameters of form factors) are optimized with respect to the experimental data, by a χ^2 tecnique, namely by minimizing the quantity:

(1,8) $$\chi^2 = \frac{1}{N} \sum_{i=1}^{N} \left(\frac{\sigma_{i\,exp} - \sigma_{i\,calc}}{\Delta \sigma_{i\,exp}}\right)^2$$

where $\sigma_{i\,exp}$, $\sigma_{i\,calc}$ are the cross sections measured and those calculated with particular choice for the parameters. Very often, physical guesses are able to reduce the amount of parameters involved.

1.2 Non Locality

If we consider also non local potentials $V(\underline{r}, \underline{r}')$, by expanding the wave function in a Taylor serie, we have the connection (Hodgson 1971):

(1,9) $$V(\underline{r},\underline{r}')\psi(\underline{r}') = V(\underline{r},\underline{r}')\sum_{L,M} \left\{\psi(\underline{r}) + (\underline{r}-\underline{r}')\dot{\psi}(\underline{r}) + \frac{(\underline{r}-\underline{r}')^2}{2!}\ddot{\psi}(\underline{r}) + \dots\right\} i^L Y_L^M(\underline{r})$$

The classical equivalent of a radial derivative is $-i\frac{p}{\hbar}$; the $\int$ r.h.s. of (1,9) is thus:

(1,10) $$\int V(\underline{r},\underline{r}')\sum_{L,M} \left\{\psi(\underline{r}) + (\underline{r}-\underline{r}')\frac{p}{i\hbar}\dot{\psi}(\underline{r}) + \dots\right\} i^L Y_L^M(\underline{r}')\, d\underline{r}' = V_0(r) + V_2(r)\, p^2 + \dots$$

The equality (1,10) accounts the fact that the odd powers of p are zero because they do not allow invariance under spatial reflection. We find that, neglecting higher order terms, the potential develops a linear dependence on energy.

The energy dependence of the local optical model previously discussed has thus two different sources: the first is of a dynamical origin, the second one is a spurious dependence introduced by the use of a local potential to describe a non local process. The existence of this spurious dependence is clearly shown by connecting the central to the imaginary part of potential with dispersion relations (Passatore 1968). This is shown in fig. 4

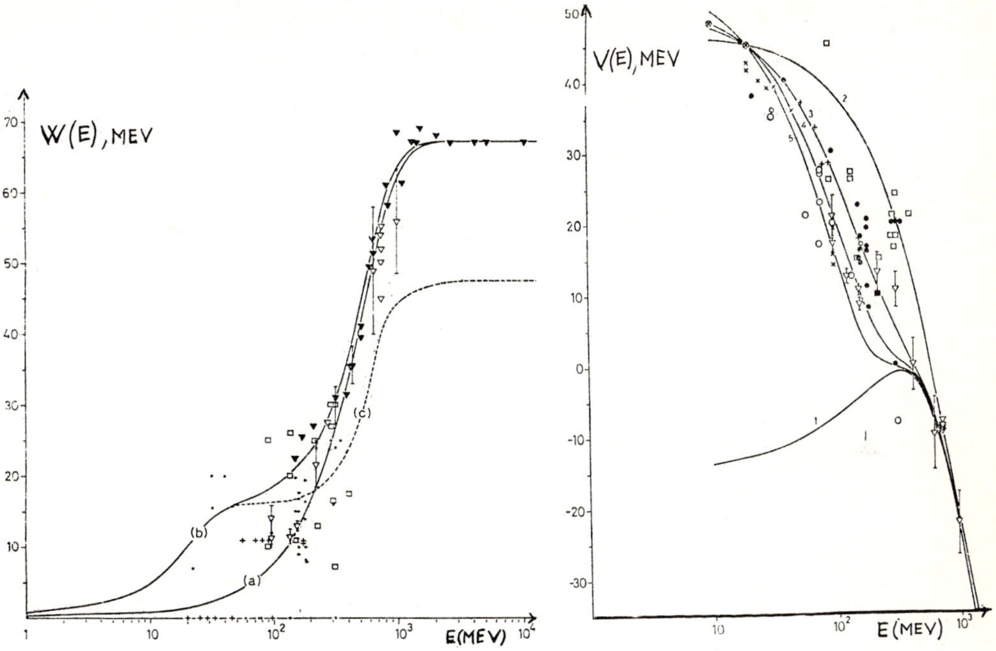

Fig. 4- (left) Imaginary depth of the Optical Potential for different incident particles. (right) Trend of the real part of the Optical Potential calculated with different assumptions. Curve labelled 1 is the result obtained by dispersion relations from the imaginary part (a) left.

An estimate of the different contributions has been given (Jeukenne 1976) as follows:

	energy dependence	
	dynamical	spurious
Real part	$\frac{\partial}{\partial E} V(E, k(E)) \simeq .12$	$\frac{\partial}{\partial k} V(E, k(E)) \frac{dk}{dE} \simeq -.42$
Imagynary part	$\frac{\partial}{\partial E} W(E, k(E)) \simeq .20$	$\frac{\partial}{\partial k} W(E, k(E)) \frac{dk}{dE} \simeq -.06$

These values are (roughly) in agreement with dispersion relations calculations.

A more sophisticated approach has been attempted by Giannini and Ricco (Giannini 1976) who were able, with the help of Perey–Buck like transformations, to correlate local to non local quantities, and to express the "true" non local potential in terms of corresponding local quantities extracted from experimental analysis.

Section 2. Nucleon nucleon Potential

The story of nucleon nucleon potential is indeed a long story. Already in early fifties Bethe estimated that more man-hours of work had been devoted to the N–N problem than to any other scientific question in the history of mankind. However, different approaches were attempted with poor success up to 1962, when Hamada and Johnston (Hamada 1962) proposed the form:

(2,1) $$V(r) = V_c(r) + V_T(r)\, S_{12} + V_{LS}(r)\, \vec{L}\cdot\vec{\sigma} + V_{LL}(r)\, L_{12}$$

where S_{12} is the tensor operator: $S_{12} = 3(\vec{\sigma}_1\cdot\hat{r})(\vec{\sigma}_2\cdot\hat{r}) - \vec{\sigma}_1\cdot\vec{\sigma}_2$

and $L_{12} = [\delta_{LJ} + (\vec{\sigma}_1\cdot\vec{\sigma}_2)]\,\vec{L}^2 - (\vec{L}\cdot\vec{S})^2$

A slight different form was proposed at the same time by the Yale group (Lassila 1962). The introduction of the term $V_{LS}(r)$ alone allows at most a linear dependence on p.
The potential (2,1) and the Yale one represented a vast improvement on all predecessors, but yet it did not correspond to a satisfactory agreement with data.
It must be stressed that the only presence of V_c and V_T accounts for the most general form of a local potential with conserves total angular momentum, parity, charge, time reversal invariance, and it is charge independent. It was to consider a proposal (Chziffra 1959) of accounting higher waves with on pion exchange contribution, namely with the potential:

(2,2) $$V^{OPEP} = \frac{g^2}{12} m_\pi c^2 \left(\frac{m_\pi}{m_N}\right)^2 \vec{\tau}_1\cdot\vec{\tau}_2 \left[\vec{\sigma}_1\cdot\vec{\sigma}_2 + S_{12}\left(1 + \frac{3}{x} + \frac{3}{x^2}\right)\right] \frac{e^{-x}}{x}$$

that settled the problem.
Here $\langle\vec{\tau}_1\cdot\vec{\tau}_2\rangle = 1(-3)$ for isospin 1(0), and $x = \mu r$, $\mu = \frac{m_\pi c}{\hbar}$. Expression (2,2) gives the most of NN interaction at distances greater than about 3 fermis.

2.1 Reid Potential

Reid (1968) considered the states with $J \leq 2$, in the energy range 0-350 Mev. He assumed an expression:

$$V(r) = V_c(r) + V_T(r) S_{12} + V_{LS}(r) \vec{L} \cdot \vec{S} \qquad (2,3)$$

for coupled states, and different forms for central potentials $V_c(r)$ in each uncoupled state (these are states with total angular momentum J=L, and moreover $^3P_0$ state).
At intermediate distances, the potential was assumed as a sum of convenient Yukawa terms $exp(-nx)/x$ with n integer. Higher waves were calculated with (2,2) after substracting terms in its tensor part to remove the bad x^{-2} and x^{-3} behaviour at small x.
In total, with an impressive sum of terms, Reid was able to handle the game so to give a remarkable accord with phase shifts data with both a soft or an hard core. He concluded that a p or p^2 dependence and/or non locality was required by data, while not affecting considerably calculations involving NN potentials. His potentials are usually interpreted to give an excellent reprodution of phase shifts. Also more recent approaches are always confined into the elastic region.

2.2. Interpretation of NN Potential

Since the Reid empirical fit to two nucleons data, the meson theoretic methods of computing nucleon nucleon potential sharpened. The need to interpret results in this frame comes from the impossibility, on a phenomenological basis alone, to get an unique expression for the potential, namely that the correct off-shell behaviour remains unknown. However, it becomes clear that, in coordinate space, NN potential may be splitted in three regions:outside about 2 fermis dominant effects are due to the one pion exchange; in the range $\sim$ 0.2-2 fermis two mesons exchanges give the bulk of the interaction, while the region $r \leq$ 0.5 fermis can only nowaday be treated phenomenologically. The meson theoretical methods can be divided into on-energy and off-energy methods, namely into the techniques of dispersion relations and into field theoretic methods, essentially founded on covariant perturbation theory (for a survay, see Erkelenz 1974). The former stem on the basic idea that the gross structure of the NN potential is given by the pole term, and are developed on the momentum space.
However, a description of the NN forces cannot be given in terms of the pole model alone. That is mainly because a pole contribution from a largely fictitious isoscalar meson (the σ)

is required to describe the intermediate range attraction of the force. The σ object with a mass about 400 to 600 Mev is considered to approximate somehow the continuum of the two pion exchange with T=0, S-wave state. Calculations have been performed, taking explicitely account of two pion exchange in the intermediate region, by folding the contribution of the phe_ nomenological amplitude $N\bar{N} \to \pi\pi$ extrapolated to the region $t \geq 4$ (Epstein 1974, Lacombe 1975, Nagel 1975, Nutt 1976). Unfortunately, notwhitstanding an impressive quantity of calculations and efforts, computed shifts usually get worse with increasing energy.

Section 3. Separable Potentials

For a central potential in momentum space let us consider (Yamaguchi 1958) the interaction between two particles with relative orbital angular momentum l through a potential:

$$(3,1) \quad v_\ell(\underline{k},\underline{k}') = \sigma_\ell \frac{k^2}{m_N} \frac{1}{2\pi^2} (2\ell+1)\, g_\ell(k)\, g_\ell(k')\, P_\ell(\hat{k}\cdot\hat{k}')$$

where P_ℓ is a Legendre polynominal, and the sign factor σ_ℓ is equal to -1(+1) for attraction (repulsion). In such case, one can solve Schroedinger equation for the outgoing scattered wave of relative motion:

$$(3,2) \quad \varphi_\ell(r) = j_\ell(kr) - \frac{2}{\pi}\int_0^\infty \frac{dK\cdot K^2\cdot j_\ell(Kr)}{K^2-k^2-i\varepsilon}\, T_\ell(K|k)$$

The off-energy-shell T matrix turns out:

$$(3,3) \quad T_\ell(k|k') = \sigma_\ell \left[g_\ell(k)\, g_\ell(k')/D_\ell^+(k')\right]$$

The asymptotic wave function and the phase shifts are related by the on energy-shell T matrix:

$$(3,4) \quad k\,T_\ell(k|k) = -\exp\left[i\,\delta_\ell(k)\right]\sin\delta_\ell(k) = \sigma_\ell\, k\left[g_\ell^2(k)/D_\ell^+(k)\right]$$

3.1 Absortive separable potentials from pion nucleon data.

Landau and Tabakin (22) have considered the appropriate relativistic Schroedinger equation (Lippmann–Schwinger equation with relativistic kinematics):

$$(3,5)\qquad T(\underline{k}',\underline{k},E)=V(\underline{k}',\underline{k})+\int\frac{d^3\underline{p}\,V(\underline{k}',\underline{p})\,T(\underline{p},\underline{k},E)}{E-E_1(p)-E_2(p)+i\varepsilon}$$

By decomposing T and V into parts which describe scattering in definite isospin and angular momentum states, they reduce (3,5) to a one dimensional integral equation for each partial wave:

$$(3,6)\qquad T_\ell(k',k,E)=V_\ell(k',k)+\frac{2}{\pi}\int_0^\infty\frac{dp\,p^2\,V_\ell(k',p)\,T_\ell(p,k,E)}{E-E_1(p)-E_2(p)+i\varepsilon}$$

and

$$(3,7)\qquad V_\ell(k',k)=\sigma_\ell\, g_{\ell}(k')\,g_{\ell}(k)$$

Equation (3,5) can be solved for the off-energy-shell T matrix

$$(3,8)\qquad T_\ell(k',k,E)=\sigma_\ell\, g_{\ell}(k')g_{\ell}(k)/D_\ell^+(E)$$

and the on-energy-shell T matrix is directly connected with (complex) phase shifts $\gamma_\ell(k)$:

$$(3,9)\qquad T_\ell(k,k,E)=\sigma_\ell\, g_{\ell}^2(k)/D_\ell^+(E(k))=-\frac{1}{2k}\,\frac{E_1(k)+E_2(k)}{E_1(k)E_2(k)}\exp[i\gamma_\ell(k)]\sin\gamma_\ell(k)$$

and

$$(3,10)\qquad g_{\ell}^2(k)=-\frac{1}{k}\,\frac{\sigma_\ell}{2E_1E_2/(E_1+E_2)}\exp[-\Delta_\ell(E(k))]\sin\gamma_\ell(k)$$

where $D_\ell^+(E)=-i\gamma_\ell(E)-\Delta_\ell(E)$, $\Delta_\ell(E)=\frac{1}{\pi}\int_0^\infty dp\left\{\frac{\gamma(p)\,p\,E(p)/E_1(p)E_2(p)}{E(p)-E(k)}-\frac{2k\gamma(k)}{p^2-k^2}\right\}$

The construction thus relies on the knowledge of the complex phase shifts γ_ℓ at all momenta. They use experimental values up to 800 Mev/c, then calculate high energy values with "Regge" phase shifts, Some of their behaviour is shown in fig. 5.

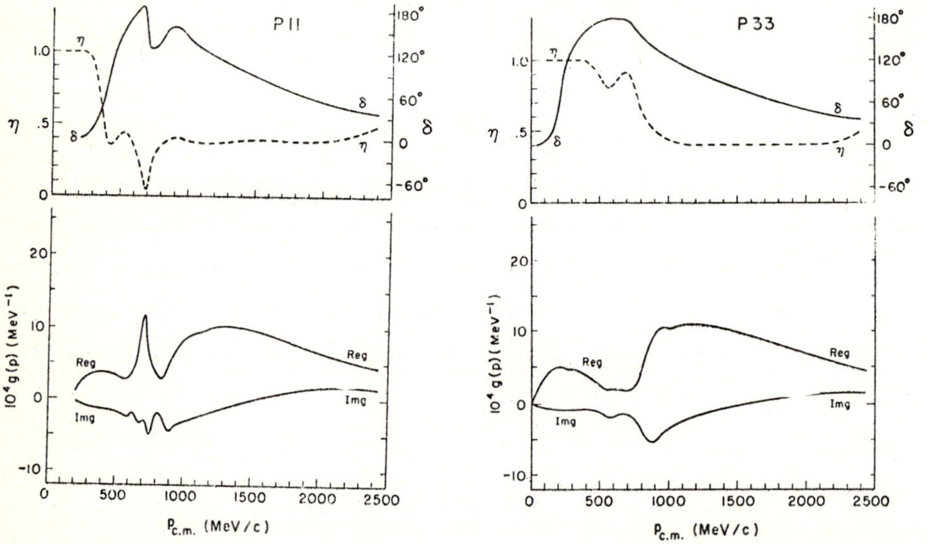

Fig. 5 - (left) Upper graphs: the π-nucleon phase shift (η, δ) for P_{11} partial wave. Lower graphs: the corresponding real and imaginary part of the separable potential function $g(p)$ calculated from the η's and δ's of the upper graphs. (right) The same as left for P_{33} partial wave.

While the method is very elegant and simple, it gives troubles as far as off-energy-shell behaviour is concerned. Indeed, it is strongly forced (eq. 3.8) and it is well known that for instance an O.P.E potential, which is admittedly reasonable at large distance, is not a separable one.

Section 4. General Approaches.

The aim of constructing a potential with general analytical tools, namely of solving the inverse problem with a general machinery,is twofold: formerly, the way offers the unique opportunity to appreciate how potentials fitting the same set of phase shifts are far from the one under the spot, and secondarity, it gives a method to build up a potential also when a search by trial and error does not work, as for instance one can guess if some phase shifts are near a resonance.

4.1 Fixed 1 potentials

The inverse problem at fixed 1 has been completely solved by Gasimov and Levitan (see Corbella 1970) for particles scattered by a spherically symmetric local potential. With the knowledge of a phase shift δ_ℓ for a fixed 1 at all energies, it is possible to construct a potential independent of E . If one however asks whether the knowledge of δ_ℓ over any finite range of energy tells us if the potential decreases asymptotically as an exponential or not, the answer must clearly be "no": a small kink at high energy is sufficient to alter radically the asymptotic behaviour of the potential (Newton 1966).

4.2 Fixed Energy Potentials

The fixed energy method has been suggested by Newton (1962) and later improved by Sabatier (1966). In its simplest formulation, it deals with an infinite system of linear equations, relating a vector of an infinite number of components a_ℓ to the input phase shifts through a known (infinite) matrix M . With these coefficients a_ℓ one can construct both wave functions $\varphi_\ell(r)$ and a central local potential analytically related to them. The problem arises that the inversion of the matrix M allows infinite solutions, so that solutions for a_ℓ are defined but for a calculable vector $\underline{v}$.

Every combination of a given solution with an arbitrary amount of this vector $\underline{v}$ is again a solution. The situation corresponds to the presence of potentials $V_0(r)$ which generate zero phase shifts (transparent potentials). It was shown that the asymptotic form of a general potential V(r) generated by the method is:

$$(4,1) \qquad V(r) \xrightarrow[r\to\infty]{} -\frac{2}{\pi}(\alpha-\beta)\, r^{-3/2} \cos\left(2r-\frac{\pi}{4}\right) + O\left(r^{\varepsilon-2}\right)$$

where α is the arbitrary amount of transparent potential introduced, β a known function of input phase shifts, and ε a positive quantity. If one chooses $\alpha=\beta$ (α-specification), the first part in the r.h.s. of (4,1) is killed, and V(r) is so generated to have an asymptotic behaviour going to zero faster than $r^{-3/2}$.

The space in which the potential is searched, is not the complete space of all possible potentials, but a restricted one. The " α -specification" picks up only one potential from an infinite set, working into a reduced space of the possible potentials.

Sabatier (1971) found that the Newton-Sabatier potential may be put in a form which shows that the above method acts as a filter in momentum space, larger and larger as the energy increases. A general construction has been formulated (Sabatier 1972) which however may not be at present so constrained to produce potentials with prescribed properties.

So far as a local potential to be introduced into a Schroedinger equation is concerned, Coudray and Cox (Coudray 1971) were able to show that a similar formulation may be sketched also for the Klein-Gordon and the Dirac equation.

In particular, for Klein-Gordon equation, potential turns out to have a form connected to the Newton-Sabatier solution V(r) by the transformation:

$$V_{KG}^{2}(r) - 2E V_{KG}(r) = V(r) \tag{4,2}$$

This fact may be interpreted as an indication that a local potential constructed with Newton-Sabatier method for a Schroedinger equation behaves as a non local one for Klein-Gordon equation.

4.3 A naive pion nucleon potential

The Newton-Sabatier method was recently (Pelosi 1978) applied to unfolding pion nucleon potential at low energies. Three energies were considered (25, 94, 143 Mev) and absorption neglected. The potential in each isospin state I, was assumed to be a linear combination:

$$V^{I}(r) = V_{PP}^{I}(r) + V_{PA}^{I}(r) \tag{4,3}$$

where the index PP(PA) correponds to a state with angular momentum parallel (antiparallel) to the spin.

Care was taken to reconstruct phase shifts at an intermediate stage. Input phase shifts differed from reconstructed ones at most in the first decimal already at $r \sim 5$.

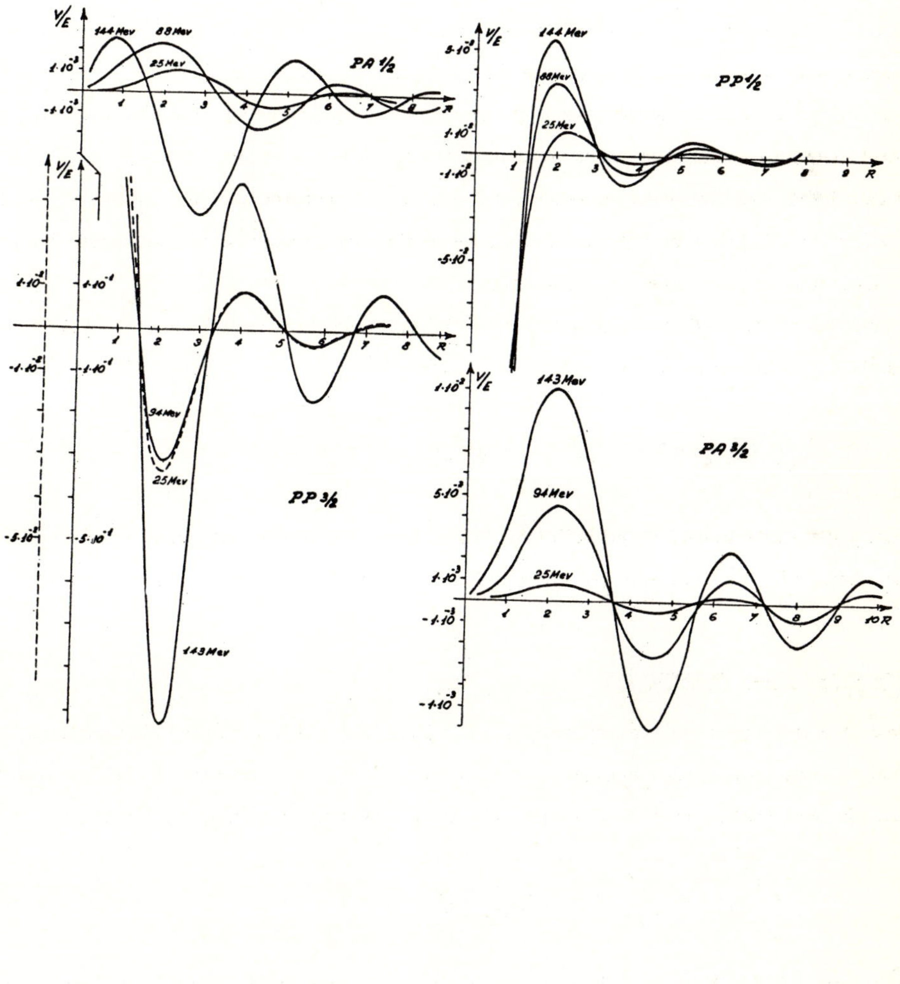

Fig. 6 Partial antiparallel (PA) and partial parallel (PP) part of the potential for π-nucleon elastic scattering in the two states of isospin as unfolded with the Newton Sabatier method.

Results are plotted in fig. 6, where the components $V^{I}_{PP}(r)$, $V^{I}_{PA}(r)$ are represented as ratios to the incident centre-of-mass energy, and the reduced radius r is the radius in fermi multiplied by the relative wave number k. The radial scale in fermi is thus more and more stretched with decreasing energy. One may see that, at small r, PP components go presumably to infinity, while this trend is scarcely connected with a kind of infinity induced by the method itself, which usually appears at smaller values of r. On the contrary, PA components in both isospin states, tend toward zero. The strenght increases regularly with increasing energy.

Oscillations are strongly present, presumably dues to either arbitrary truncation in angular momentum (which was considered as high as 40) or to the small number of phase shifts in input (only the ones experimentally determined) or to both causes. As to the second source, in a less naive search, higher waves ought to be trusted, by calculating them with some OBE model. The physics however seems to be contained only in the first few wingles, because further in r coordinate a plot of the modulus of maxima and minima versus r, in bilogarithmic scale, reveals a common trend in the different components, suggesting a behaviour independent from input data.

Oscillations were also present in a search (Coudray 1977) where, from a starting potential of a given shape, phase shifts were calculated and introduced as input data in the Newton-Sabatier method (fig. 7). From there, one may hope to learn how to handle wingles in the reconstruction of the original shape. If one succeeds in doing so, with some ingenuity, he could argue that in fig.6 some kind of a Yukawian behaviour in PP states is present.

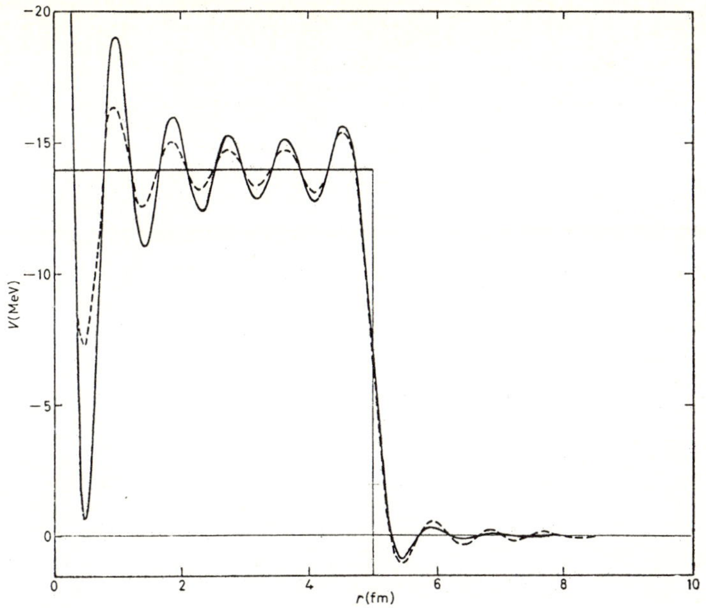

Fig. 7 Real (continuous line) and imaginary (broken line) potential reconstructed from phase shifts generated by the reported square well potential both as a real and imaginary part.

Section 5. Conclusion

5.1 General remarks

My personal opinion is that the motivation of all the game is of aesthetical origin.
Actually experimental phase shifts for two body processes are known, although with all the limitations presented in the Introduction. We know also that phase shifts may be generated with a suitable potential in a suitable equation: but what equation and what potential?
From the Optical Model we have learned how to handle a trial and error method when the analytical form of the potential is known, and so the equation. We may also presume to build up non locality, once a local potential is known. The nucleon nucleon potential reconstruction is perhaps more difficult and more simple at the same time than the corresponding meson nucleon

problem: more difficult with respect to the larger number of independent amplitudes involved, and more simple as it deals only with the complet el y elastic case, in which no absorption is present, and, in the same energy region, no resonance arises.The separable potentials afford a nicely frame, but difficulties blow-up when handling off-mass -shell quantities. We have thus to skip to more general methods if we will try to disentangling foundamental proper ties. In this respect the two orthogonal approaches at fixed 1 and fixed E should cope with the limitation due to the input data: the former is strongly dependent on the asymptotical values of phase shifts (which will be known only through theory or ingenuity) while the latter should deal also with phase shifts smaller than the measurable ones, and presumably in these latter is concealed the right off-shell behaviour. Could be the answer is hidden in a crossed search.

5.2 Aknowledgment

The author is indebted to proff. L.Zuffi and Prof. I.Iori for suggestions and criticisms, anto to Miss M.Stiffoni for carefully typing the manuscript.

References

F.A.Berends, J.C.J.M.Van Reisen, Nucl. Phys. B115, 225 (1976)

F.A.Berends, J.C.J.M.Van Reisen, Nucl. Phys. B118,53 (1977)

H.A.Bethe, Sci.An. 189, 58 (1953)

A.E.Bisson, K.A.Eberhard, R.M.Davies Phys. Rev. C1, 539 (1970)

O.D.Corbella, J.Math.Phys. 11, 1695 (1970)

C.Coudray, M.Cox, J.Math.Phys. 12, 1166 (1971)

C.C.Coudray, Lettere al N.C., 19, 319 (1977)

P.Cziffra, M.H.Mac Gregor, M.J.Moravczik, H.P.Stapp, Phys. Rev. 114, 880 (1959)

G.N.Epstein, B.M.J.Kellar, Phys. Rev. D10, 1005 (1974)

K.Erkelenz, Physics Reports C 13, 191 (1974)

H.Feshback: Nuclear Spetroscopy B, New York(1960)

M.M.Giannini, G.Ricco, Ann. Phys. 102, 458 (1976)

G.W.Greenless, G.J.Pyle, Phys. Rev. 149, 836 (1966)

T.Hamada, I.Johnston Nucl. Phys. 34, 382 (1962)

P.E.Hodgson, Nuclear Reaction and Nuclear Structure, Oxford (1971)

J.P.Jeukenne, A.Lejeune, in Nuclear Optical Model Potential Heidelberg (1976)

M.Lacombe, B.Loiseau, J.M.Richard, R.Vinh Mau, P.Pires, R. de Tourreil, Phys. Rev. D12, 1495 (1975)

R.H.Landau, F.Tabakin, Phys. Rev. 5, 2746 (1972)

K.E.Lassila, M.H.Hull jr. H.M.Ruppel, F.A.Mc Donald, G.Breit, Phys. Rev. 126, 881 (1962)

M.M.Nagels, T.A.Ryken, J.J.de Swart, Phy. Rev. D12, 744 (1975)

R.G.Newton, J.Math. Phys. 3, 75 (1962)

R.G.Newton, Scattering Theory of waves and Particles, New York (1966)

W.T.Nutt, Ann. Phys. 100, 490 (1976)

Particle Data Group, Rev. Mod. Phys. 48 (1976)

G.Passatore, Nucl. Phys. A 110, 91 (1968)

V.Pelosi, R.Cirelli, C.Marioni (to be published, 1978)

R.V.Reid jr. , Ann. Phys. 50, 411 (1968)

P.C.Sabatier, J.Math. Phys. 7, 1515 (1966)

P.C.Sabatier, P.Quyen Van Phy, Phys. Rev. D4, 127 (1971)

P.C.Sabatier, J.Math. Phys. 13, 675 (1972)

M.M.Sternheim, R.R.Silbar, Ann. Rev. Nucl. Sci. 24, 249 (1974)

Y.Yamaguchi, Y.Yamaguchi, Phys. Rev. 95, 1628 (1954)

A STUDY OF AN INVERSE PROBLEM FOR FINITE RANGE POTENTIALS

C. Coudray
Division de Physique Théorique*
Institut de Physique Nucléaire
91406 Orsay Cedex-France

We are interested in what follows by inverse problem at fixed energy in quantum mechanics. The fundamental equation is then the Schrödinger equation. This problem was solved for real central potentials by Loeffel (1968).

However in nuclear physics, one often deals with optical potentials. An optical potential may be defined as the projection of the total hamiltonian on the subspace corresponding to a well-defined channel. Such an optical potential is energy dependent, complex, non-local. A study of the form it could take if it was theoretically derived from the Schrödinger equation was made by Bertero and Dillon (1971). However, we shall in the following only consider local energy independent potentials. But these potentials will be complex.

For short-range real potentials, i.e. for potentials such as :

$$V(r) = 0 \qquad \text{for} \quad r > a .$$

Loeffel (1968) gave a method to map the results of Agranovich and Marchenko (1963) -results which concern the inverse problem for zero angular momentum- into similar results concerning the inverse problem at fixed energy. Our aim will be to show that this transformation is still possible for complex potentials ; but we need a generalization to complex potentials of the work of Agranovich and Marchenko : such a generalization had been done by Ljance (1966;1967). These studies are based upon the properties of the spectrum corresponding to the operator associated with the Schrödinger equation. So in the following study we shall first recall these properties, in the situation of a complex potential. We shall insist on the differences between real and complex potentials. Then we will introduce the Loeffel transformation. And we will show that this transformation may still be applied when the potential is complex. The last part of this paper will be a summary of a work done in collaboration with Pierre Sergent (1977). The inverse problem corresponding to finite range potentials is solved, the condition on the data

*Laboratoire associé au C.N.R.S.

to be coherent is given, the fundamental equation is derived, and the unicity of the reconstructed potential is proved.

I - Properties of the spectrum of a non self-adjoint differential operator $\mathcal{L}$.

a. Definitions

Let us recall first that λ will be called an eigenvalue of an operator $\mathcal{L}$ if there is a non-zero function f belonging to the domain of definition of $\mathcal{L}$ such that :

$$\mathcal{L} f = \lambda f. \tag{1}$$

If λ is not an eigenvalue of $\mathcal{L}$, then the operator $(\mathcal{L} - \lambda I)^{-1}$ exists. If $(\mathcal{L} - \lambda I)^{-1}$ is unbounded, then λ belongs to the continuous spectrum of $\mathcal{L}$ (the other values of λ belonging to the resolvent set of $\mathcal{L}$). We shall introduce k such as $\lambda = -k^2$. So condition (1) becomes :

$$\mathcal{L} f = -k^2 f \tag{2}$$

We define the operator $\mathcal{L}$ associated to the Schrödinger equation :

$$\left[\frac{d^2}{dr^2} - U(r) + k^2\right] f(r,k) = 0 \tag{3}$$

corresponding to a complex potential $U(r)=U_1(r)+iU_2(r)$ by imposing f and $\mathcal{L} f$ to belong to $\mathcal{L}^2(0,\infty)$, and f to vanish at the origin :

$$\mathcal{L} f \in \mathcal{L}^2(0,\infty) \qquad f(0) = 0 \quad . \tag{4}$$

This definition would lead to a self adjoint operator if $U_2(r)$ was zero. We shall in the following suppose it is not the case, but such a definition will allow us to compare the complex and the real case. In what follows, we shall always impose the following condition on U(r) :

$$\sigma_1 = \int_0^\infty e^{\varepsilon x} |U(x)| dx < \infty \tag{5}$$

b. A fundamental set of solutions

Equation (2) is a second order differential operator. So every solution of this equation may be expressed as a linear combination of two independent solutions. One may find two such solutions by defining $u(r,k)$ and $u_1(r,k)$ by the integral equations :

$$u(r,k) = \exp(irk) - \int_r^\infty \frac{\sin(r-\xi)k}{k} U(\xi)u(\xi,k)d\xi \tag{5}$$

$$u_1(r,k) = \exp(-irk) + \frac{1}{2ik}\int_a^r \exp[i(r-\xi)k]U(\xi)u_1(\xi,k)d\xi$$

$$+ \frac{1}{2ik}\int_r^\infty \exp[i(\xi-r)k]U(\xi)u_1(\xi,k)d\xi \tag{6}$$

$$a \leqslant r < \infty .$$

For $r \geqslant 0$ the first function is defined and holomorphic in the half-plane $\mathrm{Im}k > -\varepsilon/2$; the second one is defined and holomorphic in the domain $\mathrm{Im}k \geqslant 0$, $|k|>\delta$ (δ being a function of a). So the common domain of existence and holomorphy of u and u_1 is the half-plane $\mathrm{Im}k \geqslant 0$, except for a neighbourhood of the origin. In this common domain, the following equality holds :

$$W[u(r,k),u_1(r,k)] = -2ik \tag{7}$$

and this proves that one has a fundamental set.

Another fundamental set of functions may be found in the strip $|\mathrm{Im}k|>\varepsilon/2$. This strip is the common domain of existence and holomorphy of functions $u(r,k)$ and $u(r,-k)$. We have there :

$$W[u(r,k),u(r,-k)] = -2ik \qquad \mathrm{Im}k<\varepsilon/2 \quad . \tag{8}$$

We remark immediately that for $U(r)$ real, we have the identities :

$$u(r,k) = u^*(r,-k^*) \tag{9.a}$$

$$u_1(r,k) = u_1^*(r,-k^*) \tag{9.b}$$

c. The spectrum of the non-self adjoint operator .

We shall in this paragraph follow Ljance's work in the book of Naimark (Ljance 1968).

1. There is no eigenvalue of $\mathcal{L}_2$ for k real.

Let us consider a function f such as :

$$\mathcal{L} f = -k^2 f.$$

$$f(r,k) = C_1 u(r,k) + C_2 u_1(r,k).$$

$$f(r,k) \underset{r\to\infty}{\to} C_1' \sin kr + C_2' \cos kr.$$

So if k is real, $f(r,k) \notin \mathcal{L}^2(0,\infty)$.
This property holds if U(r) is real.

2. If Imk>0, the necessary and sufficient condition that $k^2 \neq 0$ be an eigenvalue of $\mathcal{L}$ is

$$\boxed{u(0,k) = 0} \tag{10}$$

We have imposed to f(r,k) to be such that f(0,k)=0. So the identity :

$$f(r,k) = \frac{u_1(0,k)u(r,k)-u(0,k)u_1(r,k)}{2ik} \tag{11}$$

is easily proved. If $\text{Im}k = k_1 > 0$

$$u(r,k) \underset{r\to\infty}{\to} e^{-rk_1} \times \text{phase}.$$

$$u_1(r,k) \underset{r\to\infty}{\to} e^{+rk_1} \times \text{phase}.$$

So, if we want that $f(r,k) \in \mathcal{L}^2(0,\infty)$, we must impose condition (10).
We shall define :

$$u(k) = u(0,k) \tag{12.a}$$

$$u_1(k) = u_1(0,k) \tag{12.b}$$

So equation 11 becomes :

$$f(r,k) = \frac{u_1(k)}{2ik} u(r,k) \tag{13}$$

equation which shows that, at an eigenvalue, the two solutions are multiples of one another.

If U(r) is real, the eigenvalues are all situated on the real axis of the k^2-plane. The demonstration is simple, and may be found in a paper by Newton (1960). Equations 11 and 9 show that for U(r) real $f^*(r,-k^*) = f(r,k)$. Let us write the corresponding eigenvalue equations :

$$\{\frac{d^2}{dr^2} - U(r) + k^2\} f(r,k) = 0$$

$$\{\frac{d^2}{dr^2} - U(r) + k^{*2}\} f^*(r,-k^*) = 0.$$

Multiplication of the first equation by $f^*(r,-k^*)$, of the second one by $f(r,k)$, then substraction of the second line from the first one provide :

$$0 = \frac{d}{dr} W[f^*(r,-k^*),f(r,k)] = -2i(\mathrm{Im}k^2) f(r,k) f^*(r,-k^*)$$

$$0 = -2i(\mathrm{Im}k^2)\int_0^\infty |f(r,k)|^2\, dr$$

and as the integral cannot be zero, necessarily $\mathrm{Im}k^2=0$.

3. The set of eigenvalues of $\mathcal{L}$ is finite.

Equation (5) show that for $\mathrm{Im}k \geqslant 0$ and $|k| \to \infty$:

$$u(k) = 1 + \mathcal{O}\left(\frac{1}{k}\right) \qquad (14)$$

As this function u(k) is holomorphic in the half-plane $\mathrm{Im}k > -\varepsilon/2$ the set of its roots is bounded and those with $\mathrm{Im}k \geqslant 0$ finite.
This property is true if U(r) is real.

4. Every real number k belongs to the continuous spectrum of $\mathcal{L}$.

The operator $\mathcal{L}^*$ corresponds to the potential $U^*(r)$. It has no eigenvalue on the k-real axis (from 1.). So the orthogonal complement of the set of values of the operator ($\mathcal{L} - \lambda I$) is empty. This proves the assertion 4.
This last assertion holds for real potentials.

5. Singular values.

Let us first define the singular values of the operator $\mathcal{L}$ by the roots of the equation $u(k)=0$ for $k\neq 0$, $\mathrm{Im}k \geqslant 0$.

If $\mathrm{Im}k > 0$ we obtain the eigenvalues.

If $\mathrm{Im}k = 0$ we obtain the spectral singularities.

To every singular value corresponds the multiplicity m_k of the root of equation $u(k) = 0$: it is called the multiplicity of the singular value. If this multiplicity is greater than 1, the functions $f^{(m)}(r,k_i) = (\frac{\partial}{\partial\lambda})^m f(r,\lambda_i)$ (with $\lambda_i=k_i^2$) belong to $\mathcal{L}^2(0,\infty)$, $(m=0, 1,\dots,m_k-1)$. They are called the principal functions of the point spectrum, or of the spectral singularity.

We remark that these types of singularities are characteristic of complex potentials. For real potentials, it is well known that the eigenvalues are simple. A demonstration of this property may be found in the paper of Newton (1960). And no spectral singularity can appear.

To show this, we may remark that, when $U(r)$ is real, Equality (9.a) holds and becomes for real k, $u^*(r,-k) = u(r,k)$. So $u^*(-k)=u(k)$. On the other hand, Equation 8 becomes for $r=0$:

$$u(k)u'(0,-k)-u(-k)u'(0,k) = -2ik \neq 0 \text{ for } k\neq 0.$$

If $u(k)=0$, then $u(-k)$ must be different from zero. Then $u^*(-k)\neq 0$, and this contradicts our hypothesis that $u(k)=0$. So a self-adjoint operator possesses no spectral singularity.

6. The number 0 cannot be a root of the equation $u(k)=0$ with multiplicity greater than unity.

To prove this assertion, first derive with respect to k Equation 8 then put $r=0$ and $k=0$. The result is that $\dot{u}(0)$ must be different from zero.

This last property holds for real potentials.

These properties of the spectrum will be sufficient for us to study the inverse problem at fixed energy for short range complex potentials. We now recall the procedure of Loeffel (1968).

II - The Loeffel transformation

We start from the Schrödinger equation (3), with conditions (4) for the function f. With these conditions, as quoted before, of U(r) is real, the differential operator $\mathcal{L} = \frac{d^2}{dr^2} - U(r)$ is a self-adjoint differential operator on the previously defined domain.
Let us write :

$$x = e^{-r} \qquad r = \log x \tag{15.a}$$

$$\nu = -ik \tag{15.b}$$

The first transformation is a bijection from the half-line $[0,+\infty[$ of the variable r onto the open interval $[1,0[$ of the variable x. The second transformation is a rotation of $-\pi/2$ in the plane of the k-variable. We now introduce :

$$f(r,k) = e^{-r/2}\, g(e^{-r},-ik) = x^{1/2}\, g(x,\nu) \tag{16}$$

$$U(r) = q(e^{-r}) = q(x) \tag{17}$$

Equation 3 may now be written as :

$$\{-\frac{d}{dx}(x^2 \frac{d}{dx}) + q(x) + \nu^2 - \frac{1}{4}\}\, g(x,\nu) = 0 \qquad x \in\,]0,1] \tag{18}$$

which is a new eigenvalue equation $\mathcal{L}_1 g = -\nu^2 g$, with the following conditions on $g(x,\nu)$:

$$g(1,\nu) = 0 \qquad \text{and} \qquad \mathcal{L}_1 g \in \mathcal{L}^2(0,1) \tag{19}$$

If U(r) is real, q(x) is real too, and $\mathcal{L}_1$ is a new self-adjoint differential operator. If now we define :

$$q(x) = x^2[V(x) - 1] \tag{20}$$

Equation (18) is nothing else than the Schrödinger equation for the potential V(x) in the open interval $[1,0[$.

Two remarks may be done at this step :
- The first one is that condition (5) on U(r) becomes :

$$\int_0^1 x^{-1-\varepsilon} |q(x)|\, dx < \infty \tag{21}$$

- The second one is that the condition on r at infinity

$$\lim_{r\to\infty} e^{-ikr}\, u(r,k) = 1 \tag{22.a}$$

becomes :

$$\lim_{x\to 0} x^{-\nu+1/2}\, \varphi(x,\nu) = 1 \tag{22.b}$$

which is nothing else than the condition for defining a regular solution at the origin.

III - Generalization to complex potentials

Loeffel had introduced this transformation for real potentials. The first step of our work has been to generalize it to complex potentials. This generalization is straightforward. All the equations of the previous paragraph may be written for complex potentials. The only difference being evidently that the differential operators become non-self adjoint when U(r) becomes complex. So their spectrum corresponds to non self-adjoint operators, as studied before.

To summarize this discussion, we may draw figures 1 and 2 which correspond to real and complex potentials, first in the k-plane, second in the ν-plane (after Loeffel's transform)

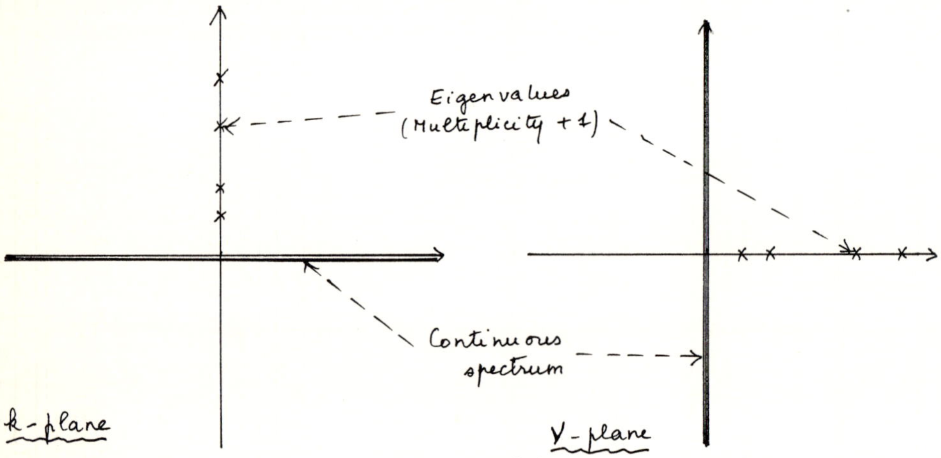

Fig.1 U(r) real

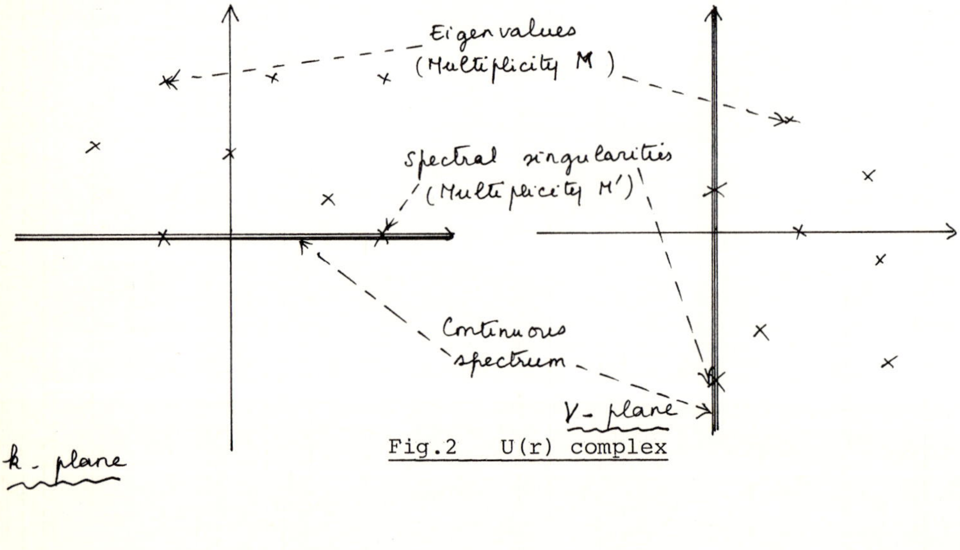

Fig.2 U(r) complex

Now we posses all elements to solve the inverse problem at fixed energy for short-range potentials (we shall only study potentials of range 1, but it obvious that a simple change of variables would lead to any range a $\neq$ 1). We summarize now our paper (1977), without giving here any rigourous demonstration, but trying to give the main ideas. (The demonstrations will be found in our paper). We have followed the method of Ljance (1967).

IV - <u>The direct problem</u>.

We must first solve the direct problems corresponding to Equation (18) with condition (21).

A fundamental set of solutions is easily found. First we have the regular solution $\varphi(x,\nu)$ defined by (22.b). For $0 < x \leqslant 1$, this function is holomorphic in the half-plane $\mathrm{Re}\nu > -\varepsilon/2$. The other solution is the transform of $u_1(r,k)$, and obeys to the relation :

$$\lim_{x \to 0} \psi(x,\nu)\ x^{\nu+1/2} = 1 \tag{23}$$

Its domain of existence and holomorphy deduces from that of $u_1(r,k)$ by a rotation of $-\pi/2$ in the k-plane : it is the right-hand half-plane, except for a neighbourhood of the origin.
The wronskian associated to the operator $\mathcal{L}_1$ is now :

$$\mathrm{Wr}[f,g] = x^2[f'(x)g(x) - f(x)g'(x)]$$

and its value is 2ν for φ and ψ in their common domain of existence and holomorphy. Another fundamental set of solutions would be $\varphi(x,\nu)$ and $\varphi(x,-\nu)$. The common domain of holomorphy of these two last functions is a strip parallel to the imaginary axis : $|\mathrm{Re}\nu| < \varepsilon/2$. Their wronskian is equally 2ν.

The transformation of the integral representation of $u(r,k)$ provides the integral representation of $\varphi(x,\nu)$:

$$\varphi(x,\nu) = x^{\nu-1/2} + \int_0^x K(x,u)\ u^{\nu-1/2}\ du \tag{24}$$

equation which shows that the Fourier transform in the k-plane is replaced by a Mellin transform in the ν-plane (this comes from the fact that $\exp ikr = (\exp(-r))^{-ik} = x^{\nu}$).
This kernel may be majored according to the law :

$$|K(x,u)| < C(xu)^{\frac{\varepsilon-1}{2}} \tag{25}$$

One finds also bounds for its first derivatives.

We must introduce the spectrum of $\mathcal{L}_1$. This spectrum has been previously studied. It contains a continuous part (on the imaginary axis), and a discrete part. The analog of Equation (10) is now :

$$w(\nu) = \varphi(1,\nu) = 0 \quad . \tag{26}$$

and the spectrum could be obtained by studying Equation 26. The function $w(\nu)$ plays a fundamental role in this problem, so we recall its principal properties. If we call ε_0 the minimum of ε and ε_1 , least distance from the imaginary axis to the non purely imaginary roots of $w(\nu)$, the properties of $w(\nu)$ write (properties I) :

- $w(\nu)$ is holomorphic when $\mathrm{Re}\nu > -\varepsilon_0$
- for each $\eta < \varepsilon_0$ one has, uniformly in the half-plane $\mathrm{Re}\nu \geqslant -\eta$:

$$w(\nu) = 1 + \mathcal{O}(\frac{1}{\nu})$$

- $w(\nu) \neq 0$ for $0 < |\mathrm{Re}\nu| < \varepsilon_0$.
- If $\mathrm{Re}\nu = 0$, $\nu \neq 0$ and $w(\nu) = 0$ then $w(-\nu) \neq 0$.
- If $w(0) = 0$ then $w'(0) \neq 0$.

Let us now define :

$$S_1(\nu) = \frac{w(-\nu)}{w(\nu)} \tag{27}$$

The properties of $S_1(\nu)$ (properties II) deduce easily from these of $w(\nu)$:

- $S_1(\nu)$ is defined and meromorphic in the strip $|\mathrm{Re}\nu| < \varepsilon_0$.
- $S_1(\nu)$ does not have non purely imaginary poles in this strip.
- For each $\eta<\varepsilon_0$, one has : $S_1(\nu) = 1+\mathcal{O}(\frac{1}{\nu})$, $|\nu|\to\infty$ uniformly in the strip $|\mathrm{Re}\nu| < \varepsilon_0$.
- $S_1(\nu)S_1(-\nu) = +1$. In particular $S_1(0) = \pm 1$.
- $w(0) = 0$ if and only if $S_1(0) = -1$.

The singularities ν_k of the function $S_1(\nu)$ are the roots of $w(\nu)$ according to Equation (27). They are all situated on the imaginary axis and have the multiplicity M_k.

We come back now to the inverse problem.

V - The inverse problem.

a. The symmetric factorization of the function S_1 (cf. Riemann problem)

Following Ljance, we shall call "problem of symmetric factorization" the following problem :

Given : - a function $S_1(\nu)$ satisfying properties II

- a set of complex numbers $\nu_1,\ldots,\nu_\gamma$ belonging to the half-plane $\mathrm{Re}\,\nu \geqslant \varepsilon_0$

- a corresponding set of natural numbers $M_1,\ldots,M_\gamma$; we require the determination of a function $w(\nu)$ possessing properties I.

The analog in the k-plane of this problem had been solved by Ljance. A simple rotation of $-\pi/2$ suffices to transpose his result in the ν-plane.

We first define the ν-index of a function $S_1(\nu)$ in the strip $|\mathrm{Re}\,\nu| < \varepsilon_0$. Let $\mathscr{C}$ be a curve in this strip running from $+i\infty$ to $-i\infty$, and having all the roots (poles) of $S_1(\nu)$ at its right (left) side. We call ν-index of $S_1(\nu)$, and denote it by $\mathrm{ind}_\nu S_1$, the increment, divided by 2π, of the argument of a continuous branch $\arg S_1(\nu)$ when ν varies on $\mathscr{C}$ from $+i\infty$ to $-i\infty$.

Theorem : The problem of symmetric factorization is solvable if and only if :

$$\mathrm{ind}_\nu\, S_1 + 2(M_1+\ldots+M_\gamma) + \frac{1}{2}\,[1-S(0)] = 0. \qquad (28)$$

Then the solution of the problem is unique. (This is a generalized Levinson theorem).

b. The method of inversion

We now study the method for deriving the potential $V(x)$. The first step will be to introduce a function $F_s(r)$, which is the Mellin transform of $S_1(\nu)$. A quoted before, it is the analog of the Fourier transform of $S(k)$. So :

$$F_s(x) = \frac{i}{2\pi}\,\mathscr{L}_2 \lim_{a\to\infty} \int_{ia+\eta}^{-ia+\eta} [S_1(\nu)-1]\,x^{\nu-1/2}d\nu \qquad 0<\eta<\varepsilon_0 \qquad (29)$$

the integral being shifted by an amount of η from the imaginary axis to avoid the spectral singularities of $\mathcal{L}_1$.
But we need also informations on the point spectrum of $\mathcal{L}_1$. To take them into account we introduce :

$$F_k(x) = \frac{i}{2\pi}\int_C \left[\frac{w_1(\nu)}{w(\nu)} - 1\right] x^{\nu-1/2}\, d\nu \qquad (30)$$

where $w_1(\nu) = \psi(1,\nu)$ and C is constituted of a parallel to the imaginary axis situated in the interval $]0,\varepsilon_0[$, and of a circle of infinite radius in the right half plane, this curve being described in the direct sense. The integrant is meromorphic inside the curve C_k and its poles are the roots ν_k of $w(\nu)$.

$$F_k(x) = -\mathrm{Res}_{\nu=\nu_k} \frac{w_1(\nu)}{w(\nu)} x^{\nu-1/2} \qquad (k=1,2,\ldots,\alpha)$$

$$F_k(x) = p_k\left(\frac{d}{d\nu}\right)[x^{\nu-1/2}]_{\nu=\nu_k} = p_k(\mathrm{Log}\ x)x^{\nu_k-1/2} = \bar{p}_k(x)x^{\nu_k-1/2} \qquad (31)$$

where $p_k(x)$ are polynomials of degree M_k-1, M_k being the multiplicity of the root ν_k ; $\bar{p}_k(x)$ are the corresponding logarithmic polynomials. The functions $\bar{p}_k(x)$ will be called the normalization polynomials of the operator $\mathcal{L}_1$. They reduce in the real case to the normalization multipliers which are constants, because all zeros of $w(\nu)$ are then simple.

So the following function includes information at once on the continuous spectrum and on its point spectrum :

$$F(x) = F_s(x) - \sum_{k=1}^{\gamma} F_k(x) \qquad (32)$$

The summation runs over all the indices corresponding to the non-purely imaginary zeros of $w(\nu)$. The next step of the work will be to introduce the fundamental equation of the inverse problem. Ljance had given this equation for the case of fixed $\ell=0$ angular momentum. This equation is the same as Marchenko's fundamental equation. It is a convolution equation.

It is easily seen that the addition property of the convolution for Fourier transforms become a multiplication property for Mellin transforms. In fact if we write :

$$f(t+v) = e^{-(\frac{t+v}{2})} F(e^{-(t+v)})$$

$$= e^{-t/2} e^{-v/2} F(e^{-t} e^{-v})$$

and if we define $u=e^{-t}$, $v=e^{-v}$, we obtain :

$$f(t+v) = uz F(uz).$$

So the fundamental equation of Ljance becomes after transformation :

$$K(x,u) - \int_0^x K(x,z)F(zu)dz = F(xu) \quad \text{for} \quad 0 < u \leqslant x \leqslant 1 \tag{33}$$

where K(x,u) was defined in Equation 24.

The problem of the unicity of the solution of Equation 33 must be solved. The last equation is of Fredholm type and one will have unicity if the corresponding homogeneous equation admits no other solution than the solution identically zero. This last property may be proved if F(x) possesses the two following properties (properties III) :

- The function $x^{1/2}F(x)$ possesses a continuous derivative such as :

$$\int_0^1 x^{-\varepsilon/2} \left| \frac{d}{dx} \left[x^{1/2} F(x) \right] \right| dx < \infty$$

- If the function $Y_x(t)\, t^{\frac{\varepsilon-1}{2}}$ is summable on $0<t<x \leqslant 1$ and if

$$Y_x(t) = \int_0^x Y_x(u)\, F(ut)du \qquad \text{for} \qquad t<x$$

then $Y_x(t) = 0$ for $t < x$.

One may show that F(x) possesses properties III, so that Equation (33) admits a unique solution.

So if we define the inverse problem data as the set of

- the function $S_1(\nu)$
- the non purely imaginary singular numbers ν_i and their corresponding normalization polynomials,

this set uniquely determine the operator $\mathcal{L}_1$.

Indeed, the function F(x) is built from the inverse problem data. Then the fundamental equation provides uniquely the kernel K(x,u). As this kernel is the kernel of the integral representation of the solution $\varphi(x,\nu)$, the following relation is easily obtained :

$$q(x) = 2x \frac{d}{dx} [x K(x,x)] . \qquad (34)$$

So the operator $\mathcal{L}_1$, and the corresponding potential $U(r)$ are obtained uniquely from the set of inverse problem data.

c. Reconstruction of $\mathcal{L}_1$ from the scattering data

The last step of our work will be to show, that given a set of coherent inverse problem data, one may build a differential operator $\mathcal{L}_1'$, this operator belonging to the class defined by condition (21). And to show that reciprocally, the inverse problem data of the operator $\mathcal{L}_1'$ are identical with the initial ones.

We shall call coherent inverse problem data a set of inverse problem data such as :

- the equality :

$$\text{ind}_\nu\, S_1(\nu) + 2(M_1+\ldots+M_\gamma) + \frac{1}{2}[1-S_1(0)] = 0$$

holds, where (M_1-1) is the degree of the polynomial p_i

- the function $F(x)$ possesses properties III.

If these conditions are realized, then we have shown that the non self adjoint operator $\mathcal{L}_1'$ obtained via the method described in the last paragraph belongs effectively to the class studied here.

The consideration of the direct problem corresponding to the operator $\mathcal{L}_1'$ shows that its scattering data are identical with the initial ones. So the following theorem may be written :

Theorem

Suppose given a function $S_1(\nu)$ possessing properties II in the half plane $\text{Re}\,\nu>-\varepsilon_0$, $\varepsilon_0>0$, numbers $\nu_1,\ldots,\nu_\gamma$ such as $\text{Im}\,\nu_i \geqslant \varepsilon_0$ $(i=1,\ldots,\gamma)$ and corresponding polynomials $p_1(x),\ldots,p_\gamma(x)$, the degree of the polynomial p_i being M_i-1. These data will be the inverse problem data of a certain non self adjoint differential operator $\mathcal{L}_1$ if they are coherent, i.e. if the two following conditions are realized :

i- $\text{ind}_\nu\, S_1 + 2(M_1+M_2+\ldots+M_\gamma) + \frac{1}{2}[1-S_1(0)] = 0$

ii- The function $F(x)$ defined by :

$$F(x) = \frac{i}{2\pi} \mathcal{L}_2 \lim_{a\to\infty} \int_{ia+\eta}^{-ia+\eta} [S_1(\nu)-1]x^{\nu-1/2}d\nu - \sum_{k=1}^{\gamma} p_k(\text{Log } x)x^{\nu_k-1/2}$$

$$0 < x \leqslant 1$$

possesses properties III for some $\varepsilon>2\varepsilon_0$. Then the fundamental equation possesses a unique solution $K(x,t)$, from which it is possible to deduce $q(x)$ satisfying the following inequality :

$$\int_0^1 x^{-1-\varepsilon}|q(x)|dx < \infty$$

and the inverse problem data of the operator $\mathcal{L}_1$ corresponding to $q(x)$ are identical with the initial ones.

VI - Conclusion

To conclude this study, we can say that we have found in the complex λ-plane a set of coherent inverse problem data, the knowledge of which being strictly equivalent to the knowledge of the differential operator $\mathcal{L}_1$, i.e. to the potential V(x). However these inverse problem data do not coincide with the scattering data, and the question which arises is the following : is it, or not, possible to deduce the inverse problem data from the experimental scattering data ? In the real case, Loeffel was able to show that this deduction was unique for short-range potentials, and he gave a method to obtain the inverse problem data from the phase-shifts. However this method has not been used until now, because it gives rise to numerical instability problems. We recall here the conclusions of Di Salvo and Viano (1976). The reconstruction on the imaginary axis of a function holomorphic in a half-plane, from the values it takes at discrete points of the real axis (the physical integer values of $\ell=\lambda-1/2$) is the worst situation for numerical estimations. An arbitrarily small perturbation in the phase-shift can give rise to an arbitrarily large change in the evaluation of $S_1(\nu)$. Suitable stabilizing constraints may be used to guarantee the continuity of the procedure, nevertheless this continuity remains very poor. One will be easily convinced that in the complex case the situation is still more complicated.

Bibliography

Agranovich Z.S. and Marchenko V.A. (1963) : The inverse problem of scattering theory, English transl., Gordon and Breach, New York.

Bertero M. and Dillon G. (1971) : An outline of scattering theory for absorptive potentials, Nuovo Cimento, 2A, 1024-1038.

Di Salvo E. and Viano G.A. (1976) : Uniqueness and stability in the inverse problem of scattering theory, Nuovo Cimento, 33B, 547-565.

Ljance V.E. (1966) : The inverse problem for a non self-adjoint operator English transl., Soviet Math. Dokl. 7, 27-30 [Dokl. Akad. Nauk.SSSR 166, 30-33].

Ljance V.E. (1967) : An analog of the inverse problem of scattering theory for a non self-adjoint operator, English transl. Math. USSR-Sbornik 1, 485-504 [Mat. Sbornik 72, 114] .

Ljance V.E. (1968) The non self-adjoint differential operator of the second order on the real axis : Appendix II of : Naimark M.A., Linear differential operators, part II, English transl. Frederick Ungar Publishing Co., New York.

Loeffel J.J. (1968) : On an inverse problem in potential scattering theory, Ann. Inst. Henri Poincaré A8, 339-447.

Newton (1960) : Analytic properties of radial wave functions, J. Math. Phys. 1, 319-347.

Sergent P. and Coudray C. : the inverse problem at fixed energy for finite range complex potentials, Preprint Orsay IPNO/TH 77-51.

ALGORITHMES POUR UN PROBLEME INVERSE DISCRET DE STURM-LIOUVILLE.

P. Morel

Université de Bordeaux I
33405 Talence / France

I - Introduction.

On considère sur $[o,\pi]$ l'opérateur différentiel de Sturm-Liouville $L(y) = -y'' + q(x)y = \lambda y$ avec les conditions aux limites $\alpha_1 y(o) + \beta_1 y'(o) = 0$, $\alpha_2 y(\pi) + \beta_2 y'(\pi) = 0$. On appelle problème inverse de valeurs propres la recherche de la fonction q connaissant le spectre de l'opérateur.

Nous voulons obtenir numériquement la fonction q ; c'est donc la version discrétisée de ce dernier problème qui nous intéresse. Après avoir introduit un pas de discrétisation, un maillage ,on obtient un problème matriciel qui légèrement généralisé s'énonce de la manière suivante. On appelle problème (P_s) la recherche d'une matrice nxn diagonale réelle $X = (x_i \delta_{ij}) \in \mathfrak{M}_{nn}$ (IR) telle que A étant une matrice n×n donnée symétrique de $\mathfrak{M}_{nn}$ (IR) le spectre: Sp (A+X) de A+X soit égal au spectre de la matrice fixée $S = (s_i \delta_{ij}) \in \mathfrak{M}_{nn}$ (IR).

On peut faire deux hypothèses ne diminuant en rien la généralité du problème traité. On peut supposer que la diagonale de la matrice A est nulle. En effet X est une solution pour A et S fixées ie. Sp(A+X)=SpS si et seulement si X-Diag A est une solution pour A et S données.
D'autre part soit $t \in$ IR tel que pour $i=1, 2, \ldots, n$ $\quad s_i > t > 0$
Si alors X est telle que :

$$Sp(A+X) = Sp((s_i+t)\,\delta_{ij})$$

alors :

$$X-(t\delta_{ij}) \text{ vérifie } Sp(A+X-(t\delta_{ij})) = Sp(S) \;;$$

en d'autres termes on peut supposer que le spectre visé est strictement positif. Nous incluons ces deux hypothèses, non restrictives, dans la formulation du problème (P_S).

2 - Des conditions nécessaires et des conditions suffisantes.

L'étude en dimension 2x2 montre immédiatement que le problème (Ps) ne possède pas toujours de solution. K. Hadeler [2], F. Laborde [4], P. Morel [6, 7] ont donné des conditions nécessaires de plus en plus précises pour que le problème

(Ps) possède des solutions. On montre dans Morel [7] que nécessairement $S = (s_i \delta_{ij})$ doit vérifier :

$$\sum_{i=1}^{n} s_i^2 - \frac{1}{n} (\Sigma s_i)^2 \geq \sum_{i,j} a_{ij} a_{ji}$$

ce qui d'une manière équivalente mettant en évidence une nécessaire séparation du spectre visé s'écrit :

$$2n \sum_{i,j} a_{ij} a_{ji} \leq \sum_{i,j} (s_i - s_j)^2$$

Dans le cas où A est symétrique $\sum_{i,j} a_{ij} a_{ji} = \sum_{i,j} a_{ij}^2 = \mathrm{tr}\, A^2 = \|A\|_S^2$; $\|A\|_s$ désignant la norme de Schur de A. Sous cette hypothèse cela permet d'affirmer que : l'application $x \to \mu(A+(x\delta_{ij}))$, où $\mu(A+x_i\delta_{ij})$ désigne le vecteur dont les composantes sont les valeurs propres de $A+(x_i\delta_{ij})$ numérotées dans l'ordre non croissant n'est surjective que si A est nulle.

Dans [6], on obtient la condition nécessaire suivante, qui est strictement plus précise que les précédentes

$$(\overset{<}{\mu}(S)|\mu(A)) \leq \sum_{i,j} a_{ij}^2 \leq (\mu(S)|\mu(A))$$

où $(\overset{<}{\mu}(S)|(\mu(A))$ désigne le produit scalaire entre les vecteurs $\overset{<}{\mu}(S)$ et $\mu(A)$ qui désignent respectivement les valeurs propres de S dans l'ordre croissant et les valeurs propres de A dans l'ordre décroissant.

Ces conditions nécessaires, ne sont suffisantes que si la dimension est inférieure ou égale à 2. Dans Morel [7] on recherche systèmatiquement une localisation de la solution ; cela pour choisir le plus correctement possible une approximation initiale lors de la mise en oeuvre d'un algorithme. Dans cet ordre d'idée citons:

$$\sum_{i=1}^{n} x_i^2 = \sum_{i=1}^{n} s_i^2 - \sum_{i\, j=1}^{n} a_{ij}^2$$

Toutes les conditions suffisantes connues expriment que le spectre visé est suffisamment séparé. Plus précisément que $d(s) = \min_{i \neq j} |s_i - s_j| > f(\lambda_1(A), \ldots, \lambda_n(A))$ où f est une fonction des valeurs propres de A.

Donnons celle de Morel [7] :

$$d(s) = \min_{i \neq j} |s_i - s_j| \geq 2^{(1-1/p)} (\sum_{i=j}^{n} |\lambda_i(A)|^p)^{1/p}$$

Elle recouvre celle de Laborde (p=+∞) et rappelle sans être pourtant identique celle de de Oliviera (p=1) et, la première connue, celle de K. Hadeler (p=2).

Notons que l'on connait également des conditions suffisantes pour un problème analogue à (P_s) mais dans lequel on ne suppose pas que A soit symétrique cf [2, 4, 5, 6, 7].

On peut adapter une démonstration de de Oliveira [1] et de Friedland [9] pour obtenir le résultat d'existence, et en quelque sorte d'unicité suivant.

<u>PROPOSITION 1.</u> - Si $d(s) = \min_{i \neq j} |s_i - s_j| > \ell(A) = \max_{i \neq j} |\lambda_i(A) - \lambda_j(A)|$ alors le problème (P_s) possède

i) $n!$ solutions

ii) une et une seule solution $X = (x_i \delta_{ij})$ vérifiant $x_1 \geq x_2 \geq \ldots\ldots \geq x_n > 0$

La démonstration est basée sur le théorème du point fixe de Brouwer.

3 - Un algorithme du type des approximations successives et un algorithme du type Newton.

Pour $n \geq 2$ toutes les conditions suffisantes assurant l'existence proviennent de l'application du théorème de Brouwer ;il est donc naturel de rechercher sous quelles conditions l'algorithme des approximations successives sera convergent. Pour montrer qu'un opérateur est une contraction il est classique d'étudier sa dérivée, ce qui entraîne à regarder la dérivabilité de $x \to \mu(A+X)$.

Si x est tel que A+X n'a que des valeurs propres simples alors en ce point $\mu(A+X)$ est de classe C^{∞}, d'après Lancaster [13] et Kato [12]. Le fait d'imposer que A+X n'ait que des valeurs propres simples est assez restrictif mais l'on peut donner des exemples où une telle situation a lieu.

Supposons que A soit symétrique, tridiagonale et que $a_{i-1,i} \neq 0$ $i=2,\ldots,n$ alors d'après Wilkinson [11] page 300,on sait que pour tout $x \in \mathbb{R}^n$ $A+(x_i\delta_{ij})$ n'aura que des valeurs propres simples. Dans ce cas $x \to \mu(A+(x_i\delta_{ij}))$ appartient à $C^{\infty}(\mathbb{R}^n)$. Notons que ce cas correspond exactement à la discrétisation de l'opérateur de Sturm-Liouville.

Supposons que $x \to \mu(A+x_i\delta_{ij})$ soit dans $C^1(\Omega)$. Notons $J(x)$ la valeur en x de la matrice jacobienne de $x \to (\mu(A+X))$; d'après Lancaster [13] on obtient :

$$J(x) = \left[\frac{\partial\mu_i(A+X)}{\partial x_j}\right]_{ij} = (u_{ji}^2)_{ij}$$

où $U=(u_{ij})_{ij}$ est la matrice orthogonale qui diagonalise la matrice symétrique A+X, ie $A+X=U.\ \mathrm{Diag}\,\mu(A+X).\ U^T$. Il est important de remarquer que $J(x)$ est une matrice doublement stochastique.

Pour les propriétés des matrices doublement stochastiques, on pourra consulter Horn [14], Hardy-Littlewood-Polya [15].

ALG 1 : <u>Un algorithme du type approximations successives</u> : Alg 1

C'est Hadeler [2] qui a obtenu les résultats les plus précis sur l'algorithme des approximations successives :

Alg 1 : $$x^{n+1} = x^n + \mu(S) - \mu\,(A+(x_i^n\delta_{ij})) \qquad n \geq 0$$

Reformulons son résultat en introduisant un coefficient de relaxation ω qui assure un meilleur comportement numérique .

<u>PROPOSITION 2.</u> - Soit A appartenant à $\mathcal{M}_{nn}(\mathbb{R})$, symétrique à diagonale nulle.

Si $\min_{i \neq j} |s_i - s_j| \geq 4 \max_i \sqrt{\sum_j a_{ij}^2}$ alors quelque soit $\omega \in]0,1]$ l'application $T : x \to x + \omega(\mu(S) - \mu(A+X))$ est k-lipschitzienne de constante $k \leq 13/18$ de la boule $B(s\ d(s)/12)$ dans elle même.

Laborde [4] a démontré également que sous l'hypothèse $\min_{i \neq j} |s_i - s_j| > 2\rho(A)$ $\rho(A)$ rayon spectral de A la solution était un point attractif (cf Ortega [16] page 383)pour les approximations successives.

De fait les conditions de Hadeler, aussi bien que celles de Laborde impliquent que sur la solution $\bar{x}$ le jacobien $J(\bar{x})$ est inversible. Cela donne en quelque sorte la limite de leur résultat car il est facile de construire des exemples pour lesquels une solution existe mais dont le jacobien en ce point n'est pas inversible.

Nous n'avons pas réussi à construire d'exemple pour lequel à la fois

$d(s) > \ell(A)$ et $J(\bar{x})$ non inversible ; mais cette conjecture semble plausible.

L'avantage majeur de cet algorithme est le fait qu'il n'utilise pas les vecteurs propres ; l'unique opération couteuse est l'extraction des valeurs propres de $A+X^n$ ce que l'on réalise par une méthode du type Q. R. avec shift.

Un autre avantage est sa tendance à conserver l'invariant important pour le problème, qu'est la trace. Appelons défaut de trace à l'itération k le nombre :

$$e_k = \sum_{i=1}^{n} x_i^k - \sum_{i=1}^{n} s_i$$

PROPOSITION 3. - Pour tout ω de $]0, 1]$ considérons l'algorithme

$$x^{n+1} = x^n + \omega(s - \mu(A+(x_i^n \delta_{ij}))) \qquad n \geqslant 0$$

alors si 1/ $e_0 = 0 \Rightarrow \forall k \geqslant 0 \quad e_k = 0$

2/ $e_0 \neq 0 \Rightarrow \lim_{k \to \infty} e_k = 0$

ALG 2 : Un algorithme du type Newton : Alg 2.

Le problème à résoudre étant essentiellement celui de la résolution d'un système non linéaire, il est naturel d'envisager l'algorithme de Newton.

PROPOSITION 4. - Supposons :

1/ qu'il existe une solution $\bar{x}$ ie $\mu(A+(\bar{x}_i \delta_{ij})) = \mu(S)$

2/ $f : x \to \mu(A+X)$ soit de classe C^1 dans un voisinage Ω de $\bar{x}$

3/ $\forall$ i=1,2,.... n $\mathcal{R}e\, \rho_i > 0$ ou ρ_i valeurs propres de $J(\bar{x})$

alors $\forall \lambda > 0$ et $\forall \omega \in]0\ 1]$ l'algorithme de Newton

$$\text{alg 2 :} \quad x^{n+1} = x^n - \omega(J(x_n)+\lambda I)^{-1}(\mu(A+(x_i^n \delta_{ij})) - \mu(S))$$

possède $\bar{x}$ comme point d'attraction.

Remarquons que s'il existe une solution $\bar{x}$ pour un spectre visé $Sp(s_i \delta_{ij})$ qui est bien séparé alors nécessairement $f : x \to \mu(A+X)$ est de classe C^1 dans un voisinage Ω de $\bar{x}$; la seule hypothèse restante est la 3° . $J(\bar{x}) = (u_{ji}^2)_{ij}$ est une matrice doublement stochastique, ce qui implique d'après le théorème de Gerchgorin que 1 est toujours la valeur de plus grand module d'une part, d'autre part que toutes les autres valeurs propres sont contenues dans la réunion pour $i=1,2,\ldots\ldots,n$ des disques centrés en u_{ii}^2 et de rayon $1-u_{ii}^2$. Tous ces disques seront contenus dans le 1/2 plan $\Re e\, z > 0$ dès que $\forall\, i=1\ 2,\ldots\ldots,n\quad u_{ii}^2 > \frac{1}{2}$. Or d'après Laborde [4] cela est réalisé si $\min |s_i - s_j| > 2\rho(A)$. D'où le corollaire

<u>COROLLAIRE 1.</u> - Si A est une matrice symétrique à diagonale nulle et si $\min |s_i - s_j| > 2\rho(A)$ alors

1/ il existe $\bar{x}$ solution de (Ps)

2/ $\bar{x}$ est un point attractif pour l'algorithme de Newton :

$$\text{Alg } 2 : \quad x^{n+1} = x^n - \omega(J(x^n)+\lambda I)^{-1} \ -(\mu(A+(x_i^n \delta_{ij}))-\mu(S)).$$

Une autre façon d'obtenir que tout les disques de centre u_{ii}^2 et de rayon $1-u_{ii}^2$ soient dans $\Re e\, z > 0$ et d'imposer que $1-u_{ii}^2 = \sum_{\substack{j=1 \\ j\neq 1}}^{n} u_{ij}^2 < \frac{1}{2}$, puisqu'ils passent tous par le point 1. En adaptant une partie de démonstration de Hadeler [2] on obtient

<u>COROLLAIRE 2.</u> - Si A est symétrique à diagonale nulle et si

$$d(s) = \min_{i\neq j} |s_i - s_j| \geqslant 2\sqrt{3}\ \max_i \sqrt{\sum_j a_{ij}^2} \quad \text{alors}$$

1/ il existe $\bar{x}$ solution de (Ps)

2/ $\bar{x}$ est un point attractif pour l'algorithme de Newton Alg 2.

On peut résumer ces deux corollaires en disant que les conditions qui assurent la convergence des approximations successives, suffisent pour entraîner la convergence de la méthode de Newton.

L'algorithme de Newton nécessite à chaque étape la connaissance de $J(x^n)$ c'est à dire de toutes les valeurs propres et de tous les vecteurs propres de $A+(x_i^n \delta_{ij})$. C'est un accroissement de la masse des calculs pour chaque itération, de fait lors des essais numériques nous nous sommes bornés à des matrices tridiagonales symétriques et nous avons employé l'algorithme du type Q. R nommé tq 12 dans Wilkinson-Reinsch [17]. Pour contre partie nous obtenons une convergence très rapide, et le fait assez surprenant que pour des approximations initiales qui sont en normes plus éloignées de la solution, que celles nécessaires à la convergence des approximations successives, nous ayons encore convergence. Ce bon comportement numérique est peut être dû au fait que l'algorithme conserve la trace, ou réduit le défaut de trace.

En effet, on a la :

PROPOSITION 5. - Pour $\omega \in]0\ 2[$ et $\lambda > 0$ considérons l'algorithme

$$x^{n+1} = x^n - \omega(J(x^n)+\lambda I)^{-1}(\mu(A+X^n)-\mu(S))$$

alors

$$e_0 = 0 \Rightarrow \forall k\ e_k = 0$$

$$e_0 \neq 0 \Rightarrow \lim_{k\to\infty} e_k = 0$$

Dans la démonstration on utilise le fait que $J(x^n)$ est une matrice doublement stochastique.

4 - Algorithmes de minimisation.

Dès que l'on sait calculer la dérivée de $x \to \mu(A+(x_i\delta_{ij}))$ il est naturel pour approximer la solution de l'équation $\mu(A+x_i\delta_{ij}) = \mu(S)$ de songer à minimiser $f(x) = \frac{1}{2}\ \|\mu(A+x_i\delta_{ij})-\mu(S)\|_2^2$.

Nous ferons l'hypothèse que A est une matrice tridiagonale à diagonale nulle telle que de plus $a_{i,i-1} \neq 0$ $i=2,3,\ldots\ldots,n$; cela pour assurer la dérivabilité de $x \to \mu(A+x_i\delta_{ij})$ en tout $x\in \mathbb{R}^n$.

La fonction $f:x\to f(x) = \frac{1}{2}\ \|\mu(A+x_i\delta_{ij}) - \mu(S)\|_2^2$ n'est pas convexe, mais elle possède de bonnes propriétés vis à vis d'une méthode de gradient. Par construction f est bornée inférieurement par zéro et il résulte d'un calcul facile que son gradient $\nabla f(x)$ en x vaut :

$$\nabla f(x) = J(x)^T \lfloor \mu(A+x_i\delta_{ij})-\mu(S)\rfloor$$

où :

$$J(x) = (u_{ji}^2)_{ij}$$

$U = (u_{ij})$ étant la matrice orthogonale qui diagonalise $A+(x_i\delta_{ij})$.

Notons également que : $\lim_{\|x\|\to+\infty} f(x) = +\infty$

Appelons Alg 3 l'algorithme de plus grande descente décrit par

$$\text{Alg 3} : x^{n+1} = x^n - \rho_n \nabla f(x^n) \qquad n \geq 0$$

Pour assurer la convergence de cet algorithme il reste à faire un choix convergent, au sens de Cea [18], du pas ρ_n.

Notons $\mu(x) = (\mu_1(x), \dots \mu_n(x))$ le vecteur de $\mathbb{R}^n$ obtenu à partir du vecteur x en renumérotant ses composantes dans l'ordre non croissant. Sur $\mathbb{R}^n \times \mathbb{R}^n$ introduisons après Hardy-Littlewood-Polya la relation $x \mathcal{R} y = x \prec y$ qui est vraie si et seulement si :

$$\forall k = 1, 2, \dots, n-1 \qquad \sum_{i=1}^{k} \mu_i(x) \leq \sum_{i=1}^{k} \mu_i(y)$$

$$\sum_{1}^{n} \mu_i(x) = \sum_{1}^{n} \mu_i(y)$$

On a alors le résultat suivant dû à Horn [14].

PROPOSITION 6. - Soit $X = (x_i \delta_{ij})$ fixée. Une condition nécessaire et suffisante pour qu'il existe une matrice réelle symétrique A à diagonale nulle telle que $\mathrm{Sp}(A+X) = (s_1, s_2, \dots, s_n)$ est que $x \prec s$.

Notons alors W_s l'ensemble des $x \in \mathbb{R}^n$ tels que $x \prec s$ Horn [14] reprenant Hardy-Littlewood-Polya montre que W_s peut encore s'écrire $W_s = \{x \in \mathbb{R}^n \mid x = Ms$ M matrice doublement stochastique$\}$ ce qui prouve que d'après un résultat de Birkoff [23] que W_s est un polyedre convexe compact dont les sommets sont les Ps ; P décrivant l'ensemble des matrices de permutation.

Considérons d'autre part l'orbite $\mathcal{O}(S)$ de $S = (s_i \delta_{ij})$, c'est à dire l'ensemble des matrices orthogonalement semblables à S.

$$\mathcal{O}(S) = \{B \in \mathcal{M}_m(\mathbb{R}) \mid B = USU^T \qquad U \text{ orthogonale}\}$$

est un ensemble compact, mais non convexe. Il est clair que si x est une solution de (Ps) alors $A + x_i \delta_{ij} \in \mathcal{O}(S)$.

Sur l'ensemble des matrices symétriques considérons le produit scalaire $(A, B) = \mathrm{tr}(AB)$ et la norme induite, dite norme de Schur $\|A\|_S^2 = \mathrm{tr}\, A^2 = \sum_{i,j=1}^{n} a_{ij}^2$.

L'ensemble des matrices symétriques est alors un espace de Hilbert qui peut se décomposer en somme directe orthogonale entre les matrices diagonales et les matrices symétriques à diagonale nulle. D'après le théorème de Wiedlant-Hoffman [21] la distance de $A+x_i\delta_{ij}$ à $\mathcal{O}(S)$ est donnée par $\|\mu(A+x_i\delta_{ij})-\mu(S)\|_2$ c'est à dire que $f(x)$ représente au facteur 1/2 près le carré de la distance de $A+X$ à $\mathcal{O}(S)$.

Notons $A+W_s$ l'ensemble convexe compact des matrices $A+x_i\delta_{ij}$ où $x\in W_s$. $A+W_s$ est contenu dans un hyperplan passant par A parallèle à l'ensemble des matrices diagonales. Il est clair que résoudre (P_s) c'est trouver un point de l'intersection $\mathcal{O}(S)\cap(A+W_s)$, et qu'une méthode constructive sera l'obtention d'une suite minimisant la distance. Gubin-Polyak-Raik [19], Pierra [20] ont développé des algorithmes de projection successives pour trouver un point de l'intersection de plusieurs convexes ; reprenons cette idée en l'adaptant.

Soit $A^k = A+X^k = A+(x_i^k\,\delta_{ij})$ une matrice de $(A+W_s)$. D'après le théorème de Wiedlandt-Hoffman [21] la distance de A^k à $\mathcal{O}(S)$ est donnée par :

$$\mathrm{dist}\,(A^k,\mathcal{O}(S)) = \min_{B\in\mathcal{O}(S)} \|A^k-B\|_s = \|\mu(A+X^k)-\mu(S)\|_2$$

Car si $A^k=A+X^k=M^k.\ \mathrm{Diag}\,\mu(A+X^k).\ M^{k^T}$, la matrice $B^k=M^k\ \mathrm{Diag}\,\mu(S)\ M^{k^T}$ réalise le minimum de la distance. En d'autres termes, on sait projeter sur $\mathcal{O}(S)$. Notons que $\mathcal{O}(S)$ n'étant pas convexe il peut exister plusieurs projections, mais notre façon de procéder en détermine une seule.

La détermination de la projection $A^{k+1}=A+X^{k+1}$ de B^k sur le convexe compact $(A+W_s)$ est particulièrement simple, d'après la proposition 6, il vient :

$$\mathop{\mathrm{Proj}}_{(A+W_s)} B^k = A+\mathrm{Diag}\,B^k$$

car $\mathrm{diag}\ B^k\in W_s$.

On appellera algorithme des projections successives ou Alg 4 l'itération des deux étapes suivantes.

a) $B^k = M^k\ \mathrm{Diag}\,\mu(S)\ M^{k^T}$ si $A^k\ A+X^k=M^k.\ \mathrm{Diag}\,(A+X^k).\ M^{k^T}$

b) $A^{k+1} = A+X^{k+1} = A+\mathrm{Diag}\,B^k.$

On a par construction la proposition suivante :

<u>PROPOSITION 7.</u> - Si $f(x) = \frac{1}{2}\ \|\mu(A+x_i\delta_{ij})-\mu(S)\|_2^2$ alors pour la suite

$\{x^k\}_1^\infty$ fournit par Alg 4 on a $f(x^{k+1}) \leqslant f(x^k)$

Explicitons le passage de X^n à X^{n+1} dans l'algorithme précédent.

On a :

$$X^{n+1} = X^n + \text{Diag}\{M^n(\text{Diag}\,\mu(S) - \text{Diag}\,\mu\,(A+X^n))\,M^{n^T}\}.$$

Cette écriture prouve que l'algorithme Alg 4, de projections successives est l'algorithme de O. Hald [9].

Multiplions à droite chaque terme par e où $e^T=(1,1,1,\ldots\ldots,1)$; il vient :

$$X^{n+1}\,e = x^{n+1} + \text{Diag}\{M^n(\text{Diag}\,\mu(S) - \text{Diag}\,\mu(A+X^n)\,M^{n^T}\}\,e$$

d'où

$$x^{n+1} = x^n + J(x^n)^T\,\{\mu(S) - \mu(A+x_i^n\delta_{ij})\} = x^n - \nabla f(x^n).$$

Il y a donc coïncidence entre l'algorithme de double projection Alg 4, l'algorithme de Hald, la méthode de plus grande descente Alg 3 avec le choix du pas $\rho_n=1$. C'est la concordance de ces trois méthodes qui va permettre de prouver la convergence.

<u>LEMME.</u> - (Hald [9] page 162). Pour la suite de matrices diagonales obtenues par l'algorithme de Hald [resp Alg 3, Alg 4] on a

$$\sum_{n>0} \|X^{n+1} - X^n\|_s^2 < +\infty$$

Il est connu que ce résultat n'implique pas à lui seul la convergence de la suite $\{x^n\}_1^\infty$. C'est notre interprétation comme méthode de gradient qui permet de conclure.

<u>PROPOSITION 8.</u> - Soit $A \in \mathcal{M}_{nn}(\mathbb{R})$ symétrique à diagonale nulle

Soit $S=(s_i\delta_{ij})$ fixée

Alors

1/ il y a coïncidence entre les trois algorithmes de Hald, des projections successives : Alg 4, de gradient à pas fixe $\rho_n=1$: Alg 3

ie. $x^{n+1} = x^n - J(x^n)^T\{\mu(A+x_i^n\delta_{ij}) - \mu(S)\}$

2/ tout point adhérant $\bar{x}$ à la suite $\{x^n\}_1^\infty$ est un point stationnaire pour $f(x) = \frac{1}{2}\|\mu(A+x_i\delta_{ij}) - \mu(S)\|_2^2$.

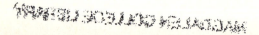

De même que l'interprétation de l'algorithme de Hald comme algorithme de descente fournit des variantes, l'interprétation géométrique suscite de même des variantes numériquement intéressantes.

De la relation

$$A+X=M \text{ Diag } \mu(S)\, M^T$$

on déduit en prenant le carré de la norme de Schur des deux membres que

$$\sum_{i=1}^{n} x_i^2 = \sum_{i=1}^{n} s_i^2 - \sum_{i,j=1}^{n} a_{ij}^2 = r^2$$

Si une solution x au problème (Ps) existe alors, nécessairement, $x \in S(o, r) \subset B(o, r)$.

posons

$$A+B(o, r)=\{M \in \mathcal{M}_{nn}(\mathbb{R}) \mid M=A+(x_i \delta_{ij}) , \sum_{i=1}^{n} x_i^2 \leq r^2\}.$$

Cet ensemble est contenu dans hyperplan parallèle à l'ensemble des matrices diagonales et passant par A.

La solution si elle existe appartient à $(A+B(o, r)) \cap \mathcal{O}(S)$; il est facile d'adapter l'algorithme des projections successives.

Soit :

$$A^k = A+X^k = M^k = M^k. \text{ Diag } \mu(A+X^k). \; M^{k^T}$$

la matrice $C^k=M^k \text{ Diag } \mu(S) M^{k^T}$ réalise le minimum de la distance entre A^k et $\mathcal{O}(S)$. La projection de C^k sur le convexe compact $(A+B(o, r))$ est facile à obtenir ; c'est $A^{k+1} = A+X^{k+1} = A+r \dfrac{\text{Diag } C^k}{\|\text{Diag } C^k\|_2}$. On itère ces deux projections successives .

L'intérêt dans cet algorithme est la conservation à priori de la norme de la solution.

Il est clair que la suite $\{x^n\}_1^\infty$ obtenue définie un algorithme de descente pour $f(x) = \frac{1}{2} \|\mu(A+x_i\delta_{ij})-\mu(S)\|_2^2$, c'est à dire $f(x^n) \geq f(x^{n+1})$.

De fait cet algorithme n'est rien d'autre que la minimisation de f(x) sous la contrainte $x \in B = B(o, r) = \{x \mid \|x\| \leq r\}$ par une méthode de gradient projeté.

On obtient alors le résultat suivant :

PROPOSITION 9. - Soit A tridiagonale symétrique à diagonale nulle telle que

$a_{i-1,i} \neq 0 \quad i=2, 3, \ldots, n$

alors si

$$K = \operatorname{Sup}_{\xi \in \mathbb{R}^n} \|f''(\xi)\| < 2$$

où

$$f(x) = \frac{1}{2} \|\mu' A + x_i \delta_{ij}) - \mu(S)\|_2^2$$

l'algorithme $x^{n+1} = \operatorname{Proj}(x^n - \nabla f(x^n))$ produit une suite dont tous les points adhérants sont des points stationnaires de f.

5 - Essais numériques.

Nous avons fait de nombreuses expériences numériques, aussi nous n'en présentons ici qu'une partie. Dans les essais suivants nous aurons $\omega = 1$ Alg 1, $\omega = 1$ et $\lambda = 0.1/(\text{n}^\circ$ de l'itération) dans Alg 2.

Nous avons également multiplié les essais pour étudier le domaine d'attraction d'une solution relativement aux divers algorithmes. L'expérience a révélé un fait assez inhabituel : l'ensemble des approximations initiales pour lesquelles l'algorithme de Newton Alg 2 converge semble plus grand que l'ensemble des approximations initiales assurant la convergence de l'algorithme des approximations successives Alg 1. Naturellement lorsque tous les deux convergent Alg 2 est bien plus rapide que Alg 1.

Pour comparer l'efficacité des divers algorithmes étudiés nous avons construit des problèmes tests à partir de la discrétisation avec un pas $h = 1/(n+1)$, de problèmes de Sturm-Liouville.

$$Ly = y'' + q(x)y = \lambda y$$

$$\alpha_1 y(o) - \beta_1 y'(o) = 0$$

$$\alpha_2 y(1) + \beta_2 y'(1) = 0$$

On approxime $y''(x_i) = y''(ih)$ par $(y_{i-1} - 2y_i + y_{i+1})/h^2$, $y'(o)$ par $(y_1 - y_o)/h$ et $y'(1)$ par $(y_{n+1} - y_n)/h$. Il est alors facile de vérifier que l'approximation de l'opérateur L est la matrice A symétrique tridiagonale ayant des -1 sur les deux codiagonales et $\{C_a + (2 + h^2 q_1), 2 + h^2 q_2, \ldots, 2 + h^2 q_{n-1}, (2 + h^2 q_n) + C_b\}$ comme diagonale, avec $C_a = -\beta_1/(\alpha_1 h + \beta_1)$ et $C_b = -\beta_2/(\alpha_2 h + \beta_2)$. La procédure est alors la suivante. On se donne une fonction q et les constantes $\alpha_1, \alpha_2, \beta_1, \beta_2$ pour n=10 on obtient des matrices 10 x 10 . On calcule alors le spectre de cette matrice par l'algorithme du type Q. R que Wilkinson-Reinsh nomme tq 12 cf (17).

Les programmes pour les calculs des jeux d'essais et les algorithmes Alg 1, Alg 2, Alg 3, Alg 4 sont écrits en Fortran et testés sur IRIS 80 de CII. Pour chaque programme la phase essentielle du calcul des valeurs propres et vecteurs propres est effectuée par le sous programme tq 12.

Nous reproduisons dans les tableaux ci-dessous deux séries d'essais ; l'une est construite à partir d'une fonction $q(x)=x(1-x)$, c'est à dire symétrique sur $[0,1]$, l'autre à partir de $q(x)=1-x$. Pour chaque série nous faisons varier les conditions aux limites.

i) $y(o) = y(1) = 0$

ii) $y'(o) = o$ et $y(1) = 0$

iii) $y(o) = y'(o)$ et $y(1) = y'(1)$

Nous donnons pour diverses itérations l'erreur relative :

$$\|\text{Spectre visé} - \text{Spectre obtenu}\|_2 \,/\, \|\text{Spectre visé}\|_2.$$

L'approximation initiale de la diagonale est pour tous les essais le spectre visé.

Tableau 1 -$q(x) = x(1-x)$; $y(o) = y(1) = 0$

N°	Alg 1	Alg 2	Alg 3	Alg 4
0	0.260	0.260	0.260	0.260
10	$0.351\ 10^{+1}$	$0.414\ 10^{-4}$	$0.196\ 10^{-1}$	$0.174\ 10^{-1}$
20	diverge	$0.373\ 10^{-5}$	$0.106\ 10^{-1}$	$0.940\ 10^{-2}$
50		$0.153\ 10^{-5}$	$0.494\ 10^{-2}$	$0.421\ 10^{-2}$
80		$0.381\ 10^{-7}$	$0.322\ 10^{-2}$	$0.272\ 10^{-2}$

Tableau 2 $q(x) = x(1-x)$; $y'(o)=0$ $y(1)=0$

N°	Alg 1	Alg 2	Alg 3	Alg 4
0	0.268	0.268	0.268	0.268
10	$0.972\ 10^{+1}$	0.257	$0.234\ 10^{-1}$	$0.209\ 10^{-1}$
20	diverge	$0.159\ 10^{-9}$	$0.160\ 10^{-1}$	$0.141\ 10^{-1}$
50		arrêt	$0.577\ 10^{-2}$	$0.463\ 10^{-2}$

Tableau 2 (suite)

N°	Alg 1	Alg 2	Alg 3	Alg 4
80			$0.150\ 10^{-2}$	$0.117\ 10^{-2}$

Tableau 3 $q(x) = x(1-x); y(o) = y'(o); y(1) = y'(1)$

N°	Alg 1	Alg 2	Alg 3	Alg 4
0	0.280	0.280	0.280	0.280
10	$0.112\ 10^{+3}$	0.179	$0.427\ 10^{-1}$	$0.408\ 10^{-1}$
20	diverge	0.128	$0.372\ 10^{-1}$	$0.336\ 10^{-1}$
50		$0.253\ 10^{-5}$	$0.115\ 10^{-1}$	$0.864\ 10^{-2}$
80		$0.669\ 10^{-7}$	$0.386\ 10^{-2}$	$0.351\ 10^{-2}$

Tableau 4 $q(x) = 1-x ; y(o) = y(1) = 0$

N°	Alg 1	Alg 2	Alg 3	Alg 4
0	0.259	0.259	0.259	0.259
10	$0.710\ 10^{+1}$	$0.282\ 10^{-4}$	$0.195\ 10^{-1}$	$0.174\ 10^{-1}$
20	diverge	$0.250\ 10^{-5}$	$0.106\ 10^{-1}$	$0.941\ 10^{-2}$
50		$0.584\ 10^{-4}$	$0.494\ 10^{-2}$	$0.422\ 10^{-2}$
80		$0.345\ 10^{-6}$	$0.322\ 10^{-2}$	$0.272\ 10^{-2}$

Tableau 5 $q(x) = 1-x ; y'(o) = 0 ; y(1) = 0$

N°	Alg 1	Alg 2	Alg 3	Alg 4
0	0.268	0.268	0.268	0.268
10	0.976	0.577	$0.235\ 10^{-1}$	$0.408\ 10^{-1}$
20	diverge	0.161	$0.161\ 10^{-1}$	$0.136\ 10^{-1}$
50		$0.196\ 10^{-1}$	$0.573\ 10^{-2}$	$0.863\ 10^{-2}$
80		$0.145\ 10^{-9}$ en N° 60	$0.386\ 10^{-2}$	$0.118\ 10^{-2}$

Tableau 6 $q(x) = 1-x$; $y(o) = y'(o)$; $y(1)=y'(1)$

N°	Alg 1	Alg 2	Alg 3	Alg 4
0	0.279	0.279	0.279	0.279
10	$0.140\ 10^{+2}$	0.314	$0.426\ 10^{-1}$	$0.208\ 10^{-1}$
20	diverge	$0.219\ 10^{-1}$	$0.371\ 10^{-1}$	$0.140\ 10^{-1}$
50		$0.134\ 10^{-5}$	$0.114\ 10^{-1}$	$0.464\ 10^{-2}$
80		$0.157\ 10^{-6}$	$0.386\ 10^{-2}$	$0.118\ 10^{-2}$

A la vue de ces résultats deux remarques au moins s'imposent. La plus importante est que pour l'approximation initiale choisie l'algorithme Alg 1 diverge à chaque fois tandis que les autres convergent ; cela corrobore une remarque déjà faite.

Le seconde remarque consiste en l'opposition entre d'une part les deux algorithmes de minimisation Alg 3 et Alg 4 et d'autre part l'algorithme de Newton Alg 2; Alg 3 et Alg 4 donnent des résultats très similaires avec un très léger avantage pour Alg 4. Mais pour ces deux méthodes la convergence bien que très régulière est aussi très lente ; on n'arrive pas en 80 itérations à dépasser le seuil d'une erreur relative en 10^{-3}, ce qui dans notre cas assure 10 pour cent d'erreur sur la plus petite composante et 0,25 pour cent d'erreur sur la plus grande composante du spectre visé. Par contre Alg 2 donne à chaque fois une erreur relative de l'ordre de 10^{-7}, ce qui assure sept chiffres caractéristiques exacts pour toutes les composantes du spectre visé.

La régularité des algorithmes 3 et 4 et la rapidité de l'algorithme de Newton incite à étudier un algorithme pour le problème inverse des valeurs propres qui aurait ces deux excellentes propriétés.

BIBLIOGRAPHIE

[1] De OLIVEIRA G. - Note on inverse characteristic problem. Numer. Math Vol 15, (1970) 339-341.

[2] HADELER K. P. - Ein inverses Eigen wert problem. Linear algebra and its appl. Vol 1, (1968), 83-101.

[3] HADELER K. P. - Newton-Verfahren für inverse Eigenwertaufgen. Num. Math. Vol 12, (1968), 35-39.

[4] LABORDE F. - Sur un problème inverse de valeurs propres. CRAS tome 268, (1969), 153-156.

[5] CHATELIN-LABORDE F. - Thèse Mathodes numériques de calcul de valeurs propres et vecteurs propres d'un opérateur linéaire. Grenoble 1971.

[6] MOREL P. - A propos d'un problème inverse de valeurs propres. CRAS tome 277, (1973), 125-128.

[7] MOREL P. - Sur le problème inverse des valeurs propres. Numer. Math 23, (1974), 83-94.

[8] HALD O. - On discrete and numerical inverse Sturin-Liouville problems, Uppsala University, Dep. of Computer Sciences, Report 42, 1972.

[9] FRIEDLAND S. - Matrices with prescribed off diagonal elements, Israel J. of Math. Vol 11, (1972), 184-189.

[10] FRIEDLAND S. - Inverse eigenvalue problems. A paraître.

[11] WILKINSON - The algebric eigenvalue problem. Oxford University Press (1965).

[12] KATO T. - Perturbation theory of linear operators. Springer Verlag (1966).

[13] LANCASTER P. - On eigenvalues of matrices dependent on a parametor. Numer. Math Vol 6, (1964), . 377-387.

[14] HORN A. - Doubly stochastic matrices and the diagonal of a rotation matrice. Amer. J. Math. Vol 76, (1954), 620-630.

[15] HARDY-LITTLEWOOD-POLYA - Inegalities, Cambridge University Press (1948).

[16] ORTEGA-RHEINBOLDT - Iterature solution of non linear equations in several variables, Academic Press (1970).

[17] WILKINSON- REINSCH - Linear algebra. Handbook for compotations. Springer Verlag.

[18] CEA J. - Optimisation, théorie et algorithmes. Dunod 1971.

[19] GUBIN-POLYAK-RAIK - The method of projections for fonding the common point of convex sets. USSR Comp. Math and Math Phys. Vol 6, (1967), 1-24.

[20] PIERRA G. - Sur le croissement de méthodes de descente. CRASS t 277, (1973) 1071-1074.

[21] WIELANDT-HOFFMAN - The variation of the spectrum of a normal matrix, Doke J. of Math Vol 20, (1953) , 37-39.

[22] GOLSTEIN A. - Constructive real analysis. Harper International Edition (1967).

Construction of Regge Amplitudes through Solution of S-Matrix Equations

P.W. Johnson‡
University of Groningen
The Netherlands

Presented at "Etudes interdisciplinaires des problèmes inverses et la rencontre de phénoménologues à Montpellier"
October, 1975

I shall describe some recent work concerning the construction of scattering amplitudes through solution of S-matrix equations. This work has been carried out in collaboration with D. Atkinson and M. Kaekebeke of the University of Groningen, R.L. Warnock of Illinois Institute of Technology, and J.S. Frederiksen of the University of Mentone (Australia). Our approach has been to carry through constructive proofs of the existence of solutions to nonlinear equations for partial wave amplitudes at physical energies and complex angular momentum. These nonlinear equations are obtained from such physical requirements as unitarity (conservation of probability), analyticity (causality), and Lorentz invariance, and the scattering amplitude which we construct must satisfy these general requirements.

I shall describe a pure S-matrix approach to nonrelativistic potential scattering in Part I, and shall discuss construction of relativistic amplitudes in Part II.

I. S-Matrix Approach to Potential Scattering

Considerable attention has been devoted to nonrelativistic scattering by central potentials of Yukawa type, for which

$$rV(r) = \frac{1}{2m}\int_4^\infty dt\ \rho(t)\ e^{-r\sqrt{t}} . \tag{1}$$

(For convenience of notation, I have set $\hbar = 1 = c$ and the range of the Yukawa force equal to 2; m isthe mass appearing in the Schrödinger equation).

‡ Permanent Address: Physics Department; Illinois Institute of Technology; Chicago, Illinois.

The scattering amplitudes corresponding to this class of potentials have been studied extensively, since the interaction is somewhat analogous to that encountered in a Lagrangian quantum field theory. In particular, the analytic structure of the scattering amplitude A(s,t) has been analyzed with care. Here the variables s and t are defined in terms of the momentum q and the scattering angle θ by the relations

$$s = 4(q^2+1) \quad \text{and} \quad t = -2q^2(1-\cos\theta), \tag{2}$$

The original sketch of the proof that A(s,t) satisfies a Mandelstam representation was given by Blankenbecler et al in Ref. 1. Bessis[2] subsequently carried through a more careful analysis to establish the Mandelstam representation for a restricted class of Yukawa potentials. As a result, one could express the scattering amplitude at complex s and t through the relation

$$A(s,t) = \frac{1}{\pi}\int_4^\infty \frac{dt'}{t'-t}\rho(t') + \sum_{r=0}^{n-1} a_k (t-t_0)^k + \frac{(t-t_0)^n}{\pi}\int_4^\infty \frac{ds'}{s'-s}\int_{16}^\infty \frac{dt'\rho(s',t')}{(t'-t_0)^n(t'-t)} \tag{3}$$

The subtraction point $t=t_0$ may be taken anywhere in the cut plane, and n, the number of subtraction constants required, is determined by the asymptotic behaviour of A(s,t) at large t; n is known to be finite as a result of the pioneering work of Regge.

The analysis of Ref. 1 is based upon the Lippmann-Schwinger equation, which may be written in momentum space as

$$T(\vec{k},\vec{k}') = V(\vec{k}-\vec{k}') + \frac{1}{(2\pi^2)}\int d\vec{k}''\ \frac{V(\vec{k}-\vec{k}'')\ T(\vec{k}'',\vec{k}')}{k''^2-k^2-i\varepsilon} \tag{4}$$

For appropriately chosen potentials $V(\vec{k})$, this is an essentially Fredholm equation for the T-matrix amplitude. However, this equation involves $T(\vec{k},\vec{k}')$ with the initial and final momenta, $\vec{k}$ and $\vec{k}'$, not necessarily equal in magnitude. This "off-shell" feature of the Lippmann-Schwinger requires that one determine the T-matrix at unphysical values, and then get the physical T-matrix by setting $|\vec{k}| = |\vec{k}'|$ in the solution. One cannot determine the physical T-matrix directly, but must solve an equation involving unphysical quantities. Of course, one may determine the scattering amplitude by direct solution of the Schrödinger equation; in so doing one gets an asymptote of the wave function after having first determined the wave function inside the interaction region. Even in a classical problem, one determines the scattering angle for a given impact parameter by calculating the trajectory of the particle in the interaction region. The object of an S-matrix theory is to calculate the scattering amplitude by solving equations involving only the physical scattering amplitude, or a stable analytic continuation of it. The equations to be solved are usually nonlinear: that is the price one apparently must pay for avoiding unphysical entities.

We have developed a pure S-matrix approach to scattering from weak potentials of Yukawa type; the details are presented in Ref. 3. We have obtained a constructive existence proof of solutions of S-matrix equations which satisfy Eq. (3) with no subtractions (n=0) and which agree with the scattering amplitudes obtained by solving the Schrödinger equation. The construction is effected by obtaining solutions of a nonlinear equation for $a(\ell,s)$, the partial-wave amplitude defined at complex angular momentum ℓ and physical s. The equation which we solve is

$$a(\ell,s) = \frac{2}{\pi(s-4)}\int_{16}^{\infty} dt\,Q_\ell\left(1+\frac{2t}{s-4}\right)\left\{ \rho(t) + \frac{1}{\pi}\int_4^{\infty}\frac{ds'}{s'-s}\,\rho(s',t)\right\}, \tag{5}$$

where

$$\rho(s,t) = \frac{1}{2i}\int_{\mathrm{Re}\,\ell=-\varepsilon} d\ell\,(2\ell+1)\,q(s)\,a(\ell,s_+)\,a(\ell,s_-)P_\ell\left(1+\frac{2t}{s-4}\right). \tag{6}$$

We have incorporated elastic unitarity in this approach, since a solution of (5)-(6) will satisfy the generalized unitarity relation

$$\frac{1}{2i}\left[\, a(\ell,s_+) - a(\ell,s_-)\,\right] = q(s)\,a(\ell,s_+)\,a(\ell,s_-), \tag{7}$$

with q(s) the usual non-relativistic phase-space factor.

It was inconvenient for us to do a direct analysis of the system (5)-(6), because of an instability inherent in it. We chose, rather, to write an equivalent nonlinear equation for the reduced partial-wave amplitude,

$$b(\ell,s) = a(\ell,s)/(s-4)^\ell, \tag{8}$$

and to write a partial-wave dispersion relation for $b(\ell,s)$.

This approach was suggested by Mandelstam[4] in the context of the relativistic problem discussed in Part II. We write the phase-space factor for this reduced amplitude as

$$q(\ell,s) = q(s)\,(s-4)^{\ell} \tag{9}$$

The equation for b which we use is

$$b(\ell,s_+) = F(b,\ell,s) = b_\beta(\ell,s) + \frac{1}{\pi}\int_4^\infty \frac{ds'}{s'-s_+}\, q(\ell,s')\, b(\ell,s'_+)\, b(\ell,s'_-)$$
$$+ \frac{2}{\pi^2}\int_4^\infty \frac{ds'}{s'-s}\int_{16}^\infty dt\, \rho(s',t)\,\{ \frac{1}{(s-4)^{\ell+1}}\, Q_\ell(1+\frac{2t}{s-4}) - \frac{1}{(s'-4)^{\ell+1}}\, Q_\ell(1+\frac{2t}{s'-4})\,\} \tag{10}$$

The Born term, b_β, serves as the inhomogeneity of the system, and is given in terms of the Yukawa density $\rho(t)$ as

$$b_\beta(\ell,s) = \frac{2}{\pi(s-4)^{\ell+1}}\int_4^\infty dt\, \rho(t)\, Q_\ell(1+\frac{2t}{s-4}). \tag{11}$$

The double-spectral function $\rho(s,t)$ is determined from b via Eq. (6).

We analyze Eq. (10) when $b(\ell,s)$ is an element of a Banach space B of functions which are analytic in ℓ for $\mathrm{Re}\,\ell > -\varepsilon$, Hölder-continuous in s for $s \geq 4$, as well as subject to the reflection property

$$b(\ell^*,s_+) = [\,b(\ell,s_-)]^*. \tag{12}$$

The functions in this space are further restricted to satisfy the bounds

$$|b(\ell,s_+)| \leq \frac{1}{(|\ell|+1)^{\nu} s^{\lambda+\mathrm{Re}\,\ell}} \quad , \mathrm{Re}\,\ell < 0 \tag{13}$$

and

$$|b(\ell,s_+)| \leq \frac{1}{(|\ell|+1)^{\nu} s^{\lambda}} \quad \frac{1}{(\sqrt{s}+2)^{\mathrm{Re}\,\ell}} \quad , \mathrm{Re}\,\ell > 0. \tag{14}$$

We required the parameters to satisfy the constraints $3/4 < \nu < 1$, $\nu + 2\lambda \leq 1$, and $0 < \varepsilon < \frac{1}{2}$. Under these conditions, we were able to apply the contraction mapping theorem to show that for weak coupling Eq. (10) has a unique solution in a restricted subset of the Banach space B.

This solution may be determined by iteration, and it agrees with that obtained by solving the Schrödinger equation at complex angular momentum.

I shall now describe our S-matrix scheme to construct the scattering amplitude when the potential has sufficiently strong attraction to form bound states. For such potentials the scattering amplitude may no longer be expressed via an unsubstracted Mandelstam representation, and the partial wave amplitude will have Regge poles which lie in the right half ℓ-plane and move with energy. For convenience of notation I shall discuss the case in which all but one of the Regge poles lie to the left of $\mathrm{Re}\ell = -\varepsilon$, so that the partial wave amplitude is written as

$$b(\ell,s) = \frac{\beta(s)}{\ell-\alpha(s)} + \bar{b}(\ell,s) \ , \qquad (15)$$

where $\bar{b}$ is analytic for $\mathrm{Re}\ell > -\varepsilon$. The Regge trajectory function $\alpha(s)$ and Regge residue function $\beta(s)$ are analytic in a cut s-plane, with the cut lying along $s \geq 4$. We make use of the feature of potential scattering that the Regge trajectory eventually turns back, so that $\mathrm{Re}\alpha(s_\pm) \leq -\varepsilon$ for $s \geq s_1$.

Let us write the Mandelstam representation in subtracted form as

$$A(s,t) = \frac{1}{\pi}\int_4^\infty \frac{dt'}{t'-t}\rho(t') + \sum_{\ell=0}^{L} \frac{C_\ell}{s-s_\ell} P_\ell(1+\frac{2t}{s_\ell-4}) + \frac{1}{\pi^2}\int_4^\infty \frac{ds'}{s'-s}\int_0^\infty \frac{dt'}{t'-t}\bar{\rho}(s',t')$$

$$-\frac{1}{2i}\int_4^{s_1} \frac{ds'}{s'-s}\left\{ (2\alpha(s'_+)+1)\beta(s'_+)(s'-4)^{\alpha(s'_+)} \frac{P_{\alpha(s'_+)}(-1-\frac{2t}{s'-4})}{\sin\pi\alpha(s'_+)} - (+\to-1)\right\} \qquad (16)$$

The coefficients C_ℓ are the residues of the bound-state poles in the scattering amplitudes, with the bound state energies s_ℓ less than 4; the function $\bar{\rho}(s,t)$ is given as

$$\bar{\rho}(s,t) = \frac{1}{2i}\int_{\mathrm{Re}\ell=-\varepsilon} d\ell(2\ell+1)\ q(s)\ a(\ell,s_+)\ a(\ell,s_-)\ P_\ell\ (1+\frac{2t}{s-4}). \qquad (17)$$

The form of the Mandelstam representation (16) does not have the single- and double- spectral components separated as in (3) above; it is more convenient in this form for our purposes.

We define a partial-wave amplitude $c(\ell,s)$ by the relation

$$a(\ell,s) = [\frac{s-4}{p(s)}]^\ell\ c(\ell,s) \ , \qquad (18)$$

where $p(s) = [\sqrt{s+a^2}+2]^2$ (19)

The appropriate phase-space factor for c is the function

$$r(\ell,s) = \left[\frac{s-4}{p(s)}\right]^{\ell} q(s) \tag{20}$$

Our approach is to write a partial-wave dispersion relation for $c(\ell,s)$, which is quite similar in form to Eq. (10), the partial-wave dispersion relation for $b(\ell,s)$. Next, we replace this partial-wave dispersion relation by an equivalent N/D equation. Finally, we write this N/D equation as a non-linear equation to be satisfied by the numeration function $n(\ell,s)$. The equation is

$$n(\ell,s) = F(n,\ell,s) = C(\ell,s) + \frac{1}{\pi}\int_4^\infty ds' \frac{C(\ell,s')-C(\ell,s)}{s'-s} r(\ell,s')\, n(\ell,s') \tag{21}$$

where

$$d(\ell,s) = 1 - \frac{1}{\pi}\int_4^\infty \frac{ds'}{s'-s} r(\ell,s')\, n(\ell,s') , \tag{22}$$

$$c(\ell,s) = \frac{n(\ell,s)}{d(\ell,s)} , \tag{23}$$

and the left-cut integral $C(\ell,s)$ is determined from $c(\ell,s)$ and the Yukawa density $\rho(t)$. We determine the Regge trajectories as zeros of the function $d(\ell,s)$; i.e,

$$d(\alpha(s),s) = 0, \tag{24}$$

We use the functions $\alpha(s)$ and $\beta(s)$ to regularize the left-cut integral $C(\ell,s)$.

Let $n_0(\ell,s)$ be a solution of (21), which lies in the Banach space B. Then one may show that F(n) maps a neighbourhood of n_0 in B into B; in fact one may apply the implicit function theorem to show that solutions of (21) vary continuously with the coupling strength of the potential, and may be generated in a stable manner. Thus, we show that this S-matrix determination of Regge trajectories in potential scattering may be cast in a form quite tractible for numerical computations. The details shall be presented in a forthcoming publication.

II. S-matrix Approach to $\pi\pi$ Scattering

The scattering amplitude for $\pi\pi$ scattering is expected to satisfy the general conditions imposed by Lorentz invariance, crossing symmetry, Mandelstam analyticity, and unitarity. The requirements of unitarity are of a direct character, in the restrictions of two-body unitarity, as well as an indirect character, in the necessity that $\pi\pi$ amplitudes be consistent with multiparticle unitarity for multiparticle amplitudes. The S-matrix approach was first proposed by Heisenberg, as a means to avoid the traumas of quantum field theory. A concrete S-matrix scheme was constructed by D. Atkinson[5], in which the requirements of two-particle unitarity, Lorentz invariance, and Mandelstam analyticity were incorporated in an explicit computational scheme. Atkinson's approach involved an inhomogeneous term, the contribution to the double-spectral function from true multiparticle states, which could be chosen with some freedom. However, the scheme of Atkinson did not permit any Regge trajectories above $\text{Re}\,\ell = 1$, for any value of s, as his equations developed unmanageable singularities whenever this situation arose.

We have developed an algorithm for construction of $\pi\pi$ amplitudes via Sommerfeld-Watson integrals, which is similar in spirit to the S-matrix approach to potential scattering. While the goal has been to incorporate reasonable Regge trajectories into such a scheme, the initial work[6] has involved an analysis of the unsubtracted equations, for which no Regge poles may be present. We shall describe the dynamical scheme in the unsubtracted case, where we treat the π mesons as isotopic spin scalars, for simplicity. The scattering amplitude is to be expressed in terms of the Mandelstam variables as

$$A(s,t) = \frac{1}{\pi^2} \int\int \frac{ds'\,dt'}{(s'-s)(t'-t)}\,\rho(s',t) + \left[\begin{array}{l}\text{cyclic permutation of}\\ s,\ t,\ \text{and}\ \mu\end{array}\right] \tag{25}$$

The dynamical entity in the approach is $a(\ell,s)$, the analytic continuation of the <u>even</u> partial waves to complex angular momentum. The function $a(\ell,s)$ is constructed as the solution of the equation

$$a(\ell,s) = \frac{4}{\pi(s-4)} \int_4^\infty dt\; Q_\ell\left(1+\frac{2t}{s-4}\right) A_t(s,t), \tag{26}$$

with

$$A_t(s,t) = \frac{1}{\pi} \int_4^\infty ds'\rho(s',t)\left[\frac{1}{s'-s} + \frac{1}{s'+s+t-4}\right] \tag{27}$$

The double-spectral function is written as

$$\rho(s,t) = \rho^{\varepsilon\ell}(s,t) + \rho^{\varepsilon\ell}(t,s) + v(s,t), \tag{28}$$

where v(s,t) is an inhomogeneous contribution to be specified at the outset, and

$$\rho^{\varepsilon\ell}(s,t) = \frac{h(s)}{4i} \int_{\mathrm{Re}\ell=-\varepsilon} d\ell(2\ell+1)\, q(s)\, a(\ell,s_+)\, a(\ell,s_-)\, P_\ell\left(1+\frac{2t}{s-4}\right) \tag{29}$$

The function h(s) is a cut-off of the two-body unitarity integral, and it has the properties $0 \leq h(s) \leq 1$, as well as

$$h(s) = \begin{cases} 1 & s \leq 16 \\ 0 & s \to \infty \end{cases} \tag{30}$$

This cut-off is quite consistent with elastic unitarity, which is required for $s \leq 16$, since it has an effect only at large s. However, the requirement of inelastic unitarity, for $s > 16$, is not automatic in this approach. The detailed proof of existence of solutions of (26)-(29) is given in Ref. 6.

The work in the unsubtracted case has served as a prelude to analysis of the more interesting situation in which sensible Regge trajectories are present. The initial phase of analysis of the latter problem has been to determine what form of Regge behaviour is consistent with crossing symmetry, analyticity, and two-body unitarity. We choose the functions $\alpha(s)$ and $\beta(s)$, which describe the Regge pole, at the outset, and write the partial-wave amplitude as

$$b(\ell,s) = \frac{\beta(s)}{\ell-\alpha(s)} + \bar{b}(\ell,s), \tag{31}$$

where the background partial-wave amplitude $\bar{b}(\ell,s)$ is analytic in ℓ for $\mathrm{Re}\,\ell > -\varepsilon$. It is easy to show that the Regge pole term maps through the appropriate equation so that we may write an equivalent equation of the form

$$\bar{b}(\ell,s) = F(\bar{b},\ell,s) \tag{32}$$

The Regge trajectory function $\alpha(s)$ is chosen so that $\alpha(0) \leq 1$, $\operatorname{Re}\alpha(s_{\pm}) < L$ for all s, and finally, $\operatorname{Re}\alpha(s_{1\pm}) = -\varepsilon$. A suitable shape for $\alpha(s)$ is shown in the figure below.

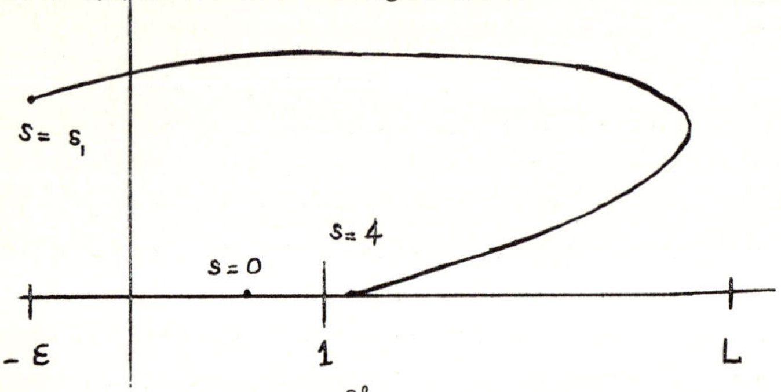

The expression (29) for $\rho^{\varepsilon\ell}(s,t)$ is no longer consistent with the Mandelstam representation; instead we must take the Sommerfeld-Watson contour to the right of all the Regge poles. Thus we obtain the modified expression

$$\rho^{\varepsilon\ell}(s,t) = \frac{h(s)}{4i}\int_{\operatorname{Re}\ell=L} d\ell\ (2\ell+1)\ q(\ell,s)\ b(\ell,s_+)\ b(\ell,s_-)(s-4)^{\ell}\ P_{\ell}(1+\frac{2t}{s-4})$$

$$= \frac{h(s)}{4i}\Big\{\int_{\operatorname{Re}\ell=-\varepsilon} d\ell\ (2\ell+1)\ q(\ell,s)\ b(\ell,s_+)\ b(\ell,s_-)\ (s-4)^{\ell}\ P_{\ell}(1+\frac{2t}{s-4}) + \pi\Theta(s_1-s)\ \mathrm{x}$$

$$[\,(2\alpha(s_+)+1\beta(s_+)(s-4)^{\alpha(s_+)}\ P_{\alpha(s_+)}(1+\frac{2t}{s-4}) - (+\rightarrow-)]\,\Big\} \qquad (33)$$

The cut-off function h(s) must be chosen as a sufficiently smooth function that obeys the requirement (30); in fact one must make the further restriction here that $h(s) = O(s^{-2L})$ as $s \rightarrow \infty$. The cut-off function h(s) was convenient, although not strictly necessary, in the unsubtracted case; but in the present context this cut-off becomes essential for the analysis. The character of the equations is altered to bear a closer relation to that of those obtained in potential scattering, as a result of this cut-off. Without the cut-off, the so-called "crossed Regge term" produces difficulties, which we have not been able to circumvent, in analysis of the equation.

The analysis of Eq. (32) for the function $\bar{b}(\ell,s)$ has been carried out in a Banach space of functions defined for $\operatorname{Re}\ell=-\varepsilon$ and $s \geq 4$. The elements of that space satisfy the constraint

$$|\bar{b}(\ell,s)| \le \frac{K}{(|\ell|+1)^\nu} s^L , \tag{34}$$

for $5/4 < \nu < 3/2$. We were able to show that, if the Regge residue function $\beta(s)$ and the overlap function $\mathbf{v}(s,t)$ are kept sufficiently small, the equation (32) has a locally unique small solution in the Banach space. From the solution of (32) we may construct a scattering amplitude satisfying a subtracted Mandelstam representation. Furthermore, the partial-wave projections of the scattering amplitude remain bounded at large s and at large physical (even integer) angular momentum ℓ. These partial-waves satisfy the elastic unitarity constraint,

$$\mathrm{Im}\, a(\ell,s) = q(s)\ |a(\ell,s)|^2 , \tag{35}$$

for physical ℓ in the elastic region $s \le 16$. However, for the inelastic region $s > 16$, we should instead satisfy the constraint of inelastic unitarity; namely

$$\frac{1-\eta^2(\ell,s)}{4q(\ell,s)} = \mathrm{Im}\, a(\ell,s) - q(s)\ |a(\ell,s)|^2 \ge 0 . \tag{36}$$

This constraint is not automatic in our S-matrix approach, since the multiparticle amplitudes themselves do not explicitly occur here. We have obtained crossing symmetry for the two-particle amplitude, at the expense of having the inelastic contribution to two-particle unitarity not manifest as a sum over intermediate states. In fact, since we have cut off the two-particle unitarity term with the function h(s), the expression which we obtain for the inelasticity is

$$\frac{1-\eta^2(\ell,s)}{4q(\ell,s)} = \#1 + \#2 = -(1-h(s))\ q(s)\ |a(\ell,s)|^2$$

$$+ \frac{2}{\pi(s-4)} \int_4^\infty dt\ Q_\ell(1+\frac{2t}{s-4})\ [\rho^{\varepsilon\ell}(t,s)+v(s,t)] . \tag{37}$$

The term #1 is definitely negative at large s, and it can be cast into a Froissart-Gribov representation similar to #2, with the t-integration beginning at t=16. As a consequence, it is possible to make term #2 uniformly larger in magnitude than term #1 for all ℓ and s.

Also, it can be arranged for term #2 to be positive, by appropriate choice of the Regge pole parameters. We can thus satisfy inelastic unitarity in the approach with a cut-off, at least when β and v are suitably small. We shall present a detailed discussion of this formulation in a forthcoming publication.

It has been established that the general requirements of crossing symmetry, two-body unitarity, and Mandelstam analyticity do permit the existence of Regge trajectories which lie to the right of $\ell = 1$ at some energies. These trajectories are allowed to rise as far as one likes, although they must eventually turn back at high energies. We can prove that suitable scattering amplitudes exist only for the physically uninteresting case of "weak coupling", when the residue function β is uniformly small. While our existence proof does not guarantee that "strong coupling" solutions exist, we would expect to be able to follow a given solution as the residue function is increased in magnitude, at least until a singularity of the system of equations is encountered.

The system (32) has the distinctive feature that the Regge pole parameters are an input, rather than an output, in the scheme. It is quite informative to establish that the general requirements of two-body unitarity and a crossing-symmetric Mandelstam representation still leave substantial freedom of choice for the Regge trajectory. However, the system (32) is not known to be stable against small variations of the Regge trajectory function α, and one might have real difficulty in using the formalism for a phenomenological analysis of $\pi\pi$ scattering. We have therefore worked upon a more ambitious undertaking, in which we generate Regge trajectories dynamically, using inelastic N/D equations to incorporate analyticity and crossing symmetry, as well as unitarity. A problem with such approaches has always been that, while the Mandelstam representation is used as ingredient to the system, one cannot be sure as to whether a solution of the equation in fact satisfies the Mandelstam representation. Also, there may be no numerically stable way of generating solutions in such a scheme. I shall describe some recent progress on these matters; the details of analysis will be given in a forthcoming publication.

The first step in obtaining a dynamical scheme is to write a partial-wave dispersion relation for a reduced amplitude $c(\ell,s)$, defined from

$$a(\ell,s) = \left[\frac{s-4}{(\sqrt{s}+2)^2}\right]^{\ell} c(\ell,s). \tag{38}$$

The two-body phase-space factor for $c(\ell,s)$ is

$$r(\ell,s) = q(s)\ h(s)\ \frac{(s-4)^{\ell}}{(\sqrt{s}+2)^{2\ell}} , \tag{39}$$

and the partial-wave dispersion relation for $c(\ell,s)$ is

$$c(\ell,s) = C_L(\ell,s) + \frac{1}{\pi}\int_4^{\infty} \frac{ds'}{s'-s}\ r(\ell,s')\ c(\ell,s'_+)\ c(\ell,s'_-) +$$

$$\frac{1}{\pi}\int_{16}^{\infty} \frac{ds'}{s'-s}\ \frac{1-\eta^2(\ell,s')}{4r(\ell,s')} . \tag{40}$$

We have incorporated a cut-off function $h(s)$ in this scheme for essentially the same reason as we did in the analysis of Eq. (32). The difficulties with the crossed Regge term are no less serious here than they were in that context.

The advantage in writing a partial-wave dispersion relation for $c(\ell,s)$ is that exponential ℓ-decrease for the partial waves, as it is required for Mandlestam analyticity, is factored out in Eq. (38). Furthermore, from the Froissart-Gribov integral one can obtain a partial-wave dispersion relation for $c(\ell,s)$ which is similar to Eq. (10) of the non-relativistic problem, in that one can avoid writing $C_L(\ell,s)$ as an explicit left-cut dispersion integral. Therefore, one can still have a well-behaved expression for $C_L(\ell,s)$, even though the exponential ℓ-decrease has been factored out of the problem. The idea is quite similar in spirit to our N/D formulation of potential scattering, except that the crossed terms make matters even more complicated here. In order to write the Frye-Warnock[7] equations equivalent to (40), we define the auxiliary quantities

$$C_1 = \frac{1}{\pi}\int_{16}^{\infty} \frac{ds'}{s'-s-i\varepsilon}\ \frac{1-\eta(\ell,s')}{2r(\ell,s')} = C_{1p} + i\ \frac{1-\eta(\ell,s)}{2r(\ell,s)} , \tag{41}$$

$$C_p(\ell,s) = C_L(\ell,s) + C_{1p}(\ell,s), \tag{42}$$

and

$$C(\ell,s) = C_L(\ell,s) + C_1(\ell,s) \tag{43}$$

We may replace the partial-wave dispersion relation (40) by the following integral equation for n, the numerator function:

$$\eta(\ell,s)\; n(\ell,s) = C_p(\ell,s) + \frac{1}{\pi}\int_4^\infty ds' \; \frac{C_p(\ell,s')-C_p(\ell,s)}{s'-s}\; r(\ell,s')\; n(\ell,s') \tag{44}$$

The denominator function d is determined from n as

$$d(\ell,s) = 1 - \frac{1}{\pi}\int_4^\infty \frac{ds'}{s'-s}\; r(\ell,s')\; n(\ell,s') \tag{45}$$

and the amplitude sarisfying the partial-wave dispersion relation (40) is obtained as

$$c(\ell,s) = C(\ell,s) + \frac{1}{\pi d(\ell,s)}\int_4^\infty \frac{ds'}{s'-s}\; C_p(\ell,s')\; r(\ell,s')\; n(\ell,s') \tag{46}$$

We obtain a dynamical scheme by using Eqs. (44)-(46), along with appropriate expressions for $\eta(\ell,s)$ and $C_p(\ell,s)$ in terms of the double spectral function $\rho^{\varepsilon\ell}$, the latter being determined from $c(\ell,s)$ via Eq. (30). For the purpose of analysis we cast our expressions into a system of equations of the form

$$[\eta, C_p] = G[\eta, C_p], \tag{47}$$

the function G being obtained from the above expressions. One may obtain a solution of (47) at weak coupling, and then follow the solutions as the coupling is increased. The coupling is increased by varying the overlap function $v(s,t)$; this latter function contributes to an inhomogeneous term in Eq. (47). The system (47) is tractible for numerical analysis, and it permits the generation of Regge trajectories by changing the function $v(s,t)$, which plays the same role as the potential in the nonrelativistic formulation. The Regge trajectories are determined as zeros of the function $d(\ell,s)$, which are determined from (45) and controlled at each stage of the analysis.

While the relativistic N/D S-matrix equations do provide a sensible framework for numerical generation of realistic amplitudes, there are several points at which the approach could run into difficulties. We shall call attention to these touchy points, which serve to illustrate the limitations of this scheme, as well as of S-matrix schemes in general.

The Frye-Warnock equation (44), with C_p and η fixed, is a Fredholm integral equation for $n(\ell,s)$, unless the function $\eta(\ell,s)$ develops one or more zeros in its domain of definition. The Fredholm character of that equation is an essential property in our approach, and we cannot tolerate any such zeros. It is possible, also, that we come upon a unit eigenvalue of the integral equation (44), so that it would be ambiguous or impossible to obtain the solution of the equation. For that matter, we could come upon singularities of the Fréchet derivative of Eq. (47). These difficulties may appear as the strength of the coupling to true many-particle states is increased, although they are distinct in character subtleties caused by the Regge poles themselves. Thus, one would expect to be able to produce sensible Regge trajectories without encountering these difficulties by a suitably careful choice of the overlap function $v(s,t)$.

We have analyzed this dynamical scheme under the assumption that the (dynamically generated) Regge trajectories do not collide with one another in the right half ℓ-plane. In fact, we have made the restrictive simplifying assumption that the trajectory functions $\alpha(s)$ are univalent (schlicht) over the portion of the cut s-plane such that $\mathrm{Re}\,\alpha(s) > -\varepsilon$. This condition is known to be satisfied in ordinary potential scattering, but we know of no proof of it in a relativistic theory. In addition, we cannot guarantee the positivity condition (38) in this N/D approach, since the sign of term #2 in (37) cannot be determined a priori. It is an essentially numerical question as to which overlap functions will give solutions subject to these constraints.

The crossing-unitarity problem is known to be subject to the CDD ambiguity[8], and one expects this ambiguity to be present in the equations described here. The N/D equations allow one to obtain solutions in a particular CDD class, whereas the direct analysis of Eq. (32) does not take this ambiguity into account properly. Thus, one requires the latter more sophisticated approach to do a proper phenomenology with CDD poles.

The solutions of these dynamical schemes, which have exact crossing symmetry and two-particle unitarity, must satisfy the Froissart bound, so that $\alpha(0) \leq 1$ for all Regge trajectories. We would like to know how close we can come to saturating the Froissart bound in this approach. It would also be very interesting to understand how the constraint of multiple particle unitarity (indirectly) affects the two-particle amplitudes. The question of the existence of Regge cuts is closely related to this latter point. We have been able to avoid Regge cuts in the current formulation, but there are no multiple particle amplitudes in the equations. In particular, the cut-off function h(s) might lead to inconsistencies with multiple particle unitarity.

This completes the summary of our recent work on construction of two-particle amplitudes using S-matrix equations. I would like to thank Professor P. Sabatier and other sponsors of these conferences for inviting me to Montpellier to discuss our results.

(Footnotes)

1. Blankenbecler, Goldberger, Khuri, and Treiman, Ann Phys (N.Y.) 10, 62 (1960)
2. J. Bessis, J. Math Phys 6, 637 (1965)
3. J.S. Frederiksen, P.W. Johnson, and R.L. Warnock, J. Math Phys 16, 1886 (1975).
4. S. Mandelstam, Ann Phys (N.Y.) 21, 302 (1963)
5. D. Atkinson, Nucl Phys B7, 375 (1968); B8, 377 (1968); B13, 415 (1969); B23 397 (1970). Reference to other work is given in these articles.
6. D. Atkinson and J.S. Frederiksen, Commun Math Phys 40, 55 (1975); J.S. Frederiksen, Commun Math Phys 43, 1 (1975).
7. G. Frye and R.L. Warnock, Phys Rev 130, 478 (1963).
8. D. Atkinson and R.L. Warnock, Phys Rev 188, 2098 (1969).

GEL'FAND-LEVITAN THEORY OF THE INVERSE PROBLEM FOR SINGULAR POTENTIALS

K. CHADAN

Laboratoire de Physique Théorique et Hautes Energies*

Bât. 211, Université de Paris XI

91405 ORSAY Cédex France

Abstract.

The Gel'fand-Levitan theory with two potentials is generalized to the case where the first potential, assumed to be known, is singular and repulsive at the origin. It is shown that the only modifications required are redefinitions of the regular solution and the Jost function. Some of the proofs are only sketched here, and will be given in more detail in a separate paper.

Introduction.

The purpose of this paper is to generalize the Gel'fand-Levitan formalism for nonrelativistic radial Schrödinger equation to the case where the potential is singular and repulsive at the origin. It is well-known now [1 , 2] that the Gel'fand-Levitan integral equation of the

*Laboratoire associé au C.N.R.S.

inverse scattering problem was first developped for potentials satisfying the integrability condition

$$\int_0^\infty r \,|V(r)|\, dr < \infty \qquad (1)$$

In what follows, we shall assume that

$$r\,V(r) \in L^1(a,\infty)\ , \quad \forall\, a > 0 \qquad (2)$$

and call such potentials short range.

It has been shown recently [3] that the Gel'fand-Levitan method applies without any modification to a class of short range potentials for which $r\,V(r)$ is not necessarily absolutely integrable at the origin. However, these potentials are such that, if we define

$$W(r) = -\int_r^\infty V(t)\, dt \qquad (3a)$$

then

$$W(r) \in L^1(0,a)\ , \quad \forall\, a > 0 \qquad (3b)$$

$$\lim_{r \to 0} r\, W(r) = 0 \qquad (3c)$$

It was in fact shown that, for these potentials, the scattering theory formulated in terms of Jost functions applies exactly as for potentials satisfying the old condition (1). In what follows, we shall call the W-class the class of potentials satisfying (2), (3b), and (3c). It is easy to verify that the W-class contains all the potentials satisfying (1).

In the present paper, we shall generalize the Gel'fand-Levitan theory to the case of short range potentials which are truly singular at the origin, i.e. they are outside of the W-class.

These short-range singular potentials can be classified in three classes. The first one is made of potentials for which the singularity at the origin is repulsive. Examples are

$$V(r) \underset{r \to 0}{\sim} g\, r^{-n} , \quad n > 2 , \quad g > 0 \qquad (4)$$

$$V(r) \underset{r \to 0}{\sim} g\, e^{ar^{-n}} , \quad n, a, g > 0 \qquad (5)$$

etc. The second class contains potentials which are attractive at the origin. Finally, the third class is made of very singular and violently oscillating potentials, like the one devised by Pearson [4]. For the second and the third class, it is well-known that there are several difficulties either with the definition of the Hamiltonian as a unique self-adjoint operator, or with the asymptotic completeness of wave operators [5,6]. We shall therefore consider only the first class, i.e. singular repulsive potentials, for which the scattering theory is well-defined [5,6,7]. We shall assume that the potential is of the form

$$V(r) = V_s(r) + V_1(r) \tag{6a}$$

where $V_s(r)$ is singular repulsive :

$$\lim_{r \to 0} r^2 V_s(r) = \infty \tag{6b}$$

and $V_1(r)$ is regular :

$$V_1(r) \in \text{W-class} \tag{6c}$$

It is well-known that for singular repulsive potentials the S-matrix exists and is unitary, which means that the phase-shift $\delta(k)$ is a real continuous function of the momentum k [5,8]. Moreover, it can be shown that the S-matrix is obtained as the limit of the S-matrix for the regularized potential. We think here of regularizing procedures like

$$g\, r^{-n} = \lim_{\varepsilon \downarrow 0} g\, (r+\varepsilon)^{-n} \quad , \tag{7}$$

$$g\, r^{-n} = \lim_{\varepsilon \downarrow 0} g\, r^{-n}\, \theta(r-\varepsilon) \tag{8}$$

or any other regularizing procedure. When the potential

is regularized at the origin, we have of course the usual decomposition [9]

$$S_\epsilon(k) = F_\epsilon(k)\, F_\epsilon^{-1}(-k) \tag{9}$$

where F_ϵ is the Jost function. When we let $\epsilon \downarrow 0$, we recover the S-matrix of the singular potential. The difference between the regular W-class and the truly singular repulsive potentials is that, at the limit $\epsilon = 0$, F_ϵ tends uniformly to the Jost function of the limiting potential in the case of potentials of W-class, whereas F_ϵ tends to infinity in the case of singular potentials. In other words, whereas in the case of regular potentials the numerator and the denominator of (9) have separately finite limits for all (real) values of k, it is only the ratio as a whole which has a finite limit in the case of singular potentials [8].

Another difference is that, in the regular case of W-class, the phase-shift tends to naught at $k = \infty$ (at high energies the potential is negligible compared to the kinetic energy), whereas for singular repulsive potentials the phase-shift tends to $-\infty$. As example, we have, for the potential (4), the high energy behaviour [8]

$$\delta(k) \underset{k \to \infty}{=} -\alpha_n\, k^{(n-2)/n} + \cdot\cdot,\quad \alpha_n > 0 \tag{10}$$

This fact, and the fact that $F_\epsilon(k)$ does not have a finite limit for such potentials when $k = \infty$ are intimately related, as it will be shown below. Before showing this, let us notice that the most rapid growth of the phase shift at large energies is linear, corresponding to $n = \infty$ in (4) or (5), i.e. a hard core, which is the most singular case.

The Jost Function.

For regular potentials satisfying (1), or belonging to the W-class, the Jost function can be shown to have the following integral representation [1,2]

$$F(k) = \exp \left[\int_0^\infty \gamma(t)\, e^{ikt}\, dt \right] \qquad (11a)$$

where

$$\gamma(t) \in L^1(0, \infty) \qquad (11b)$$

The phase shift being minus the phase of the Jost function [9], we have (k real!)

$$\delta(k) = - \int_0^\infty \gamma(t) \sin kt\, dt \qquad (12)$$

and

$$|F(k)| = \exp \left[\int_0^\infty \gamma(t) \cos kt\, dt \right] \qquad (13)$$

These representations show that when $k = \infty$, the phase shift is zero, whereas the modulus of the Jost function is 1.

Consider now the singular case. When we regularize the potential at the origin, we obtain for its Jost function $F_\epsilon(k)$ a representation of the form (11a), with $\gamma_\epsilon(t) \in L^1(0,\infty)$. Suppose now we let $\epsilon \downarrow 0$. As we said before, F_ϵ does not have a limit. However, the S-matrix has a limit :

$$\lim_{\epsilon \downarrow 0} S_\epsilon(k) = \lim \exp[[2i\ \delta_\epsilon(k)] = e^{2i\,\delta(k)} \qquad (14)$$

According to (12), this means that the limiting function $\gamma(t)$ is such that

$$\Gamma(t) = -\int_t^\infty \gamma(u)\, du \qquad (15)$$

is negative near $t = 0$, and is there singular, but weaker than t^{-1}. Indeed, before taking the limit $\epsilon = 0$, we can write (12) as

$$\delta_\epsilon(k) = k\int_0^\infty \Gamma_\epsilon(t)\cos kt\, dt \qquad (16)$$

This shows, according to standard theorems on the asymptotic behaviour of Fourier sine and cosine transforms [10], that the limiting function $\Gamma(t)$ must satisfy the above conditions if we demand that the high energy behaviour of the phase shift should be of the type (10), i.e. negative and with a behaviour less singular than k. Notice that if $\Gamma(0)$ is finite, the phase shift does not become infinite when k becomes very large. Therefore,

$$F_\epsilon(0) = \exp\left[\int_0^\infty \gamma_\epsilon(t)\, dt \right] \qquad (18)$$

becomes infinite when $\epsilon \downarrow 0$. However,

$$|\tilde{F}_\epsilon(k)| \equiv |F_\epsilon(k)|\, F_\epsilon^{-1}(0) = \exp\left[\int_0^\infty \gamma_\epsilon(t)\, (\cos kt - 1)\, dt\right] = \exp\left[k \int_0^\infty \Gamma_\epsilon(t)\, \sin kt \, dt \right] \qquad (19)$$

has a finite limite, $|\tilde{F}(k)|$, when $\epsilon \downarrow 0$. Again, using the same asymptotic theorems as before [10], we see now that this limiting function has the asymptotic form

$$|\tilde{F}(k)| \simeq \exp\left[- \alpha\, k^{1-\eta} \right] \qquad (20)$$

when k is large, wit $\alpha > 0$, and $0 < \eta < 1$.

We call this function the modified Jost function. It is given by

$$\widetilde{F}(k) = \exp\left[- ik \int_0^\infty \Gamma(t)\, e^{ikt}\, dt \right] \qquad (21)$$

As in the regular case, the phase shift is given now by minus the phase of the modified Jost function :

$$\delta(k) = k \int_0^\infty \Gamma(t) \cos kt\, dt \qquad (22)$$

The regular solution of the Schrödinger Equation.

For the regularized potential, the regular solution of the Schrödinger equation satisfying $\varphi(0) = 0$, $\varphi'(0) = 1$, is given by [9]*

$$\varphi_\epsilon(k,r) = (2ik)^{-1} \left[F_\epsilon(-k)\, f_\epsilon(k,r) - F_\epsilon(k)\, f_\epsilon(-k,r) \right] \qquad (23)$$

where f_ϵ is the Jost solution. In fact, since the Jost solution is defined only by its asymptotic behaviour at $r = \infty$, it does not depend on ϵ except at very short distances. For all finite values of r , the Jost solution of the true potential is the same as that of the regularized potential. Therefore, according to the analysis of

---

* We consider for simplicity the S-wave here. However, the analysis is quite general and applies to all waves.

the previous section, the limit of the "renormalized" wave function

$$\tilde{\varphi}_{\epsilon}(k,r) = F_{\epsilon}^{-1}(0)\ \varphi_{\epsilon}(k,r) \qquad (24)$$

$$r > 0$$

exists as $\epsilon \downarrow 0$, and defines our regular solution for the true potential :

$$\tilde{\varphi}(k,r) = (2ik)^{-1}[\ \tilde{F}(-k)\ f(k,r) - \tilde{F}(k)\ f(-k,r)\ ] \qquad (25)$$

exactly as in the regular case. However, here, obviously

$$\tilde{\varphi}'(k,0) = \lim_{\epsilon \downarrow 0} F_{\epsilon}^{-1}(0)\ \varphi_{\epsilon}'(k,0) = 0 \qquad (26)$$

as expected, since for singular repulsive potentials one has $\varphi(0) = \varphi'(0) = \cdots = 0$

To see exactly how $\tilde{\varphi}$ is normalized, we have to look at a point $\neq 0$. We choose $r = \infty$. For r large enough, we have

$$\tilde{\varphi} \simeq k^{-1} \mid \tilde{F}(k) \mid \sin (kr + \delta) \qquad (27)$$

Using now the fact that, because of (19),

$$\tilde{F}(0) = \lim_{\varepsilon \downarrow 0} \tilde{F}_{\varepsilon} (0) \equiv 1 \qquad (28)$$

we have, for large values of r,

$$\tilde{\varphi}(0,r) \simeq r - a_0 \qquad (29)$$

where a_0 is the scattering length. This gives the exact normalization of the regular solution of the radial equation at zero energy. Once this is given, we have, for other values of energy,

$$\tilde{\varphi} (k,r) = \tilde{\varphi} (0,r) - k^2 \int_0^r G_v(r,r') \tilde{\varphi}(k,r') \, dr' \qquad (30)$$

where G_v is the appropriate total "Green's function" at zero energy [8,9].

Consider now the completeness of the solutions of the radial equation. For the regularized potential, we have [1,2,9], assuming for simplicity that there are no bound states,

$$(2/\pi) \int_0^\infty \varphi_\varepsilon(k,r) \; \varphi_\varepsilon(k,r') \mid F_\varepsilon(k) \mid^{-2} k^2 \, dk = \delta(r-r') \qquad (31)$$

Dividing φ_ϵ and F_ϵ by $F_\epsilon(0)$, and taking the limit, the above relation becomes

$$(2/\pi) \int_0^\infty \tilde{\varphi}(k,r)\, \tilde{\varphi}(k,r')\, |\tilde{F}(k)|^{-2}\, k^2\, dk = \delta(r-r') \qquad (32)$$

There is therefore no formal change in going from (31) to (32), except that all quantities should be replaced by modified ones.

<u>The Inverse Problem</u>.

We consider now the case of a potential of the form (6a), satisfying the conditions (6b) and (6c), and would like to generalize the usual Gel'fand-Levitan theory with two regular potentials to this case. Let us regularize first the singular part of (6a). We then have, in obvious notations, the Povzner-Levitan representation for the regular solution of the total potential :

$$\varphi_\epsilon(k,r) = \varphi_{s\epsilon}(k,r) + \int_0^r K_\epsilon(r,r')\, \varphi_{s\epsilon}(k,r')\, dr' \qquad (33)$$

Dividing now both sides by $F_{s\epsilon}(0)$, where $F_{s\epsilon}$ is the Jost function of the singular potential $V_{s\epsilon}$, and letting $\epsilon \downarrow 0$, one obtains

$$\tilde{\varphi}(k,r) = \tilde{\varphi}_s(k,r) + \int_0^r K(r,r')\, \tilde{\varphi}_s(k,r')\, dr' \qquad (34)$$

From this, and the completeness relations (32) both for $\tilde{\varphi}$ and $\tilde{\varphi}_s$, we find, as usual [1,2], the integral equation

$$K(r,t) + G(r,t) = \int_0^r K(r,s)\, G(s,t)\, ds \qquad (35a)$$

$$G(r,t) = (2/\pi) \int_0^\infty \tilde{\varphi}_s(k,r)\, \tilde{\varphi}_s(k,t) \left[\, |\tilde{F}(k)|^{-2} - |\tilde{F}_s(k)|^{-2} \right] k^2 \, dk \qquad (35b)$$

To see the convergence of the integral in (35b), we have first, by using (25), (20), (10) and the fact that the large k behaviour of the Jost solution is given by $f \sim e^{ikr}$, that

$$\tilde{\varphi}(k,r) \simeq k^{-1} \, |\tilde{F}(k)| \sin (kr - \alpha k^{\beta}) \qquad (36)$$

with $\beta < 1$. Since the large k behaviour of F and F_s are the same (remember that in (6a) V_1 is regular, and therefore it can be neglected compared to the kinetic energy plus the singular part of the potential when k is large), one can show the convergence of the integral in the same way as for the regular case.

When bound states are present, (35b) has to be modified in the well-known way to include their contributions.

How to solve the inverse problem.

We assume the phase shift to be given for all $k \geq 0$, and that it can be written

$$\delta(k) = \delta_s(k) + \delta_1(k) \tag{37}$$

where $\delta_s(k)$, the singular part, has an asymptotic behaviour similar to (10), whereas $\delta_1(\infty) = 0$. We assume that the corresponding singular potential V_s and its solution $\tilde{\varphi}_s$, as well as its Jost function $\tilde{F}_s$, to be explicitely known. From the total phase shift δ, we calculate also $\tilde{F}$ via (22) and (21). From these, we can now calculate the kernel G of the generalized Gel'fand-Levitan equation (35a). The potential V_1 is now given by

$$V_1(r) = -2\,(d/dr)\,K(r,r) \tag{38}$$

In résumé, the inverse problem with two potentials, where the known part is singular, is formally identical to the case where both potentials are regular, except that one must use the modified solutions $\tilde{\varphi}$ and the modified Jost function $\tilde{F}$, etc, everywhere.

REFERENCES.

1) FADDEEV, L.D. (1963), J. Math. Phys. 4, 72-104

2) CHADAN, K., and SABATIER, P.C.(1977) : Inverse Problems in Quantum Scattering Theory, Springer-Verlag.

3) BAETEMAN, M.L., and CHADAN, K. (1976), Nucl. Phys. A255, 34-49 ; (1977) , Ann. Inst. Henri Poincaré, AXXIV 1-16.

4) PEARSON, D.B. (1975), Commun. Math. Phys. 40, 125-146.

5) SIMON, B. (1971) : Quantum Mechanics for Hamiltonians defined as Quadratic Forms, Princeton University Press. See also the recent treatise by REED, M., and SIMON, B., vol . II, III, IV.

6) AMREIN, W.O., JAUCH, J.M., and SINHA, K.B. (1978) : Scattering Theory in Quantum Mechanics, W.A. Benjamin, Inc.

7) See reference 4 where a good list of papers treating scattering by singular potentials is given.

8) FRANK, W.M., LAND, D.J., and SPECTOR, R.M. (1971), Rev. Mod. Phys., 43, 36-98.

9) NEWTON, R.G. (1966) : Scattering Theory of Waves and Particles, McGRAW-HILL.

10) TITCHMARSH, E.C. (1948) : Theory of Fourier Integrals, pp. 172 ff.

List of Contributions to the meetings of the R.C.P. 264 with references to the papers in which they have been published since, when they were a first communication of new results

1 9 7 2*

P.C. SABATIER - Introduction à l'étude de la R.C.P. des problèmes physiques dans les problèmes inverses.

B. LEMAIRE - Problèmes mathématiques d'identification.

J.P. ZOLESIO - Problèmes d'identification de domaines.

G. JOBERT - Problème inverse de la géophysique interne (Sismologie).

M. SOURIAU - Méthodes matricielles pour l'inversion des sismogrammes.

Ch. HEMON - Géophysique interne - Prospection pétrolière par les méthodes sismiques.

M. FEIX - Le problème inverse en physique des plasmas.

M. NAVET - Problème inverse en hydrogéologie.

F. LEFEUVRE - L'inversion dans l'analyse des champs naturels d'ondes électromagnétiques T.B.F. dans la magnétosphère.

P. DESCHAMPS - Problème inverse particulier en géophysique externe.

D. GAUTIER - Problème inverse en Astrophysique.

M. DAIGNIERES - Problème inverse en Magnétotellurique dans le domaine temporel.

R.G. NEWTON - The inverse problem in quantum mechanics.

A. DEGASPERIS - On the inverse problem for Schrödinger's wave equation. The asymptotic expansion method.

C. COUDRAY - Problème inverse à énergie fixée. Applications de la méthode de Newton.

J. DUCRUIX - Application de la méthode de Bachus et Gilbert à la magnétotellurique.

V. COURTILLOT - Localisation des sources d'anomalies du champ géomagnétique.

P.C. SABATIER - A propos de l'équation de Fredholm de première espèce.

G. JOBERT - Une remarque sur la méthode des rais à trois dimensions.

*All the lectures of 1972 have been published in a report of "Laboratoire de Physique Mathématique" Université des Sciences et Techniques du Languedoc 34060 MONTPELLIER Cedex

J. LENOBLE - Etude d'une atmosphère planétaire.

M. JAULENT - Un problème inverse en mécanique quantique. Application à des problèmes inverses dans les milieux absorbants.

1 9 7 3

K. CHADAN - Les propriétés de la fonction de Jost des potentiels coupés, et le problème inverse.
(Published as a report of Laboratoire de Physique Théorique et Hautes Energies - Orsay)

J.J. LOEFFEL - Méthodes numériques dans un problème inverse de mécanique quantique.
(Published in : A. HEINIGER, J.J. LOEFFEL "Une classe d'algorithmes pour le problème inverse de la diffusion", Rapport de la Société Suisse de Physique Helvetica Physica Acta 46, 462-464 (1973).

J. DUCRUIX - Un problème inverse non linéaire de géophysique.
(Published in Thèse de 3e cycle, Institut de Physique du Globe, Paris).

J.L. LE MOUEL - Un problème inverse linéaire de géophysique.
(Published in : J. DUCRUIX, J.L. LE MOUEL, V. COURTILLOT "Une méthode simple d'évaluation de la correction topographique dans les problèmes de flux de chaleur", C.R. Académie des Sciences Paris 278 B, 841-843 (1974).

P. LOUIS, M. DAIGNIERES - Etude de l'analyse des données en magnétotellurique.

P.C. SABATIER - Problème inverse sismique et problème inverse quantique.
(Published in : P.C. SABATIER "Problème direct et inverse de diffraction d'une onde élastique par une zone perturbée sphérique, C.R. de l'Académie des Sciences Paris, 278 B, 545-547 (1974).

G. JOBERT - Problème inverse sismique.

J.M. DROUFFE - Détermination de l'amplitude de diffusion à partir de la section efficace.
(Published in : H. CORNILLE, J.M. DROUFFE "Phase-Shift ambiguities for spinles $>$ and $<_{max} \leq 4$ elastic scattering", Il Nuovo Cimento 20 A, 3, 401-436 (1974).

M. CUER - Ambiguités semiclassiques.
(Published in : Thèse de Doctorat de 3e Cycle, Université des Sciences et Techniques du Languedoc, Montpellier).

D. BURETTE - Méthode de résolution des problèmes inverses. Application à la détermination d'un mécanisme de compensation isostatique.
(Published in : Mémoire de DEA, Institut de Physique du Globe, Paris).

1 9 7 4

P.C. SABATIER - Les problèmes inverses.

P. MOREL - Algorithme pour le problème inverse des valeurs propres.

J.L. VIDAL - Interprétation des intensités de raies formées dans des milieux optiquement minces.

Y. BIRAUD - Une méthode de déconvolution positive.

I. MIODEK - Un problème inverse pour un plasma.
(Published in "Inverse problem for complex $r^{\hat{a}}$ analytic potentials of finite range "J. Math. Phys. 17, 2, 168-173 (1976).

K. CHADAN - Méthode de Guelfand-Levitan et problèmes des ondes solitaires.

V.E. COURTILLOT - Quelques applications du formalisme des problèmes inverses à la résolution de problèmes de potentiel.
(Published in "V. COURTILLOT, J. DUCRUIX, J.L. LE MOUEL "A solution of some inverse problems in geomagnetism and gravimetry, J. of Geophysical Research, 79, 32, 4933-4940 (1974).

M. JAULENT - Le problème inverse pour l'équation de Schrödinger à une dimension avec un potentiel dépendant de l'énergie.
(Published in : M. JAULENT, C. JEAN "The inverse problem for the one dimensional Schrödinger equations Ann. Inst. Henri Poincaré - XXV, 2, 105-118 and 119-137(1976).

M. DAIGNIERES - Application du problème inverse à des calculs d'induction électromagnétique.

1 9 7 5

P.C. SABATIER - Contraintes de positivité dans les problèmes inverses linéaires.
(Published in : P.C. SABATIER "Positivity constraints in linear inverse problems : I. General Theory II Applications" Geophysical J.R. Astr. Soc. 48, 415-469 (1977)).

J.L. VIDAL - Détermination des paramètres physiques des nébuleuses gazeuses à partir des observations.
(Published in : Thèse de Doctorat d'Etat).

J.L. GERVAIS - Solitons (I).

M. JAULENT - Problèmes inverses de la diffusion en milieux absorbants.
(Published in : M. JAULENT "Inverse scattering problems in absorbing media", J. Math. Phys. 17, 7 1351-1360 (1976).

G. MAHOUX - Problème inverse $F|^2| \to f$ unitaire.
(Published in ATKINSON D. MAHOUX G. and YNDURAIN F.J., Nucl. Phys. B 98 521-532 (1975)).

J.L. GERVAIS - Solitons (II).

B. ROUSSELET - Etude de la régularité de valeurs propres par rapport à des déformations bilipschitziennes du domaine géométrique.
(Published in : Compt. Rend. Acad. Sc. Paris 283 A, 507-509 (1976).

D. GAUTIER - Problème inverse du transfert radiatif.

M. DELANNOY - Analyse de champs d'ondes aléatoires.

D. ATKINSON - Compacité de la dérivée de Fréchet de l'opérateur permettant de construire l'amplitude de diffusion à partir de la section efficace.
(Published in : ATKINSON D., KAEBEKEKE L.P. and de ROO M. J. Math. Phys. 16, 685-693 (1975).

M. CUER - Analyse semi-classique du problème inverse à énergie fixée en mécanique quantique. Applications aux ambiguités du modèle optique.
(to be published in Annals of Physics).

K. CHADAN - Problème inverse des potentiels singuliers oscillants.
(Published in BAETEMAN M.L. and CHADAN K. Nucl. Phys. A 255 34-49 (1975).

P.W. JOHNSON - Construction of Regge amplitudes through solution of S-matrix equations.
(Published in the present book).

J.L. GERVAIS - Solitons (III).

P. WEIDELT - Problème inverse de l'électromagnétisme.

A. BAMBERGER et P. LAILLY - Problèmes inverses traités à l'Institut Français du Pétrole.
(Published in "A. BAMBERGER, G. CHAVENT, P. LAILY, Etude mathématique et numérique d'un problème inverse pour l'équation des ondes à une dimension. Rapport n° 14 de l'Ecole Polytechnique (Centre de Mathématiques Appliquées) Janvier 1977.

1 9 7 6

P.C. SABATIER - Problèmes inverses linéaires et contraintes, et applications en géophysique.
(Published in J. Geophys. 43, 115-137 (1977)).

P. MOREL - Diverses données spectrales pour le problème inverse discret de Sturm-Liouville.
(Published in present book).
- Algorithmes pour un problème inverse discret de Sturm-Liouville.
(Published in present book).

I. MIODEK - Les approches de Lax et d'Ablowitz dans le problème des solitons.
(Publ. in J. Math. Phys. 19, 19-31 (1978).

A. MARTIN - Bornes suffisantes pour garantir l'inexistence d'un spectre discret.
(Published in : GLASER V., MARTIN A., GROSSE H., THIRRING W., in studies in mathematical Physics. Ed. by E.H. LIEB, B. SIMON and A.S. WIGHTMAN - Princeton Univ. Press.)

B.L.N. KENNETT - Ray theoretical inverse methods in geophysic.
(Published in present book).

M. CUER - Les potentiels transparents en étude classique et semiclassique.
(Published in J. Math. Phys. 18, 658-669 (1977).

J. DUCRUIX - Continuation du potentiel de gravité.

SAFON - Applications de la programmation linéaire en géophysique.
(Published in SAFON C., VASSEUR G., CUER M., Geophysics 42, 1215-1229 (1977).

B. KARLSSON - Inverse method for off-shell continuation of the scattering amplitude in quantum mechanics.
(Published in the present book).

M. CADILHAC - Problème inverse électromagnétique.

K. CHADAN - Problèmes inverses à une dimension.

M. JAULENT - Applications du problème inverse à une dimension.

1 9 7 7

D. ATKINSON - Analytic extrapolations and inverse problem.
(Published in the present book).

M. CADILHAC, A. ROGER - Etude numérique d'une méthode de régularisation de Tichonov appliquée à un problème inverse en optique.
(Published in the present book).

F. CALOGERO F., DE GASPERIS A. - Spectral transform and nonlinear evolution equations.
(Published in the present book).

J. CEA - Quelques méthodes sur la recherche d'un domaine optimal.
(Published in the present book).

K. CHADAN - Solution du problème inverse à l fixé avec les vrais potentials singuliers.
(Published in the present book).

H. CORNILLE - Inversion-like integral equations.
(Published in the present book).

C. COUDRAY - A study of an inverse problem for finite range potentials.
(Published in the present book).

V. COURTILLOT, J. DUCRUIX, J.L. LE MOUVEL - Inverse methods applied to continuation problems in geophysics.
(Published in the present book).

M. CUER - Application of linear programming to the inverse gravity of magnetic problems basic numeral techniques.
(Published in the present book).

C. DE MOL, M. BERTERO, G.A. VIANO - On the regularization of linear inverse problems in Fourier optics.
(Published in the present book).

M. FEIX, J.R. BURGAN, E. FIJALKOW, J. GUTIERRES, A. MUNIER - Utilisation de groupes de transformation pour la résolution des équations aux dérivées partielles.
(Published in the present book).

D. GAUTIER - Problèmes inverses de transfert radiatif.

D. JACKSON - Linear inverse problems with a priori data.
(Published in the present book).

M. JAULENT - Problèmes inverses dans les milieux absorbants.

G. JOBERT, A. CISTERNAS - On the inverse problem of local seismic fois.
(Published in the present book).

F. LAMBERT - Equations de Yang-Mills et problème inverse.

J.P. LEFEBVRE - Application à l'acoustique des algorithmes de Gelfand Levitan Jost Kohn".

A. MARTEN - Inversion de l'équation de transfert radiatif dans les atmosphères terrestres et planétaires.

I. MIODEK - What you always want to know about the application of the inverse problems to nonlinear equations".
(Published in the present book).

V. PELOSI - Survey of the phenomenological approach to the inverse problem in elementary particles scattering.

A. ROGER - Determination of the index profile of a dielectric plate from scattering data.

P.C. SABATIER - Problèmes inverses de la terre traités en problèmes inverses de collision. (Spectral and scattering inverse problems, to be published in the J. of Math. Physics).

N. SCOTT, A. CHEDIN - Problèmes ou méthodes inverses pour la géophysique extérieure.

G. TURCHETTI, C. SAGRETTI - Stieltjes functions and approximate solutions of an inverse problem.
(Published in the present book).

G. VASSEUR, V. BARTHES - An inverse problem for electromagnetic prospection.
(Published in the present book).